D1141231

ATOMIC COLLISIONS GROUP

$V2703(2)$

Advances in
ATOMIC AND MOLECULAR PHYSICS

VOLUME 6

CONTRIBUTORS TO THIS VOLUME

J. N. BARDSLEY

D. R. BATES

M. A. BIONDI

YUKIKAZU ITIKAWA

A. S. KAUFMAN

A. E. KINGSTON

T. R. MARRERO

E. A. MASON

MANUEL ROTENBERG

KAZUO TAKAYANAGI

ADVANCES IN

ATOMIC
AND MOLECULAR
PHYSICS

Edited by

D. R. Bates

DEPARTMENT OF APPLIED MATHEMATICS
THE QUEEN'S UNIVERSITY OF BELFAST
BELFAST, NORTHERN IRELAND

Immanuel Esterman

DEPARTMENT OF PHYSICS
THE TECHNION
ISRAEL INSTITUTE OF TECHNOLOGY
HAIFA, ISRAEL

VOLUME 6

 1970

ACADEMIC PRESS New York London

ACADEMIC PRESS, INC.
111 Fifth Avenue, New York, New York 10003

United Kingdom Edition published by
ACADEMIC PRESS, INC. (LONDON) LTD.
Berkeley Square House, London W1X 6BA

LIBRARY OF CONGRESS CATALOG CARD NUMBER 65-18423

PRINTED IN THE UNITED STATES OF AMERICA

Contents

Dissociative Recombination

J. N. Bardsley and M. A. Biondi

Analysis of the Velocity Field in Plasmas from the Doppler Broadening of Spectral Emission Lines

A. S. Kaufman

The Rotational Excitation of Molecules by Slow Electrons

Kazuo Takayanagi and Yukikazu Itikawa

The Diffusion of Atoms and Molecules

E. A. Mason and T. R. Marrero

Theory and Application of Sturmian Functions

Manuel Rotenberg

Use of Classical Mechanics in the Treatment of Collisions between Massive Systems

D. R. Bates and A. E. Kingston

List of Contributors

Numbers in parentheses indicate the pages on which the authors' contributions begin.

J. N. BARDSLEY, Department of Physics, University of Pittsburgh, Pittsburgh, Pennsylvania (1)

D. R. BATES, Department of Applied Mathematics and Theoretical Physics, The Queen's University, Belfast, Northern Ireland (269)

M. A. BIONDI, Department of Physics, University of Pittsburgh, Pittsburgh, Pennsylvania (1)

YUKIKAZU ITIKAWA, Institute of Space and Aeronautical Science, University of Tokyo, Komaba, Meguro-ku, Tokyo, Japan (105)

A. S. KAUFMAN,* Bar-Ilan University, Ramat-Gan, Israel and The Hebrew University of Jerusalem, Israel (59)

A. E. KINGSTON, Department of Applied Mathematics and Theoretical Physics, The Queen's University, Belfast, Northern Ireland (269)

T. R. MARRERO,† Brown University, Providence, Rhode Island (155)

E. A. MASON, Brown University, Providence, Rhode Island (155)

MANUEL ROTENBERG, Department of Applied Physics and Institute for Pure and Applied Physical Science, University of California, San Diego, La Jolla, California (233)

KAZUO TAKAYANAGI, Institute of Space and Aeronautical Science, University of Tokyo, Komaba, Meguro-ku, Tokyo, Japan (105)

---

* Present address: Department of Physics, The Hebrew University of Jerusalem, Israel.

† Present address: Department of Chemical Engineering, University of Missouri, Columbia, Missouri.

Contents of Previous Volumes

DISSOCIATIVE RECOMBINATION

J. N. BARDSLEY and M. A. BIONDI

Department of Physics,
University of Pittsburgh,
Pittsburgh, Pennsylvania

I. Introduction

The capture of an electron by a positive ion (recombination) requires that the electron go from a free (positive energy) to a bound (negative energy) state and thus, in some manner, give up energy. The energy may be given to a third body (for example, at an encounter with a neutral atom or another electron) or it may be radiated away as a result of a free-bound transition. Three-body encounters become probable only at rather high concentrations of neutral molecules or of electrons, while the radiative process occurs with small probability during an electron–ion collision and consequently leads to a very small recombination rate. By way of contrast, the highly efficient process of

1

dissociative recombination may be thought of as one in which the positive ion (molecular in character) carries along its own "third body" to act as an energy removing agent. The process may be represented schematically by the reaction,

$$AB^+ + e \rightleftharpoons AB'' \rightarrow A + B. \tag{1}$$

In a collision of an electron with a molecular ion AB^+ a temporary state of the neutral molecule AB'' is formed. This state is unstable, since an electron can be reemitted through autoionization. However if dissociation of the state AB'' is possible, then the electron can be trapped and two neutral atoms can be formed, often in excited states. Dissociative recombination can clearly also occur in polyatomic molecules, but most of our discussion will be concerned with diatomic molecules.

The initial interest in this reaction arose from studies of the upper atmosphere. The observations of the oxygen green line in the night sky and in aurora were attributed by Kaplan (1931) to the formation of the excited atom $O(^1S)$ through the dissociative recombination of electrons with O_2^+ ions. Probably the most important feature of the reaction is its influence on the charge density in the ionosphere. In the absence of any ionization (e.g., night time) the rate of change of the electron density n_e is controlled by the rate of electron–ion recombination α_e and ion–ion recombination α_i according to the equation

$$dn_e/dt = -(\alpha_e + \lambda\alpha_i)n_e^2 = -\alpha_{\text{eff}} n_e^2. \tag{2}$$

Here λ is the ratio of negative ions to electrons, and α_{eff} is introduced as an effective recombination coefficient. In the earliest theoretical studies of the reactions in the upper atmosphere (Massey, 1937; Bates *et al.*, 1939), it was assumed that dissociative recombination could be neglected and that the electron–ion recombination occurred through radiative transitions from continuous to bound states of the electron–ion system. This radiative recombination process has a rate of the order of 10^{-12} cm^3/sec at ionospheric temperatures, which was compared with the "measured" effective recombination rates in the E region of around 10^{-8} cm^3/sec (see e.g., Bates and Massey, 1946). Massey (1937) and Bates *et al.* (1939) were forced to conclude that the ratio of negative ions to electrons in the E region is of the order of 1000 and in the F region is about 20. These high ratios were also necessary to explain the charge density indicated by the magnetic field studies of Chapman (1931). Further study showed such high ratios to be unacceptable (Bates and Massey, 1946, 1947) and led to the suggestion that dissociative recombination should be examined more closely.

The first experimental evidence that the rate of dissociative recombination might be as high as 10^{-7} cm^3/sec arose from microwave studies of electron loss processes in afterglows (Biondi and Brown, 1949b). In neon, the expected ambipolar diffusion loss was overwhelmed by a 2-body electron–ion recombination loss. Accurate determinations of the recombination coefficient yielded values of $\sim 2 \times 10^{-7}$ cm^3/sec at 300°K, some 10^5 times the expected radiative recombination rate. The significance of the data on other molecules given in the same paper is uncertain because of the lack of knowledge concerning the identity of the ions involved in the recombination.

Unfortunately, results of studies of ambipolar diffusion in helium (Biondi and Brown, 1949a) were published some 6 months before the neon results. In the helium studies, a rather poorly determined recombination correction to the diffusion data was given (the inferred value, $\alpha \sim 10^{-8}$ cm^3/sec, was obtained from too small a range of electron density decay to be accurate), and it was the helium rather than the neon data that was explained in terms of dissociative recombination (Bates, 1950).

At the time of these studies the presence of molecular ions in rare gas afterglows had not been firmly established, but in 1952 Phelps and Brown attached a mass spectrometer to a helium afterglow apparatus and demonstrated the formation of He_2^+ ions. The use of the mass spectrometer is now acknowledged to be essential for reliable afterglow studies, since a measurement of a recombination coefficient is of little value if one does not know the species to which the measurement relates.

After the demonstration of high recombination rates, a series of experiments were performed at Pittsburgh to show that dissociative recombination is responsible. These experiments will be described in Section II. The experimental techniques for measuring recombination rates were analyzed carefully by several groups, and some of the pitfalls of the experiments were realized. Thus in the last few years it has been possible to obtain reliable data on dissociative recombination, and the recombination rates for many ions have been obtained. Initially most experiments were carried out near room temperature, but the most recent work has been concerned with the temperature dependence of the recombination rate. As a result of the measurements it appears that most molecular ions (a notable exception is He_2^+) have recombination coefficients greater than 10^{-7} cm^3/sec at room temperature, and for the heavy rare gases (krypton and xenon) and for most polyatomics, the coefficient seems to be of order 10^{-6} cm^3/sec.

The analysis of the atmospheric data and the measurements of Biondi and Brown led Bates (1950) to reexamine theoretically the dissociative recombination reaction. The previous estimate of a small rate was based on the argument that the reaction involves the exchange of energy from electronic to nuclear motion and this necessitates the breakdown of the Born–Oppenheimer separation of electronic and nuclear motion. The picture of the reaction given by Bates (1950) showed that it is not necessary to invoke this breakdown;

instead, he suggested that the recombination process should be considered in two parts. First the intermediate state AB″ is formed through configuration interaction. The incident electron excites a target electron, and itself falls into an unoccupied orbital of the molecule. The potential curve controlling the motion of the nuclei is then that of the neutral state AB″ and not of the initial ionic state. Thus the initial electron capture can occur solely through exchange of energy between the incident and a target electron, and after the capture the nuclei may be forced apart if the new potential curve is repulsive, as shown in Fig. 1.

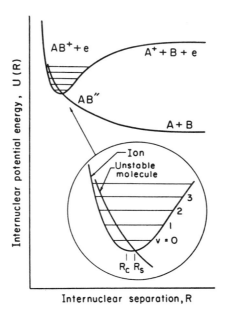

FIG. 1. Schematic representation of the potential energy curves involved in the direct dissociative recombination process. In this simplified example only one potential curve for the molecular ion AB⁺ (with an electron at rest at infinity) and one for the unstable molecule AB″ are shown. Inset: magnified view of the potential curves in the curve-crossing region.

There are two features of this process that are not found in atoms and that arise whenever dissociation can occur in the state AB″. First, the initial capture can take place for electrons of any energy. In order that capture occur without the transfer of energy between electronic and nuclear motion it is necessary only that the difference between the potential curves AB⁺ and AB″ be equal to the energy brought in by the incident electron. For thermal collisions with potential curves as shown in Fig. 1, this will always be possible at some internuclear distance R_c. This nuclear separation R_c will be referred to as the capture point. Second, dissociation provides a very efficient mechanism for trapping the added electron. The lifetimes of autoionizing states against electron emission are typically of the order of 10^{-14} sec. This must be compared with the stabilization time—the time taken by the nuclei to separate from the capture point R_c past the crossover or stabilization point R_s. Using the Born–Oppenheimer picture, autoionization is forbidden for separations larger than R_s, since the neutral curve lies below the ionic curve, and the emission of negative energy electrons is impossible. This stabilization time is of the order 10^{-16} to 10^{-15} sec, and so it is extremely likely that the state AB" will dissociate into neutral atoms rather than reionize by emitting an electron. For electron recombination with atomic ions at moderate densities there is no stabilization mechanism of comparable efficiency, and for this reason the recombination rate for molecular ions is normally several orders of magnitude greater than that for atomic ions.

Although reliable *ab initio* calculations of dissociative recombination rates have not yet been performed, it can be shown that the model suggested by Bates is consistent with the observed recombination rates. The model also leads to predictions concerning the temperature dependence, which can be compared with experiments. These points will be discussed in Section III. Also in that section we will discuss modifications of the simple model which may be important for some molecules.

II. Experimental Studies

A. Methods of Measurement

The methods of experimentally studying dissociative recombination may be broadly classifiied as plasma-afterglow and electron–ion beam techniques. To obtain quantitative determinations of the recombination coefficient for Reaction (1) it is necessary to measure the electron and/or positive ion concentration to a high degree of accuracy. In the plasma-afterglow measurements, Langmuir probe (Mott-Smith and Langmuir, 1926), microwave (Biondi, 1951a), and charge collection techniques (Young and St. John, 1966) have been used to determine the electron and/or ion concentrations in the volume of the plasma. While microwave techniques yield accurate spatially averaged electron concentrations in a straightforward manner, only a few of the probe measurements have met the required conditions of operation sufficiently well to determine local electron densities reliably (Smith and Goodall, 1968). Also, up to the present, only a semiquantitative relationship has been established between volume electron and ion concentrations and the charge collected on the parallel plates of an ionization chamber containing the plasma.

The analysis of data from all of these plasma-afterglow experiments proceeds from a consideration of the appropriate particle continuity equation. For the electrons, at each point in space we have

$$\partial n_e/\partial t = \sum_i P_i - \sum_j L_j - \nabla \cdot \mathbf{\Gamma}_e, \tag{3}$$

where n_e is the electron concentration, the P_i and L_j are the rates of processes leading to production and loss of electrons, respectively, and $\mathbf{\Gamma}_e$ is the electron particle current (e.g., resulting from ambipolar diffusion when concentration gradients are present). Electron production by such processes as photoionization or microwave breakdown may be deliberately used in the plasma generation phase; in addition, during the afterglow (after external ionization sources have been removed) metastable atoms may lead to electron production by Penning ionization or by metastable–metastable collisions; all of these contribute to the P_i terms.

In addition to the dissociative recombination loss, which is given by

$$L_{\text{d.r.}} = \alpha n_e n_+, \tag{4}$$

where α is the 2-body dissociative recombination coefficient and n_+ is the concentration of the recombining molecular ion, other volume loss processes may contribute. At high electron concentrations, 3-body collisional-radiative recombination (Bates *et al.*, 1962), which may be schematically represented by

$$A^+ + e + e \rightarrow A^* + e, \tag{5}$$

may be of importance. When electronegative gases are involved in the studies, electron attachment to neutrals to form negative ions may be a significant loss process. With the possibility of so many processes contributing to the growth or decay of electron concentration, it is necessary to choose conditions under which one, or at most a few, processes are of importance in order to obtain quantitative determinations of the dissociative recombination coefficients.

Under optimum experimental conditions a single positive ion (molecular in character) is present in the plasma, and the afterglow electron concentration decay is controlled principally by dissociative recombination. However, the spatial distribution of the electrons is affected by ambipolar diffusion to the boundaries. In this case, $\mathbf{\Gamma}_e = -D_a \nabla n_e$, where D_a is the ambipolar diffusion coefficient of electrons and ions, and Eq. (3) simplifies to

$$\partial n_e(\mathbf{r}, t)/\partial t \simeq -\alpha n_e^2(\mathbf{r}, t) + D_a \nabla^2 n_e(\mathbf{r}, t), \tag{6}$$

where we have used the quasineutrality of a plasma to set $n_e \simeq n_+$. If, at a given point in the plasma the ambipolar diffusion term is very much smaller than the recombination term (high neutral atom concentration and/or large plasma vessel), we may neglect the second term on the right and the solution of Eq. (6) becomes the well-known "recombination solution,"

$$1/n_e(\mathbf{r}, t) = [1/n_e(\mathbf{r}, 0)] + \alpha t. \tag{7}$$

In view of the form of Eq. (7), most 2-body recombination measurements present the data in the form of plots of $1/n_e$ vs. t.

Langmuir probes, which can sample the electron concentration at a point in the plasma, can therefore be used to determine the recombination coefficient directly from the slopes of reciprocal electron concentration versus time curves, provided that diffusion loss is negligible and that the probe data yield accurate *absolute* electron density values.

By way of contrast, microwave determinations of electron concentration [for example, by measuring the shift in resonant frequency of a cavity surrounding the plasma afterglow (Biondi, 1951a)] provide a weighted average of the electron concentration, $\bar{n}_{\mu w}$, called the "microwave-averaged" electron concentration and defined by

$$\bar{n}_{\mu w}(t) \equiv \int_{\text{vol}} n_e(\mathbf{r}, t) E^2(\mathbf{r})\, dV \Big/ \int_{\text{vol}} E^2(\mathbf{r})\, dV, \tag{8}$$

where $E(\mathbf{r})$ is the probing microwave field amplitude in the cavity and the integrations are carried out over the volume of the cavity. Thus, in order to obtain values of α from the measured afterglow decays of $\bar{n}_{\mu w}(t)$, it is necessary to solve Eq. (6) with known values of D_a treating α as a parameter to generate curves of $\bar{n}_{\mu w}(t)$ for comparison with the data. From best fits of the measurements to the computer solutions one determines the dissociative recombination coefficients to a high degree ($\pm 5\%$) of accuracy (Frommhold *et al.*, 1968).

In many earlier microwave afterglow studies it was tacitly assumed that, by inference from Eq. (7), the slopes of the measured $1/\bar{n}_{\mu w}$ versus time curves were equal to α. This assumption is in error, since it is a straightforward matter to show (Kasner and Biondi, 1965) that spatial averaging of a nonuniform electron concentration adds a coefficient in front of the α term in the counterpart to Eq. (7), and therefore the slopes of $1/\bar{n}_{\mu w}$ versus time plots do not directly yield the recombination coefficient. Computer solutions of Eq. (6) indicate that ambipolar diffusion to the boundaries results in a nonuniform spatial distribution of the electrons (even when the overall diffusive loss to the boundaries is small compared to volume recombination loss) whose form changes with time during the afterglow. The required correction factors to relate the slopes of the $1/\bar{n}_{\mu w}$ curves to α have been obtained for plasma containers in the form of infinite cylinders and spheres (Gray and Kerr, 1962) and for finite cylinders, rectangular parallelepipeds, and certain "one-dimensional" geometries (Frommhold and Biondi, 1968).

Charge collection experiments require a similar correction for the spatial nonuniformity of the electron concentration within the region swept by the collection plates. However, up to the present, recombination analyses have been based on the assumption of a uniform charge density in the plasma volume; this assumption may lead to rather large errors (possibly, a factor of 2) in recombination coefficients determinations.

The electron–ion beam studies, of which the recent merged beam measurements (Hagen, 1968; Theard, 1969) will serve as representative examples, have their own special problems in the measurement of charged particle concentrations and spatial distributions. To obtain quantitative determinations of the recombination coefficients it is necessary to define the merged region of electron and ion beams, to probe across the beams to determine their lateral profiles, and to relate currents and beam velocities to absolute concentrations of ions and electrons. In addition, there may be a broad distribution of relative velocities between ions and electrons, rather than the single, well-defined value which was to have been one of the advantages of this method of measurement.

A simplified block diagram of a recent afterglow apparatus (Mehr and Biondi, 1969) employing microwaves for plasma generation, electron density determinations, and controlled electron heating, together with a differentially pumped quadrupole mass spectrometer to sample the ions diffusing to the cavity walls, is shown in Fig. 2. The dual mode (high Q for electron density

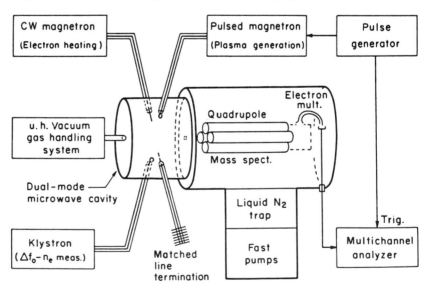

FIG. 2. Simplified diagram of a microwave afterglow/mass spectrometer apparatus employing microwave heating of the electrons (after Mehr and Biondi, 1969).

measurement, low Q for controlled electron heating) resonant cavity is at the center. Gas samples of very high purity (for example, ultrapure neon gas is obtained from evaporation of liquid neon) are continuously introduced to the cavity by means of an ultrahigh vacuum gas handling system and pumped away after effusing through a small hole (0.04 cm diam) into the mass spectrometer. Because of the small ion signals from the cavity into the mass spectrometer, pulse counting and multichannel analyzer signal averaging techniques are used to determine the decay of ion wall current during the afterglow.

The microwave measurement of average electron density depends on the detuning of a resonant cavity as a result of the dielectric effect of an electron gas when the applied microwave frequency is large compared to the electron's collision frequency. The controlled electron heating makes use of the fact that the microwaves interact strongly with the electrons but almost negligibly with the more massive ions. Thus, application of a constant amplitude microwave field to the cavity causes the electrons to reach a temperature (their energy distribution remains essentially Maxwellian) in excess of the neutral atom and ion temperature. In order to maintain a constant microwave heating field, a very low-Q ($\leqslant 10$) mode is used to essentially eliminate the effect of the electron density decay on the cavity response to the incident heating energy.

A somewhat similar afterglow apparatus (Sayers, 1956) employing a Langmuir probe to measure electron densities and temperatures and an *rf* mass spectrometer to sample the ions reaching the walls has also been used in recombination studies. The electron temperature, although found to be elevated above the ambient, cannot be altered in any controlled fashion in these studies.

A second controlled heating plasma-afterglow experiment, shown schematically in Fig. 3, involves shock heating of an rf ionized gas sample, followed by downstream measurements of the plasma decay as it flows past a series of double probes (Fox and Hobson, 1966; Cunningham and Hobson,

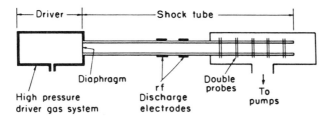

FIG. 3. Simplified diagram of an afterglow apparatus employing shock heating of the gas and double probe determinations of ion concentrations (after Cunningham and Hobson, 1969a).

1969a). The ion concentrations are deduced from the currents to the probes when a "saturation" voltage difference is applied (McLaren and Hobson, 1968). It is argued that, following shock heating of the ionized gas, the kinetic temperatures of electrons and ions quickly reach that of the neutral gas, and it is assumed that the molecular ions achieve a Boltzmann distribution of their vibrational states at the ion kinetic temperature before the downstream recombination determinations begin.

The third method by which dissociative recombination can be studied over a very wide range of relative velocities between ions and electrons is the so-called "merged beam" method. The central features of one such beam apparatus (Theard, 1969) are indicated schematically in Fig. 4. A discharge ion

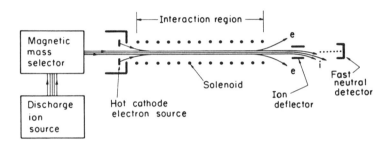

FIG. 4. Schematic representation of a merged electron–ion beam apparatus (after Theard, 1969).

source followed by a magnetic mass selector is used to select the desired ion which is then accelerated to between 4 and 50 keV to form an ion beam. This beam then passes along the center line of a long solenoid. Just outside of the solenoid in the magnetic fringing region a circular cathode emits electrons which are accelerated to energies between ~ 0.1 and 1 eV to match the velocity of the ion beam. The converging magnetic field lines bring the electrons into the ion beam (which is essentially undeflected by the magnetic field) where, for the length of the solenoid, they interact (e.g., recombine) with the ions. At the end of the solenoid the fringing magnetic field causes them to diverge from the ion beam and interaction terminates. By controlling the electron beam accelerating voltage relative velocities between electrons and ions ranging from very small to very large values can, in principle, be attained. In practice, spiraling of the electron beam in the confining magnetic field limits the minimum relative velocity that can be achieved and leads to rather broader distributions of relative velocities than originally had been anticipated. The recombination events are determined by deflecting the ion beam after it leaves the interaction region and detecting the undeviated fast neutral atoms produced by the recombination events.

B. Establishing the Recombination Mechanism

As noted in the Introduction, although the dissociative capture process was proposed to explain rapid recombination between atmospheric molecular ions and electrons in the ionosphere, it was first discovered in the laboratory (Biondi and Brown, 1949b) in neon afterglow studies. At the time, the existence of noble gas molecular ions such as Ne_2^+ and Ar_2^+ was not well established, and the occurrence of an electron loss by recombination with positive ions at some 10^5 times the radiative recombination rate was completely unforeseen. However, the electron decay data at $300°K$ appeared very much like curve A in Fig. 5. The large range of electron densities over which $1/\bar{n}_{\mu w}$ increased linearly with time $(f \sim 10)^1$ provided strong evidence for the predominance of 2-body recombination, and the slope of the curve yielded a value 2.1×10^{-7} cm^3/sec. [This slope has recently been corrected for diffusion effects, yielding a value $\alpha(Ne_2^+) = 1.7 \times 10^{-7}$ cm^3/sec, in excellent agreement with results of recent studies.] Similar electron decay studies in argon also yielded a large $(>10^{-7}$ cm^3/sec) recombination coefficient.

During the same period, studies of neon and helium afterglows (Holt et al., 1950; Johnson et al., 1950) revealed that the electron loss was accompanied by persistent optical emission from the afterglows. These radiations (atomic line spectra in the case of neon, molecular bands in the case of helium) were attributed to excited atom production by the recombination of Ne^+ and He_2^+ ions, respectively, with electrons.

Following the suggestion that the large recombination loss in the laboratory studies resulted from the dissociative process (Bates, 1950), attempts were made to determine experimentally whether or not this was indeed the process responsible. Two distinguishing features of dissociative recombination, Reaction (1), are (a) that it requires the presence of molecular ions and (b) that the neutral atoms (or molecules) produced receive a kinetic energy of dissociation. Neither radiative recombination nor the later discovered 3-body recombination (involving a second electron as the energy removing agent in the electron–ion capture process) is expected to exhibit appreciably different rates for molecular ions as opposed to atomic ions. Also, neither process leads to production of atoms or molecules with kinetic energies in excess of thermal.

The first experimental test of the dissociative hypothesis (that is, of the required presence of molecular ions) was carried out in noble gas afterglows with a microwave apparatus which did not employ mass analysis of the ions. Two experiments (Biondi, 1951b, 1963) were carried out, one under conditions where molecular Ar_2^+ ions were expected to be present, the other where

---

[1] The quantity f is defined as the ratio of electron concentrations over which the increase in $1/\bar{n}_{\mu w}$ with time remains linear to within 1%.

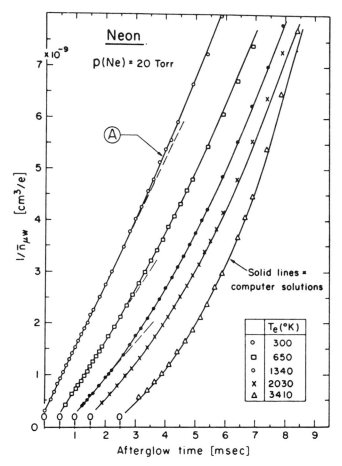

FIG. 5. "Recombination plots" of electron loss in neon as a function of electron temperature (after Frommhold *et al.*, 1968). The reciprocals of the "microwave-averaged" electron densities are plotted as a function of time during the afterglow. The solid lines represent computer solutions of the electron continuity equation.

only atomic Ar^+ ions were expected. At moderate pressures (>10 Torr) in pure argon, the afterglow should contain predominately Ar_2^+ as positive ions, since associative ionization,

$$Ar^* + Ar \rightarrow Ar_2^+ + e, \qquad (9)$$

from highly excited argon atoms Ar^* should lead to substantial formation of Ar_2^+ during the microwave discharge, and 3-body conversion of any Ar^+ ions by the reaction,

$$Ar^+ + 2\,Ar \rightarrow Ar_2^+ + Ar, \qquad (10)$$

should rather rapidly (in ~ 100 μsec) eliminate atomic ions during the after-glow. Under these conditions the measured electron concentration decays in argon at 300°K resembled the $1/\bar{n}_{\mu w}$ versus afterglow time curves shown for neon in Fig. 5, with a value $\alpha(Ar_2^+) = 6 \times 10^{-7}$ cm^3/sec obtained from the slope of the curve.

However, when a mixture of 1 part argon in 1000 parts helium was used, the form of the electron decay changed radically. In a microwave discharge afterglow in such a mixture it was expected that Penning ionization by helium metastables,

$$He^M + Ar \rightarrow He + Ar^+ + e, \tag{11}$$

would create a plasma consisting principally of Ar^+ as the positive ion. In this case, the small concentration of argon atoms reduces the rate of 3-body conversion of Ar^+ to Ar_2^+ to a point where, during the measuring interval of the afterglow, the Ar_2^+ concentration remains very small compared to that of Ar^+. In this case, at total helium–argon pressures between 1.5 and 7 Torr, the electron decay during the afterglow followed the ambipolar diffusion form obtained by solving the continuity equation [Eq. (6)] with $\alpha = 0$. The rate of electron decay varied inversely with gas pressure, as expected for ambipolar diffusion control, and the value $D_a p = 910$ cm^2/sec-Torr (where D_a is the ambipolar diffusion coefficient and p is the He pressure) determined from the measurements is consistent with the value expected for Ar^+ ions in helium. From detailed analysis of uncertainties in the data it was possible to show that, in the helium–argon studies, the recombination coefficient was at least 1000 times smaller than in the pure argon studies. Thus, the large recombination loss was clearly associated with the presence of molecular ions.

The next step was to show that the recombination process was dissociative in character by somehow detecting the kinetic energy of dissociation of the atoms produced by the recombination process. Inasmuch as Reaction (1) may lead to formation of excited atoms which subsequently radiate, under appropriate conditions, the emitted line radiation should show Doppler shifts resulting from the dissociation kinetic energy. Competing processes which might mask this effect are collisional slowing of the fast excited atoms and, even more likely, excitation transfer collisions with the ambient gas atoms according to the reaction.

$$A_f^* + A_s \rightarrow A_f + A_s^*, \tag{12}$$

where the subscripts f and s denote fast and slow atoms, respectively. In either case a thermal Doppler line would be emitted, even though the excited atoms had been produced with substantial dissociation kinetic energy. In order to avoid such effects, one works at the lowest feasible concentration of neutral atoms and with excited states having a large radiative transition probability.

If the excited atoms are produced by Reaction (1) with a single dissociation energy E_D, it is a straightforward matter to show (Rogers and Biondi, 1964) that the Doppler shifts lead to a broad, flat-topped line, whose half width is proportional to $E_D^{1/2}$ (see Fig. 6B). The search for spectral line broadening was first undertaken using a photoelectric recording Fabry–Perot interferometer (Jacquinot and Dufour, 1948; Biondi, 1956) to study helium afterglows. Helium has the required short radiative lifetime, as well as a simple afterglow spectrum, and at the time the studies were undertaken (ca. 1952) the fact was not appreciated that as a result of unfavorable curve crossings, thermal electrons might not be able to recombine with He_2^+ ions. A slight broadening of the late afterglow $\lambda5876$ ($3^3D \rightarrow 2^3P$) line relative to the discharge and

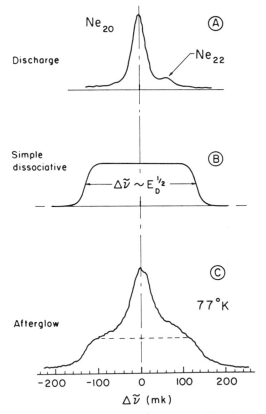

FIG. 6. Spectral line shapes of the $2p_1$–$1s_2$ ($\lambda5852$) transition in neon: (A) Tracing of chart record of the line profile during the microwave discharge at 77°K and 1 Torr pressure showing the contributions of the two isotopes. (B) Theoretical line profile resulting from the radiation of atoms produced with a single dissociation energy, E_D. (C) Tracing of observed line profile during the afterglow at 77°K and 0.5 Torr pressure, showing a thermal core sitting atop a dissociative pedestal.

early afterglow lines was systematically detected, and no origin other than dissociation kinetic energy could be suggested.

As a result of the inconclusive results in helium, spectral line shapes in pure neon afterglows were studied using a Fabry–Perot etalon of considerably improved resolution (Connor and Biondi, 1965; Frommhold and Biondi, 1969). It was observed that, while the $\lambda 5852$ ($2p_1 \rightarrow 1s_2$) neon line exhibited a shape and width appropriate to excited atoms at the ambient temperature during the discharge (see Fig. 6A), the afterglow line profile was evidently composed of a very broad, flat pedestal (as in Fig. 6B) surmounted by a thermal core (see Fig. 6C). This line profile is expected when fast excited atoms are produced by dissociative recombination and a fraction of them transfer their excitation to ambient gas atoms before radiating. In support of this interpretation was the fact that, as the neutral neon concentration was increased, the intensity in the thermal core increased in direct proportion, leading to inference of an excitation transfer cross section of $\sim 1 \times 10^{-15} \text{cm}^2$ for atoms moving with relative speeds of 2.5×10^5 cm/sec.

Thus was established the dissociative origin of the excited neon atoms produced by recombination. These studies were extended to some 21 other neon and to 5 argon transitions of the ($2p_n \rightarrow 1s_m$) series (Frommhold and Biondi, 1969), and in all cases line broadening during the afterglow was detected. In addition, the studies yielded information concerning the stability of the Ne_2^+ molecular ion. [It should be noted, however, that very recent studies using a Fabry–Perot interferometer of somewhat poorer resolving power (Edwin and Turner, 1969) have failed to detect dissociative broadening in krypton afterglows, perhaps owing to the excitation transfer effects mentioned earlier.]

As a final point concerning the mechanism of the electron–ion recombination, we note that the recent afterglow studies that employ mass analysis invariably detect *molecular* ions as the principal afterglow ions when a large recombination loss occurs, Thus, it is Ne_2^+, N_2^+, O_2^+, and NO^+ ions and not Ne^+, N^+, and O^+ ions that lead to recombination coefficients $\gtrsim 10^{-7}$ cm^3/sec.

C. Recombination Measurements near Thermal Energies

The bulk of the measurements of dissociative recombination coefficients have been limited to the temperature range $\sim 100°$ to $\sim 700°$K and have been carried out under conditions when it was expected that $T_e \simeq T_i = T_{gas}$. In addition, although a differentially pumped mass spectrometer was added to an early microwave afterglow apparatus (Phelps and Brown, 1952), most afterglow studies have not included until recently mass identification of the ions undergoing recombination. In the noble gases, where at room temperature and above *diatomic* molecular ions are the most complex expected, lack of mass identification is a less serious deficiency than in molecular (e.g., atmospheric) gases such as N_2 and O_2. Here it is found that more complex ions such as N_3^+, N_4^+, O_3^+, and O_4^+ are often important afterglow ions, thus complicating recombination coefficient determinations. For this reason many of the early laboratory determinations of "atmospheric ion" recombination coefficients probably do not refer to the desired ions, i.e., N_2^+ and O_2^+, as assumed.

1. Noble Gases

Let us first consider the recombination coefficient determinations in the noble gases He, Ne, Ar, Kr, and Xe. It is relatively straightforward to obtain very pure gas samples of helium and neon; on the other hand, it is difficult to remove all traces of nitrogen from argon samples, and krypton and xenon samples each contain some admixture of the other gas.

As was noted in the Introduction, the recombination coefficient determinations in helium (Biondi and Brown, 1949a) were rather poorly done compared to those in neon and in argon, since diffusion loss was relatively high and it was difficult to attain "recombination control" of the afterglow. Thus the linear range of the $1/\bar{n}_{\mu w}$ versus time curves at helium pressures of 20 Torr extended over only a factor $f \sim 3$ in electron density, and the inference of a coefficient $\alpha(He_2^+) \sim 10^{-8}$ cm^3/sec was therefore quite suspect. Subsequently, very extensive unpublished studies of electron concentration and optical radiation decays in helium by D. E. Kerr and co-workers failed to show any evidence for a recombination coefficient substantially in excess of $\sim 10^{-9}$ cm^3/sec.

Later microwave afterglow studies (Chen *et al.*, 1961) at 30 Torr helium pressure yielded a coefficient $\sim 10^{-8}$ cm^3/sec from $1/\bar{n}_{\mu w}$ data with a linear range, $f \sim 4$; however, the measurements were carried out at electron densities in the range of 10^{11} cm^{-3}, for which the 3-body collisional-radiative recombination process, Reaction (5), yields an effective 2-body coefficient, $\alpha_{eff} \sim 10^{-8}$ cm^3/sec. Collisional-radiative recombination, however, leads to a form of $1/\bar{n}_{\mu w}$ versus time curve which is concave downward rather than linear; therefore we cannot account for the results of Chen *et al.* completely in terms of this process. Very recently Berlande *et al.* (1970) have shown that it is possible to account for the magnitude and temporal form of Chen *et al.*'s electron decays in terms of neutral-stabilized recombination; however, the effective 2-body coefficient should increase with neutral atom concentration, contrary to the observation of an α *independent* of pressure from 15 to 30 Torr.

More recently, microwave interferometer afterglow studies of cataphoretically purified helium (Thomas *et al.*, 1966) at electron densities in the range of $\sim 3 \times 10^{11}$ cm^{-3} yielded rather good $1/\bar{n}_{\mu w}$ versus time curves (linear ranges, $f \sim 15$). The slopes of the decay curves are independent of gas pressure over the measured range 5–43 Torr, yielding a value $\alpha(\mathrm{He}_2^+) \simeq 6 \times 10^{-9}$ cm^3/sec at 300°K. While such a behavior is consistent with a 2-body process (e.g., dissociative recombination), the effect of 3-body collisional-radiative recombination should have made itself felt at these high electron densities.

Arguing against a dissociative recombination coefficient approaching 10^{-8} cm^3/sec in helium are several static and flowing afterglow studies (Collins and Robertson, 1964, 1965; Niles and Robertson, 1964; Ferguson et al., 1965; and Collins and Hurt, 1969) in which the afterglow radiation production has been correlated with the presence or absence of He$^+$ and He$_2^+$. Instead of finding a correlation between afterglow *atomic* emissions and the inferred variation of He$_2^+$, as would be expected if substantial dissociative recombination were present, they find, in general, that the *molecular* band radiation correlates with the inferred He$_2^+$ variations. These observations are consistent with the collisional-radiative recombination mechanism, where He$_2^*$ excited states are produced from He$_2^+$ ions and He* from He$^+$. At the large electron concentrations used in these studies (10^{11}–10^{12} cm^{-3}), it is not clear what upper bound can be set on the dissociative recombination rate that may be present.

The very recent microwave afterglow studies of neutral-stabilized recombination in helium (Berlande et al., 1970) have been used to determine the 2-body coefficient by extrapolation to zero pressure (where the rate of the 3-body process vanishes. A value $\alpha = (5 \pm 2) \times 10^{-10}$ cm^3/sec is obtained at 300°K.

In spite of the conflicting results concerning whether or not there is some dissociative recombination in helium, one fact is clear. All experimenters agree that the coefficient must be less than $\sim 10^{-8}$ cm^3/sec; therefore, He$_2^+$ presents a somewhat unique situation, since, as we shall see, all other well-investigated cases lead to values of the dissociative recombination coefficient in the 10^{-7}–10^{-6} cm^3/sec range at 300°K. Possible origins of this anomalously small rate for He$_2^+$, such as lack of suitable potential curve crossings (Mulliken, 1964), are discussed in Section III.

Let us now consider the studies of neon, which probably represents the most extensively investigated and best documented case of dissociative recombination. As noted earlier, dissociative recombination was first encountered during the microwave afterglow studies of neon (Biondi and Brown, 1949b). These data, when corrected for diffusion effects by solution of the continuity equation, Eq. (6), (Frommhold et al., 1968), yield a value $\alpha(\mathrm{Ne}_2^+) = 1.7 \times 10^{-7}$ cm^3/sec at 300°K. Subsequent measurements (Oskam, 1958; Biondi, 1963; Oskam and Mittelstadt, 1963; Connor and Biondi, 1965; Hess, 1965; Kasner, 1968; Frommhold et al., 1968; Philbrick et al., 1969) in neon at 300°K have all yielded linear $1/\bar{n}_{\mu w}$ versus time curves over large ranges of electron concentrations (refer to curve A of Fig. 5 for an example). When the data are analyzed either by comparing the measured decays to the computer solutions of the continuity equation or by applying correction factors for the effect of diffusion on the recombination loss determination (Gray and Kerr, 1962; Frommhold and Biondi, 1968) values for $\alpha(Ne_2^+)$ ranging from 1.7 to 2.0×10^{-7} cm³/sec are obtained. From the most accurate of these studies, it appears that, at 300°K $\alpha(Ne_2^+) = (1.8 \pm 0.1) \times 10^{-7}$ cm³/sec. This recombination coefficient can be positively assigned to Ne_2^+ since, in two of the studies (Kasner, 1968 and Philbrick, *et al.*, 1969), differentially pumped mass spectrometers correlated the decay of Ne_2^+ current to the walls of the plasma containers with the volume loss of electrons by recombination.

The early studies of argon (Biondi and Brown, 1949b), when corrected for diffusion effects, yielded a value of the recombination coefficient between 2 and 3×10^{-7} cm³/sec at 300°K. As shall be seen shortly, this value is suspiciously close to $\alpha(N_2^+)$ and substantially smaller than results of more recent studies carried out with purer argon gas samples, where values of $\alpha(Ar_2^+) = \sim 7 \times 10^{-7}$ cm³/sec (Biondi, 1963, after diffusion correction), $(6.7 \pm 0.5) \times 10^{-7}$ (Oskam and Mittelstadt, 1963), and $(8.5 \pm 0.8) \times 10^{-7}$ (Mehr and Biondi, 1968) were obtained at 300°K. In some of these studies the linear ranges of the $1/\bar{n}_{\mu w}$ versus time curves were large, $f > 30$, permitting highly accurate determinations of the coefficient. Unfortunately, mass identification of the ions undergoing recombination in argon has not been provided (of the noble gases, only Ne_2^+ in the neon studies has been so identified). Thus, our inference that the more recent studies in argon refer to Ar_2^+, rather than to an impurity ion such as N_2^+, is based on gas purity considerations.

The situation in krypton is severely complicated by the possible presence of significant amounts of xenon in the gas samples, making the identity of the ion under study difficult to specify. The most careful determination of the recombination loss in krypton afterglows (Oskam and Mittelstadt, 1963) yields a value $\alpha(Kr_2^+) = (1.2 \pm 0.1) \times 10^{-6}$ at 300°K in cataphoretically purified gas samples. The linear range of the $1/\bar{n}_{\mu w}$ data was remarkably large ($f \sim 600$ at 45 Torr krypton pressure). This value supports less accurate estimates obtained in krypton samples containing enough xenon to affect the afterglow decays; i.e., values of 0.6–1.2×10^{-6} cm³/sec (Richardson, 1952) and $\leqslant 1.1 \times 10^{-6}$ cm³/sec (Lennon and Sexton, 1959). In Richardson's studies, the early "krypton" afterglow, during which the quoted value of α was determined, was marked by atomic krypton radiation, the later afterglow by predominantly *xenon* line radiation.

In xenon, the best determinations (Oskam and Mittelstadt, 1963) of $\alpha(Xe_2^+)$ were obtained from microwave afterglow studies of gas samples containing

$<1 : 10^5$ krypton impurity. At 300°K the value $\alpha(Xe_2^+) = (1.4 \pm 0.1) \times 10^{-6}$ cm³/sec was obtained from $1/\bar{n}_{\mu w}$ data with $f \sim 250$. This value is consistent with less accurate determinations (Richardson, 1952) from the late afterglow in "krypton" (only *xenon* line radiation was emitted) where values in the range 1.2–2.1×10^{-6} cm³/sec were obtained. Other afterglow studies in xenon (Lennon and Sexton, 1959) yielded a value $\alpha(Xe_2^+) = 2.3 \times 10^{-6}$ cm³/ sec at 10 Torr with $f \sim 20$. In view of the superior recombination control of the afterglow in the studies by Oskam and Mittelstadt, we regard their value as the most accurate one.

To summarize the situation in the noble gases, we note that at 300°K there appears to be a monotonic increase in the values of the recombination co-efficients with increasing mass number proceeding from the value 1.8×10^{-7} for Ne_2^+ to 1.4×10^{-6} cm³/sec for $Xe_2^!$. As will be seen in Section III B, one might expect such a trend on the basis of weaker repulsive potential curves for the heavier noble gas molecular states.

The laboratory measurements have been shown to refer to the dissociative recombination process only in neon and argon, where the kinetic energy of the dissociating atoms produced by the recombination has been detected in spectral line profile studies. However, by inference, the same process occurs in krypton and xenon, since at pressures where Kr_2^+ and Xe_2^+ should be the principal ions, the recombination-produced afterglow radiation consists of lines from excited Kr* and Xe* *atoms*. The anomalous case in the noble gases is helium, where the dissociative recombination coefficient is almost certainly substantially less than 10^{-8} cm³/sec. As will be seen in Section III, the probable explanation of this small value lies in unfavorable crossings between the He_2^+ potential curve and the repulsive He_2^* states which lead to disso-ciation.

2. Molecular Gases

Few of the early determinations of recombination coefficients in molecular gases such as nitrogen and oxygen have stood the test of time, chiefly as a result of the intrusion of unsuspected, more complex ions in the studies. We shall, therefore, emphasize those recent studies where mass analysis of the ions undergoing recombination has been employed, and it has been possible to identify the ion responsible for a given recombination rate.

a. Nitrogen. A microwave afterglow apparatus employing a differentially pumped *rf* mass spectrometer (Kasner and Biondi, 1965) was used to deter-mine $\alpha(N_2^+)$. It was found necessary to use rather low pressures of nitrogen ($<10^{-2}$ Torr) to avoid appreciable concentrations of N_3^+ and N_4^+ ions. In order to achieve recombination control of the afterglow, diffusion loss was inhibited by addition of a substantial pressure (~ 20 Torr) of neon, which offered the additional advantage of resulting in production of the N_2^+ ions in their ground electronic and a low ($v \lesssim 4$) vibrational state as a result of Penning ionization by metastables,

$$Ne^M + N_2 \rightarrow Ne + N_2^+ + e. \tag{13}$$

Over a wide range of experimental conditions the N_2^+ ion current diffusing to the wall (and into the mass spectrometer) followed closely ("tracked") the decay of electron concentration in the volume. All other afterglow ion (e.g., N_3^+ and N_4^+) wall currents were very small compared to N_2^+. From $1/\bar{n}_{\mu w}$ versus time data which exhibited f values > 15, a value $\alpha(N_2^+) = (2.9 \pm 0.3) \times 10^{-7}$ cm^3/sec was obtained at 300°K. Later studies (Kasner, 1967) refined this value to $(2.7 \pm 0.3) \times 10^{-7}$ cm^3/sec.

In the only other study at room temperature employing mass analysis (Mehr and Biondi, 1969), the observed intrusion of somewhat larger amounts of the minority ion N_3^+ may have affected the N_2^+ recombination determinations. From data showed excellent recombination control and fit the computer solution of Eq. (6), a value $\alpha(N_2^+) = (1.8^{+0.4}_{-0.2}) \times 10^{-7}$ cm^3/sec was obtained at 300°K. Here, while the N_2^+ wall current was more than an order of magnitude larger than that of the next most abundant ion N_3^+, it rather imperfectly tracked the volume decay of electron concentration; hence rather large uncertainties ($\sim +20\%$, -10%) were assigned to the recombination coefficient.

At higher partial pressures of nitrogen ($>10^{-1}$ Torr), the dimer ion N_4^+ (or $N_2 \cdot N_2^+$) together with some N_3^+ was found to be the principal afterglow ion (Kasner and Biondi, 1965), and a recombination coefficient at 300°K of $\alpha(N_4^+) \sim 1$–2×10^{-6} cm^3/sec was obtained from $1/\bar{n}_{\mu w}$ curves with linear ranges $f \sim 6$–12. This value, obtained for mass-identified ions, is close to the value 1.4×10^{-6} cm^3/sec obtained by Biondi and Brown (1949b) at nitrogen pressures of 2–5 Torr and to the value 2×10^{-6} cm^3/sec (Hackam, 1965) at pressures $\geqslant 10$ Torr, suggesting that these latter measurements also referred to N_4^+ ions.

b. Oxygen. While there have been a number of recent determinations of recombination loss in oxygen, only three investigations have identified the ions under study by mass analysis. In addition, there is some likelihood that, in the various investigations, the O_2^+ ions are in different electronic states. If, as has been done in the studies of nitrogen, one relies on formation of O_2^+ ions in "dilute" oxygen–neon mixtures through Penning ionization of O_2 by neon metastables, then energetically it is possible to form the O_2^+ ions not only in their ground electronic state ($X\,^2\Pi_g$) but also in the excited ($a\,^4\Pi_u$) state. If, however, one uses triple mixtures composed of $\sim 10^{-3}$ to 10^{-2} Torr oxygen, ~ 1 Torr krypton, and ~ 20 Torr neon (Kasner and Biondi, 1968), then the likely O_2^+ production sequence involves Penning ionization of krypton by neon metastables, followed by the charge transfer reaction $Kr^+ + O_2$. In this case, the energetics of the reaction require that the O_2^+ ions be in their ground electronic and one of the lower ($v \leqslant 8$) vibrational states.

At 295°K, using the afterglow–mass spectrometer apparatus mentioned earlier, Kasner and Biondi (1968) obtained a value $\alpha(O_2^+) = (2.2 \pm 0.2) \times 10^{-7}$ cm^3/sec from fits of computer solutions of Eq. (2.4) to the $1/\bar{n}_{\mu w}$ versus time data. The O_2^+ wall current tracked the volume electron density decay very well in these studies. To avoid complicating effects from negative ion accumulation, the apparatus was operated in a "single pulse-afterglow" mode.

The value of $\alpha(O_2^+)$ was insensitive to the partial pressure of O_2 in the triple mixture (O_2–Kr–Ne) studies but exhibited noticeable variations at the lower partial pressures ($< 2 \times 10^{-3}$ Torr) in the binary mixture (O_2–Ne) studies. As we shall see shortly, at elevated gas temperatures there appear to be systematic differences in the values of $\alpha(O_2^+)$ for ground state ions (triple mixture studies) and for ions in both the ground and first excited state (binary mixture studies).

In the second microwave afterglow study employing mass analysis, Mehr and Biondi (1969) obtained a value $\alpha(O_2^+) = (1.95 \pm 0.2) \times 10^{-7}$ cm^3/sec at 300°K from electron density decay data which accurately followed the predicted forms obtained by computer solution of the continuity equation. In addition, the O_2^+ wall current tracked the volume electron density decay. Here also the apparatus was operated in the "single pulse-afterglow" mode. These studies were carried out in binary mixtures (6×10^{-3} Torr oxygen and 10 Torr neon) and therefore apply to O_2^+ ions which may be in either the (a $^4\Pi_u$) excited or the (X $^2\Pi_g$) ground state.

Recent afterglow studies using Langmuir probe techniques to measure the electron density at the center of the plasma (Smith and Goodall, 1968) have been carried out in helium–oxygen mixtures. Although earlier afterglow ion–molecule reaction studies by this group had included mass analysis of the reacting ions, the recombination studies did not. Instead the experimental operating conditions were chosen to duplicate those in which O_2^+ was the dominant ion in the ion–molecule studies. From $1/n_e$ versus time curves with very large linear ranges ($f > 50$), Smith and Goodall obtained the value $\alpha(O_2^+) = (2.1 \pm 0.3) \times 10^{-7}$ cm^3/sec at a gas temperature of 295°K (and an electron temperature somewhat higher), in good agreement with the results of the afterglow recombination studies employing mass analysis. Since it is thought that the O_2^+ ions are formed by the $O^+ + O_2$ charge-transfer–atom-rearrangement reaction, the O^+ having first been formed in dissociative charge transfer of He^+ on O_2, it is not clear in what electronic or vibrational state the O_2^+ ions exist in these studies.

Mentzoni (1965) carried out microwave afterglow studies in pure oxygen at ~ 2 Torr without mass analysis of the ions and found a recombination coefficient, $\alpha \simeq 2 \times 10^{-7}$ cm^3/sec at 300°K, suggesting that O_2^+ remains the dominant afterglow ion at rather high concentrations of oxygen.

In the case of the aeronomically important O_2^+ ion, it is of considerable interest to determine the partial recombination coefficients for production of excited oxygen atoms in the ^1S ($\lambda 5577$ radiating) and in the ^1D ($\lambda 6300$ radiating) states. This is a rather formidable task, since both "radiating" states are metastable, with lifetimes of ~ 1 and ~ 100 sec respectively. By use of optical spectrographic techniques employing pulse counting and a multichannel analyzer for signal storage and averaging, Zipf (1967) has measured the absolute intensities of $\lambda 5577$ and $\lambda 6300$ radiation coming from a microwave afterglow apparatus. By relating the decay of intensity to the measured decay of electron density, he obtains values for the partial recombination coefficients for production of given excited states; namely, $\alpha(^1S) = 2.1 \times 10^{-8}$ cm^3/sec and $\alpha(^1D) = 1.9 \times 10^{-7}$ cm^3/sec at 300°K.

Finally, as in the case of nitrogen, it has been possible to determine the recombination coefficient for the dimer ion O_4^+ (or $O_2 \cdot O_2^+$). At low temperatures ($\sim 200°$K) substantial amounts of the ion are formed, even at rather small oxygen concentrations; thus, Kasner and Biondi (1968) were able to determine an effective recombination coefficient for O_2^+ and O_4^+ ion mixtures and from the variation of α_{eff} with the O_4^+–O_2^+ concentration ratio obtained a value $\alpha(O_4^+) \simeq 2.3 \times 10^{-6}$ cm^3/sec at 205°K.

c. Nitric Oxide. As a result of the aeronomical importance of the NO^+ ion, there have been numerous attempts to determine accurately its dissociative recombination coefficient. Gunton and Shaw (1965) carried out one of the first definitive determinations using a microwave afterglow apparatus with a differentially pumped mass spectrometer (which, however, could only be operated at low gas pressures where recombination was not the controlling electron loss process). Photoionization of NO gas (1–6 mTorr pressure) contained in larger amounts (pressures to ~ 130 Torr) of neon acting as a buffer gas created the plasma, which then decayed principally as a result of electron–ion recombination. As in the oxygen studies, the effect of negative ion accumulation on the plasma decay was avoided by obtaining electron density decay data from a single pulse and afterglow. Since at low neon buffer gas pressures the only *significant* afterglow ion detected by the mass spectrometer was NO^+, Gunton and Shaw assumed NO^+ would remain the principal ion even at neon pressures in excess of 100 Torr. As we shall see later, this assumption may be valid at 300°K but fails as a result of significant dimer ion ($NO \cdot NO^+$) formation at lower temperatures ($\sim 200°$K). From $1/\bar{n}_{\mu w}$ versus time data which exhibited moderate linear ranges (f values from 3 to 10) a value $\alpha(NO^+) = (4.6^{+0.5}_{-1.3}) \times 10^{-7}$ cm^3/sec was obtained at 298°K.

A second microwave-afterglow study of NO^+ recombination, using a differentially pumped quadrupole mass spectrometer operable at the pressures where recombination controlled the electron and ion loss, was carried out by Weller and Biondi (1968). Lyman α radiation from a pulsed hydrogen flash lamp photoionized NO gas (~ 10 mTorr) in neon buffer gas (3–7 Torr). The measured $1/\bar{n}_{\mu w}$ curves followed the predicted forms obtained by computer solution of Eq. (6) and the wall current of NO^+ ions (the only significant afterglow ion) accurately tracked the volume electron concentration decay. Weller and Biondi obtained the value $\alpha(NO^+) = (4.1^{+0.3}_{-0.2}) \times 10^{-7}$ cm^3/sec at 300°K, in good agreement with the results of Gunton and Shaw.

An interesting determination of $\alpha(NO^+)$ from studies where, although no mass analysis of the ions was employed, the mode of ion generation suggests that the ions under study were indeed NO^+, was carried out by Young and St. John (1966). Chemionization in mixtures containing N and O atoms was used to produce the ($NO^+ + e$) plasma. A parallel plate ionization chamber using pulsed charged collection was employed to determine the equilibrium value of the plasma density (in which case production by chemionization is balanced by dissociative recombination loss) and to determine the rate of approach to equilibrium following application of an ion sweepout pulse. From a knowledge of the ion drift velocity (mobility) the currents collected by the plates can be related to the NO^+ concentration in the volume, provided that at each instant the form of the spatial distribution of the ions in the volume is known. On the basis of the rather artificial assumption of a uniform spatial distribution of the ions, Young and St. John obtained the value $\alpha(NO^+) = (5 \pm 2) \times 10^{-7}$ cm^3/sec at 300°K. As in the microwave afterglow studies, this value should be corrected (revised downward) to account for the effects of diffusion on the spatial distribution of the plasma; however, the uncorrected value is in satisfactory agreement with the more accurately determined values from photoionized plasma-afterglow studies.

As in the cases of nitrogen and oxygen, in nitric oxide it has been possible to determine the recombination coefficient of the dimer ion ($NO \cdot NO^+$). By operating at higher than normal pressures of NO (~ 100 mTorr) and of neon (~ 15 Torr), Weller and Biondi (1968) were able to enhance the concentration of the dimer ion until it was the principal afterglow ion at 300°K. From a study of the variation of the effective recombination coefficient with the relative concentrations of ($NO \cdot NO^+$) and NO^+ they determined a value $\alpha[(NO \cdot NO^+)] \simeq 1.7 \times 10^{-6}$ cm^3/sec at 300°K. Thus, in all three cases— nitrogen, oxygen, and nitric oxide—the dimer ions exhibit very large recombination coefficients, from 1 to 2×10^{-6} cm^3/sec, while the monomer ions' values are in the range 2–4 $\times 10^{-7}$ cm^3/sec at 300°K (see the discussion in Section III, E).

d. Carbon Dioxide. As a result of analyses of the atmospheres of Mars and Venus, which showed CO_2 to be a principal constituent in both cases, there has been considerable interest in the rate of dissociative recombination of CO_2^+ ions with electrons for use in model calculations of the ionospheres of these planets. Weller and Biondi (1967), therefore, modified their microwave-afterglow–mass spectrometer apparatus by replacing the photoionizing source with a pulsed microwave source for plasma generation in CO_2–neon mixtures. Penning ionization of CO_2 by neon metastables should produce the CO_2^+ ions in their ground electronic ($\tilde{X}\ ^2\Pi_g$) state and vibrational quenching should reduce them to their ground vibrational state before recombination occurs. In mixtures of 0.25 mTorr CO_2 in 10 Torr of neon, Weller and Biondi obtained $1/\bar{n}_{\mu w}$ versus time curves which were linear over a factor $f \gtrsim 10$, and from computer fits of Eq. (6) to the data deduced the value $\alpha(CO_2^+) = (3.8 \pm 0.5) \times 10^{-7}$ cm^3/sec at 300°K. The CO_2^+ wall current tracked the electron decay in the early afterglow, but in later times ($t > 3$ msec) appreciable currents of O_2^+ ions appeared and the CO_2^+ wall current decayed somewhat faster than the volume electron concentration. The effect of the O_2^+ ions was taken into account in the determination of $\alpha(CO_2^+)$. Measurements at lower temperatures encountered some difficulties in cutoff of the ion wall current signals early in the afterglow, but a comparable value, $\alpha \sim 4 \times 10^{-7}$ cm^3/sec, was obtained at 210°K.

   e. Hydrogen. Inasmuch as H_2^+ represents the ion which offers the best possibility for *ab initio* calculations of the dissociative recombination process, experimental determinations of $\alpha(H_2^+)$ would be of great value in providing a test of such calculations. Unfortunately, measurements in hydrogen afterglows carried out to date almost certainly refer to one or another of the ions H^+, H_3^+, and H_5^+. Following the original, rather unsatisfactory (small f value), determinations of electron loss in hydrogen (Biondi and Brown, 1949b; Richardson and Holt, 1951) which led to values $\alpha \gtrsim 2 \times 10^{-6}$ cm^3/sec at 300°K, Varnerin (1951) obtained a recombination coefficient that varied from a value of 3×10^{-7} cm^3/sec at low hydrogen pressures (1–3 Torr) to a value of $\sim 2.4 \times 10^{-6}$ cm^3/sec at pressures exceeding ~ 30 Torr. The electron density decays were determined from measurements of the complex conductivity of the plasma enclosed in a shorted waveguide, and the $1/\bar{n}_{\mu w}$ versus time curves were linear for electron densities ranging between 2×10^{10} and 1×10^9 cm^{-3}. The changing recombination coefficient with pressure probably reflected the changing concentrations of H_3^+ and H_5^+ as principal afterglow ions. Mass spectrometer studies of hydrogen glow discharges (Yamane, 1968) have shown that at low energies the fast reaction $H_2^+ + H_2 \rightarrow H_3^+ + H$ quickly removes H_2^+ ions produced in a primary ionization process and at higher pressures the 3-body reaction $H_3^+ + 2H_2 \rightarrow H_5^+ + H_2$ causes conversion of the H_3^+ ions (Saporoschenko, 1965a).

   Subsequently, Persson and Brown (1955) used intense microwave discharges in a microwave afterglow study of hydrogen and found that higher mode ambipolar diffusion loss accounted fully for the observed electron concentration decays. The measured ambipolar diffusion coefficient yielded a reduced ion mobility, $\mu_0 = 16.6$ cm^2/V sec which is in excellent agreement with the measured mobility of mass-identified H$^+$ ions (Albritton et al., 1968; Saporoschenko, 1965b). Thus, in these studies H$^+$ was evidently the ion produced by the intense discharge, and therefore dissociative recombination loss was excluded.

In any event, it appears that none of the studies to date have achieved conditions where H$_2^+$ is the principal afterglow ion. Studies, are therefore, contemplated in which H$_2^+$ generation by Penning ionization of H$_2$ by helium or neon metastables will be used in a microwave afterglow–mass spectrometer apparatus to provide determinations of the desired $\alpha(\mathrm{H}_2^+)$ values.

D. Recombination Measurements as a Function of Electron Temperature

Considerable insight into the mechanism of the recombination process is afforded by studies of its dependence on electron energy. In addition, ionospheric conditions involve electron temperatures elevated above the ion and neutral temperatures; therefore duplication of these conditions in laboratory studies is desirable. As noted in Section II, A, independent control of the electron energy can be attained either by use of microwave heating in the afterglow experiments or by varying the relative velocity between electron and ion beams in the merged beam experiments. The results of microwave electron heating experiments are summarized in Figs. 7–11, while the merged beam data are given in Fig. 9.

1. Noble Gases

Two studies of neon using microwave heating of the electrons in microwave afterglow apparatus have been carried out; the first (Frommhold et al., 1968) made use of a nonresonant waveguide heating mode to permit precise calculations of the electron temperature from measurements of the microwave heating field intensity. Inasmuch as this apparatus did not contain a mass spectrometer for ion identification, a second apparatus (Philbrick et al., 1969) employing a low-Q, resonant cavity heating mode and a quadrupole mass spectrometer was employed in similar studies. In both cases the afterglow electron density decays in pure neon (6–20 Torr pressure) accurately followed the predicted forms obtained by computer solution of Eq. (6)—see Fig. 5. In addition, in the apparatus with the mass spectrometer, Ne$_2^+$ was the only significant afterglow ion, and its wall current tracked the volume electron concentration decay very well at $T_e = 300°K$, though less perfectly at high electron temperatures ($T_e > 2000°K$).

The results of these two studies, extending over the range $300°K \leqslant T_e \leqslant 11,000°K$ (with $T_i = T_{gas} = 300°K$), are indicated by the solid line in Fig. 7.

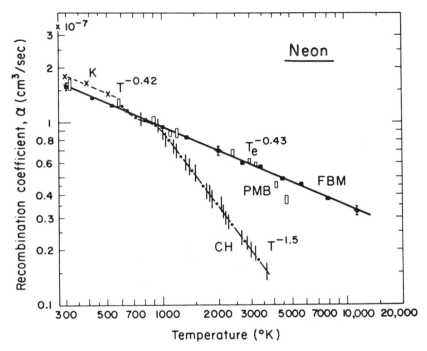

FIG. 7. Measured variation of the recombination coefficient, $\alpha(Ne_2^+)$, with temperature. The symbol T refers to variations when $T_e = T_i = T_{gas}$, while the symbol T_e refers to variations of T_e with $T_i = T_{gas} = 300°K$. The results of Kasner (*K*), Frommhold, Biondi, and Mehr (*FBM*), Philbrick, Mehr, and Biondi (*PMB*), and Cunningham and Hobson (*CH*) are shown.

It will be seen that $\alpha(Ne_2^+)$ varies approximately as $T_e^{-0.43}$, which is rather close to the theoretically expected variation, $T_e^{-1/2}$, for the direct dissociative process (see Section III for a discussion).

A similar study of the electron temperature dependence of $\alpha(Ar_2^+)$ was carried out by Mehr and Biondi (1968) using the apparatus of Frommhold *et al.* (1968). At argon pressures of ~ 20 Torr, the afterglow was recombination-controlled even at the elevated electron temperatures, and the fits of $1/\bar{n}_{\mu w}$ versus time data to computer solutions of the continuity equation yielded the values of $\alpha(Ar_2^+)$ as a function of electron temperature shown by the solid line in Fig. 8. It will be seen that a simple variation as $T_e^{-0.67}$ adequately describes the observations. This variation is somewhat more rapid than the

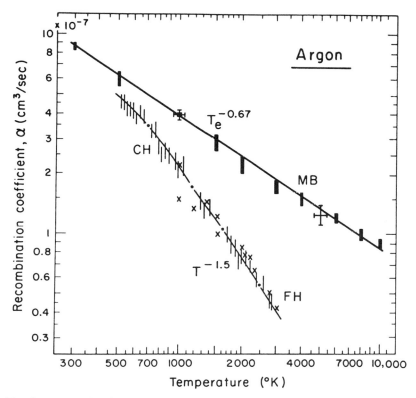

FIG. 8. Measured variation of the recombination coefficient, $\alpha(\text{Ar}_2^+)$, with temperature. The results of Mehr and Biondi (*MB*), Cunningham and Hobson (*CH*), and Fox and Hobson (*FH*) are shown.

$T_e^{-1/2}$ dependence suggested by simplified theories; possible reasons for this difference are discussed in Section III.

2. Atmospheric Gases

In addition to the determinations for the noble gas ions Ne_2^+ and Ar_2^+, the variation of α with T_e for mass-identified N_2^+ and O_2^+ ions has been studied using microwave heating-afterglow techniques. Also, merged beam experiments have provided preliminary determinations of the variation of $\alpha(\text{N}_2^+)$ with relative velocity between electron and ion.

The variation of $\alpha(\text{N}_2^+)$ with T_e has been studied by Mehr and Biondi (1969) in a microwave-afterglow–mass spectrometer apparatus employing a microwave heating mode to raise the electron temperature above that of the ions and gas molecules, which remained at the ambient temperature (300°K).

In nitrogen–neon mixtures of the same compositions which had permitted Kasner and Biondi (1965) and Kasner (1967) to study the recombination of N_2^+ ions without interference from other ions, Mehr and Biondi were unable to achieve conditions which completely eliminated N_3^+ and N_4^+ as afterglow ions, although they were reduced to minority (<5%) levels. As a result, tracking of the volume electron decay by the N_2^+ wall current was imperfect and a rather larger-than-normal error limit (+20%, −10%) was assigned to the $\alpha(N_2^+)$ values obtained from the analysis of the $1/\bar{n}_{\mu w}$ versus time curves (which fit the computer solutions of the continuity equation very well). The results of Mehr and Biondi's determinations of $\alpha(N_2^+)$ over the range $300°K \leqslant T_e \leqslant 5000°K$ are shown by the solid line in Fig. 9. It will be seen that a variation according to the simple power law $T_e^{-0.39}$ provides a satisfactory fit to the data.

In an earlier study using Langmuir probes to measure the electron concentration and electron temperature at the center of a plasma-afterglow and an rf mass spectrometer to sample the ions, Sayers (1956) had obtained a prelim-

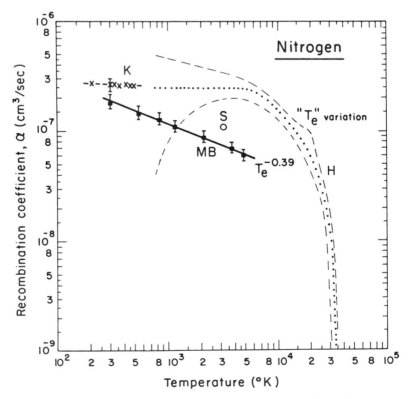

FIG. 9. Measured variation of the recombination coefficient, $\alpha(N_2^+)$, with temperature. The results of Kasner (K), Mehr and Biondi (MB), Hagen (H), and Sayers (S) are shown.

inary value $\alpha(N_2^+) = 1.1 \times 10^{-7}$ cm^3/sec at $T_e = 3200°$K in nitrogen–helium mixture studies. However, no details of the measurement are given and no further report of this work has been forthcoming; thus it is difficult to assess the accuracy of the determination.

Although results of merged beam studies are still in their preliminary stages, the very large electron energy ranges covered in the measurements makes the findings of interest. Hagen (1968) found that, as the relative velocity between electron and N_2^+ ion beams was varied, $\alpha(N_2^+)$ started at a value $(2.5^{+2.5}_{-2.0}) \times 10^{-7}$ cm^3/sec at 0.1 eV relative energy and varied little up to ~ 0.7 eV, then decreased more and more rapidly as energy was increased to ~ 5 eV.

Although the distribution of relative velocities between electrons and ions is hardly Maxwellian, we can place these results in Fig. 9 by defining an effective electron temperature $(T_e)_{\text{eff}}$ such that $(kT_e)_{\text{eff}} = \frac{2}{3}E_{\text{rel}}$, where E_{rel} is the (mean) relative energy between electrons and ions. In view of the extremely large uncertainty (a factor of 10) in the absolute values of $\alpha(N_2^+)$ Hagen's results, shown by the dotted line in Fig. 9, are of more interest for the indicated *rapid* variation of the relative values with energy.

In a similar vein, very preliminary determinations of the *relative* values of $\alpha(N_2^+)$ as a function of the energy between the electron and ion beams have been reported by Theard (1969). The results are uncertain because of difficulties in determining the exact interaction path and velocities of the electrons as they spiral through the ion beam; however, Theard finds that the relative values of $\alpha(N_2^+)$ decrease approximately as E_{rel}^{-1} between energies of 0.3 and 1 eV and then more rapidly (faster than E_{rel}^{-2}) as the relative energy approaches 3 eV. Thus, a rather rapid variation of α with electron energy at higher energies is also indicated by this merging beam experiment.

The advantage offered by the merging beam techniques of recombination coefficient determinations over a wide energy range is presently offset by difficulties in determining the exact path length and charged particle densities along the interaction track and in determining the distribution of relative velocities at a given relative energy. Further, the state (electronic and vibrational) of the N_2^+ ions emerging from the ion source is not well controlled (or specified); thus considerable additional work is required to make the results of merged beam studies quantitative.

Returning to the microwave-afterglow measurements, Mehr and Biondi (1969) also determined the variation of $\alpha(O_2^+)$ with T_e over the range 300°K $\leqslant T_e \leqslant 5000°$K in binary (O$_2$–Ne) mixtures; thus, as noted earlier, the O$_2^+$ ions may be in either the excited or the ground electronic state. In these studies, the O$_2^+$ ion wall currents tracked the volume electron decay, so the determinations of $\alpha(O_2^+)$ should be highly accurate. The results are shown by the solid line in Fig. 10, which represents a simple power law variation as $T_e^{-0.70}$ to 1200°K, then as $T_e^{-0.56}$. Here, as in the cases of Ne$_2^+$, Ar$_2^+$, and N_2^+, the

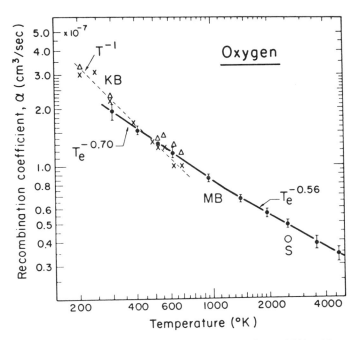

FIG. 10. Measured variation of the recombination coefficient, $\alpha(O_2^+)$, with temperature. The results of Kasner and Biondi (*KB*), Mehr and Biondi (*MB*), and Sayers (*S*) are shown.

observed variation of $\alpha(O_2^+)$ with T_e is close to that predicted by the simple theory of the direct dissociative process.

Sayers (1956) reported a preliminary value $\alpha(O_2^+) = 4 \times 10^{-8}$ at $T_e = 2500°K$ from studies in oxygen–helium mixtures using a Langmuir probe–mass spectrometer apparatus, in satisfactory agreement with Mehr and Biondi's results.

E. RECOMBINATION MEASUREMENTS AT ELEVATED ION TEMPERATURES

In the dissociative recombination process represented schematically by the potential curves in Fig. 1, it will be seen that the overlap between the initial state of the system ($AB^+ + e$) and the dissociating state AB'' depends strongly on the vibrational state of the ion. Thus, in addition to the dependence on electron energy just discussed, the recombination rate should depend on the vibrational temperature of the ions.

This feature of the dissociative process has been investigated in two ways; (a) by varying the temperature of the gas sample under study by changing the temperature of the walls surrounding the gas, in which case, under appropriate afterglow conditions, $T_e = T_i = T_{gas} = T_{wall}$ and (b) by shock heating a sample of ambient (300°K) ionized gas and observing the plasma decay while the heated sample is in the relatively "quiet" region between the shock front and the contact surface with the driver gas. In the former case (thermally controlled environment) it is reasonable to expect that the population of the various vibrational levels of the ions is given by a Boltzmann distribution characterized by a vibrational temperature T_v, such that $T_v = T_i$, where T_i is the translational temperature of the ions. In the latter case (shock-heated gas), although T_i quickly follows T_{gas}, there is some question as to whether T_v (i.e., the vibrational state distribution of the ions) has sufficient time to follow. In the thermally controlled environment experiments the range of T_e and T_i values has been limited to $\sim 200°$ to $\sim 700°K$; a maximum of $\sim 1000°K$ is at present the upper limit one can hope to attain by this method. The shock-heating studies have extended from a lower limit of $\sim 550°K$ to a maximum of $\sim 3500°K$.

1. Noble Gases

Kasner (1968) varied the gas temperature over the range 295°–503°K and found a variation of $\alpha(Ne_2^+)$ as $T^{-(0.42 \pm 0.4)}$ (dashed curve in Fig. 7). The wall currents of Ne_2^+ ions tracked the volume electron decays and the $1/\bar{n}_{\mu w}$ versus time data accurately followed the solutions of Eq. (6), leading to high confidence in the experimental values of α. It is interesting to note that this variation, when $T_e = T_i = T_{gas}$, is essentially identical to that obtained by Frommhold et al. (1968), when T_e is varied and T_i and T_{gas} remain at 300°K. Such a behavior is explicable if the Ne_2^+ vibrational spacings are sufficient that the ions essentially remain in the $v = 0$ state up to temperatures of 500°K, and therefore the decrease in $\alpha(Ne_2^+)$ is due solely to changes in the electron energy.

The elevated ion temperature studies of neon have been extended from 580° to 3500°K in measurements in shock-heated neon (Cunningham and Hobson, 1969a). A series of double probes was used to determine the decay of ion concentration as the shock-heated, ionized neon sample flowed down the tube in the region between the shock front and the contact surface. Only limited linear $1/n_i$ decay ranges were attained ($f \lesssim 5$), and there is some question as to the accuracy of the absolute ion density values obtained from double probe measurements; thus, the absolute values of $\alpha(Ne_2^+)$ may contain very substantial uncertainties. The results of Cunningham and Hobson are shown (together with their estimates of errors) by the dot-dashed curve in Fig. 7. It will be seen that their values join satisfactorily to those of Kasner and at first follow the variation with T_e noted by Frommhold et al., then fall away more rapidly above $\sim 700°K$. Such a behavior is qualitatively to be expected if the recombination coefficient decreases with increasing vibrational quantum number (as for a favorable curve crossing case of the type illustrated in Fig. 1). However, as will be seen in Section III, C, the extreme variation observed would require that the recombination coefficient be *zero* for ions in states with $v > 0$ (O'Malley, 1969a).

Chen (1969) has carried out an afterglow experiment employing a microwave interferometer for electron density determinations, a time-of-flight mass spectrometer for ion identification, a triple probe for electron temperature determinations, and a thin-film thermometer for gas temperature measurements. He analyzes the electron decays following an intense discharge created by a capacitor bank (no examples of the data are given,) extrapolates to $T_e = 300°K$ and finds the variation of α due to changes of T_{gas} alone (and presumably, therefore, of the vibrational states of the ions). Over the range $420°K \leq T_{gas} \leq 1500°K$ in neon, he infers *no* variation of $\alpha(Ne_2^+)$, in contradiction to the previously cited results, where above $\sim 900°K$ the effect of changing vibrational populations apparently causes a more rapid decrease in α than results from electron temperature effects alone. Inasmuch as only the end results of Chen's analyses (i.e., α vs. T_e and α vs. T_{gas}) are presented, it is not possible to assess the source of the discrepancy.

In the case of argon, Cunningham and Hobson (1969a) have carried out measurements in their shock-tube apparatus over the temperature range $530°$–$3000°K$. The same questions concerning the accuracy of the inferred values that were raised for the case of neon apply to the determinations of $\alpha(Ar_2^+)$, so that there may be more substantial errors in the absolute values than indicated by the authors. Their results are shown by the dot-dashed curve in Fig. 8, together with the results of similar, earlier studies by Fox and Hobson (1966). As in the case of neon the low temperature ($\sim 600°K$) variation of $\alpha(Ar_2^+)$ roughly parallels the variation with T_e noted by Mehr and Biondi, then decreases more rapidly at higher temperatures. Again the behavior is qualitatively in accord with excitation of some of the ions into vibrational states with $v > 0$, where the recombination coefficient is smaller than for the $v = 0$ state (see discussion in Section III).

The analysis by Chen (1969) of his argon data leads him to the diametrically opposite conclusion, namely, that at low electron temperatures the recombination of vibrationally excited Ar_2^+ ions is *more* rapid than of ions in the $v = 0$ state. Again insufficient details of the measurements are given to determine the reason for the discrepancy.

2. Atmospheric Gases

In the case of $\alpha(N_2^+)$ the only elevated ion temperature study involving mass-identified ions is that of Kasner (1967), who made accurate determinations over the temperature range $205°$–$480°K$. Surprisingly, he found essentially *no* variation of $\alpha(N_2^+)$ over this range (see dashed line in Fig. 9). In view of the excellent N_2^+ ion tracking of the electron decays and the substantial linear ranges ($f > 10$) of the $1/\bar{n}_{\mu w}$ versus time data, the deduced values of $\alpha(N_2^+)$ should be highly accurate.

Kasner's result cannot be reconciled with that of Mehr and Biondi (T_e variation), since the substantial vibrational spacing of the N_2^+ ion does not admit significant thermal excitation above $v = 0$ over the range 205°–480°K. Consequently, both studies should have yielded the same temperature dependence over the common temperature range (as, indeed, was the case in neon). The only resolution of the contradiction appears to be to make full use of the quoted experimental uncertainties ($\pm 10\%$ in Kasner's results, -10 to $+20\%$ in Mehr and Biondi's) and to conclude that the low temperature ($< 500°K$) variation of $\alpha(N_2^+)$ is actually intermediate between the extremes represented by the two studies.

The dependence of $\alpha(O_2^+)$ on ion temperature has been determined for mass-identified ions by Kasner and Biondi (1968). They obtained two slightly different variations of $\alpha(O_2^+)$ over the temperature range 205°–690°K for ions which were confined to their ground electronic state ($X\ ^2\Pi_g$) (x symbols in Fig. 10) and for ions in a mixture of ground and first excited (a $^4\Pi_u$) states (open triangles in Fig. 10). The $1/\bar{n}_{\mu w}$ data and O_2^+ wall current tracking of the electron decays were sufficiently accurate that the slight differences in the $\alpha(O_2^+)$ values for the two cases are probably significant.

It is interesting to note that Kasner and Biondi's results for the binary O_2–Ne mixtures (open triangles—mixed ion states) agree satisfactorily with Mehr and Biondi's observed variation of $\alpha(O_2^+)$ with T_e in similar binary mixtures over the common temperature range (300°–690°K). Thus, when the temperature range is insufficient to cause significant depopulation of the $v = 0$ state of the ion, the variation of $\alpha(O_2^+)$ is the same whether T_e and T_i are covaried or T_e alone is varied.

Afterglow studies by Smith and Goodall (1968) in oxygen–helium mixtures and by Mentzoni (1965) in pure oxygen (neither of which provided for mass identification of the ions under study) suggest a somewhat slower decrease in the recombination coefficient (approximately as $T^{-0.4}$) from 300°–900°K. However, problems with changing ion composition became apparent in Smith and Goodall's studies at low temperatures (180°K), where an anomalously large and variable value of α strongly suggests the presence of appreciable (and variable) concentrations of the dimer ion ($O_2 \cdot O_2^+$) as noted in the low temperature studies of Kasner and Biondi. Thus, we regard the studies without mass analysis as inherently less reliable in determining the energy variations of the recombination coefficients.

The variation of $\alpha(NO^+)$ with gas temperature has been determined quantitatively up to moderate temperatures ($\sim 450°K$) but has only been estimated at high temperatures (up to $\sim 5000°$K). The results of microwave afterglow studies of photoionized NO plasmas by Gunton and Shaw (1965) and by Weller and Biondi (1968) over the temperature range $200°$–$450°$K are shown by the solid line in Fig. 11. (Gunton and Shaw's low temperature point is

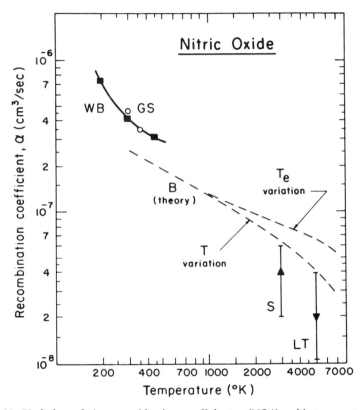

FIG. 11. Variation of the recombination coefficient, $\alpha(NO^+)$, with temperature. The results of measurements of Gunton and Shaw (*GS*), open circles; Weller and Biondi (*WB*), solid squares; Stein *et al.* (*S*); and Lin and Tease (*LT*) are shown, together with theoretical calculations of Bardsley (*B*).

omitted because of dimer ion effects.) At higher temperatures there are only microwave measurements of electron density decays in shock-heated air (Stein *et al.*, 1964), where it is presumed that NO^+ is the principal afterglow ion. A value $\alpha \sim (4 \pm 2) \times 10^{-8}$ cm^3/sec is obtained at $T \sim 2900°$K from the studies. Finally, from a complicated analysis of the ionization levels attained behind shock waves in air, Lin and Tease (1963) conclude that recombination loss with a coefficient of the order of 2×10^{-8} cm^3/sec (presumably associated with NO^+ ions) must be present at $5000°$K. It will be seen that these high temperature values fall considerably below Bardsley's theoretical curve indicated by the dashed line in Fig. 11. As discussed in Section III, inasmuch as Bardsley's calculations apply to only two of the four states involved in the direct dissociative process, his values represent a lower bound on the total coefficient $\alpha(NO^+)$, suggesting that the coefficients inferred from the high temperature measurements may be seriously in error.

III. Theory

A. GENERAL DISCUSSION

Following the model discussed in the Introduction we consider dissociative recombination as the formation and dissociation of an intermediate state AB″. Of the many electronic states of AB only a few can be important in the reaction. For the recombination of thermal electrons it is necessary that the potential energy curve for AB″ come close to that of the ion AB^+ and preferably cross it within the range of nuclear vibrations. During most of our discussion we will assume that the potential curves of AB^+ and AB″ do cross (as in Fig. 1). States belonging to Rydberg series which converge to the first ionization limit do not exhibit this crossing property. Suitable states AB″ can be formed by raising two electrons from the ground state orbitals, or by exciting an inner electron.

An intermediate state of this type is formed by the transfer of energy from the incident electron to one of the target electrons. It will be shown later that the formation is most likely to occur when the nuclear separation is close to the value R_c, at which the difference between the potential curves for AB″ and AB^+ is equal to the incident energy ε. At this point electron capture can occur without any simultaneous transfer of energy into nuclear motion. After the formation of the state AB″, two things may happen. The repulsive force between the nuclei causes them to move apart and they may dissociate, trapping the extra electron. The recombination process is then complete. Alternatively an electron may be emitted through autoionization, the reverse of the process by which the electron was captured. The probability of autoionization rapidly decreases once the nuclei are further apart than the separation R_s at which the two potential curves cross. For this reason the crossover point is often also called the stabilization point.

In calculating the contribution of an intermediate state AB″ to the recombination cross section it is often helpful to write

$$\sigma(\varepsilon) = \sigma_{cap}(\varepsilon)S(\varepsilon). \tag{14}$$

Here $\sigma_{cap}(\varepsilon)$ represents the cross section for formation of the state AB″, and the survival factor $S(\varepsilon)$ is the probability that this state will decay by dissociation rather than by autoionization. The separation of the cross section into two independent factors describing the formation and dissociation of the intermediate state cannot be fully justified (Bardsley, 1968, 1970) but is extremely useful for a qualitative analysis of the recombination process.

We will assume that the intermediate state is formed by configuration interaction, with a target electron being excited by the incident electron which drops into an unoccupied molecular orbital. If we denote the electronic matrix element governing the configuration interaction by $V(R)$, we can express the capture cross section as

$$\sigma_{\rm cap}(\varepsilon) = (2\pi^3/\hbar m \varepsilon)(r/2)|\langle \zeta_{\rm AB+}(R)|V(R)|\zeta_{\rm AB''}(R)\rangle|^2, \qquad (15)$$

with

$$V(R) = \langle \psi_{\rm AB+}(r, R)\phi_\varepsilon(r, R)|{\rm H}_{\rm el}(r, R)|\psi_{\rm AB''}(r, R)\rangle. \qquad (16)$$

Here we use $\psi(r, R)$ to denote electronic wave functions for the molecule and $\zeta(R)$ to represent nuclear functions. The wave function describing the incident electron is $\phi_\varepsilon(r, R)$. The factor r in Eq. (15) represents the ratio of the multiplicities of the intermediate state AB″ to the initial state AB⁺. It is divided by two to allow for the two possible orientations of the spin of the incident electron.

The alternative mechanism for electron capture is through the breakdown of the Born–Oppenheimer approximation, that is, through the transfer of energy from electronic to nuclear motion. Discussion of this latter interaction will be postponed until Section III, E.

Equation (15) for the capture cross section can be simplified by taking note of the forms of the nuclear wave functions $\zeta_{\rm AB+}(R)$ and $\zeta_{\rm AB''}(R)$. The vibrational motion of the initial ion AB⁺ can be represented by a harmonic oscillator wave function as shown in Fig. 12. The form of $\zeta_{\rm AB''}(R)$ is also illustrated in this figure. The classical turning point, at which the internuclear potential $U_{\rm AB''}(R)$ is equal to the total energy E, is denoted by R_E. $\zeta_{\rm AB''}(R)$ is largest near R_E, oscillates rapidly for $R > R_E$, and decreases exponentially as R is decreased from R_E. The wavelength of the oscillations depends on the magnitude of the slope of the potential energy curve $U_{\rm AB''}(R)$. For the atmospheric gases this slope is of the order of 0.5 a.u. (25 eV/Å). With such values it is a good approximation to replace $\zeta_{\rm AB''}(R)$ in Eq. (15) by a delta function, multiplied by a constant to achieve the correct normalization

$$\zeta_{\rm AB''}(R) = (1/U')^{1/2}\,\delta(R - R_c). \qquad (17)$$

Here U' is the slope of the potential curve of AB″ at the singularity. The position of the singularity must be chosen to be close to the classical turning

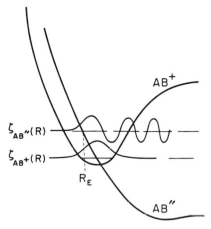

Fig. 12. Hypothetical potential energy curves and associated wave functions for one of the states of the molecular ion AB^+ and of the unstable molecule AB'' involved in the *direct* dissociative recombination process.

point. Following the concept of the R centroid used by spectroscopists we choose R_c as the point where

$$U_{AB''}(R_c) - U_{AB^+}(R_c) = \varepsilon. \qquad (18)$$

With this approximation for $\zeta_{AB''}(R)$ the capture cross section becomes

$$\sigma_{cap}(\varepsilon) = (2\pi^3/\hbar m\varepsilon)(r/2)(1/U')|V(R_c)|^2|\zeta_{AB^+}(R_c)|^2. \qquad (19)$$

This equation is often modified by introducing the capture width

$$\Gamma_c = \Gamma(R_c) = (2\pi/\hbar)|V(R_c)|^2, \qquad (20)$$

so that

$$\sigma_{cap}(\varepsilon) = (\pi^2/m\varepsilon)(r/2)(\Gamma_c/U')|\zeta_{AB^+}(R_c)|^2. \qquad (21)$$

This expression demonstrates that the intermediate states that are important are those for which R_c is within the range of the nuclear vibrations in the initial ionic state. For thermal electrons R_c is very close to the stabilization point R_s, justifying our earlier assertion that R_s should lie within the range of the nuclear vibrations. Equation (21) also shows that the determination of the capture cross section involves the calculation of three quantities, the capture width Γ_c, the potential slope U', and the position of the capture point R_c. The methods that have been used in the calculation of these quantities will be described in Sections III, B and D.

The calculation of the survival factor is in general very difficult. In the study of the analogous process of the dissociative attachment of electrons to neutral

molecules for incident energies of a few electron volts, the following expression
has been shown to be extremely useful (Bardsley and Mandl, 1968)

$$S(\varepsilon) = \exp\left\{-\int [\Gamma(R)/\hbar]\, dt\right\}. \tag{22}$$

For each value of R, an autoionization width $\Gamma(R)$ is defined, so that the
probability of electron emission per unit time is $\Gamma(R)/\hbar$. For the recombination
of ions in their ground electronic state the decay width is equal to the capture
width as defined in Eq. (20). The limits on the integration over time in Eq.
(22) are from the formation of the state AB″ at R_c to the stabilization at R_s,
after which autoionization is assumed to be negligible.

The derivation of Eq. (22) is based on two assumptions. The introduction
of time depends on the use of classical concepts of motion for the nuclei.
This seems justifiable because of the relatively large mass of nuclei; however
this assumption can be easily avoided if necessary. A more serious limitation
is the assumption that at each nuclear separation a definite rate of ionization
can be defined. This would be possible only if the autoionization rate is
governed solely by the electronic transition probability. It is also dependent,
however, on the nuclear wave functions. After emission of an electron, the
nuclear wave function must be one of the eigenfunctions $\zeta_{AB^+}(R)$. If the total
energy of the system is such that only one or two final vibrational states are
allowed, then this may cause a significant constraint on the emission of an
electron. The result of such considerations is that the autoionization must be
represented by a nonlocal operator, and the derivation of a simple and
helpful expression for the survival factor does not seem possible. Numerical
techniques must then be used to calculate the effects of autoionization
(Bardsley 1968, 1970).

Although Eq. (22) with a local $\Gamma(R)$ is not strictly applicable to recombination at thermal energies, it is useful in enabling us to estimate the importance
of autoionization. The time taken for the nuclei to move from the capture
point R_c to the stabilization point R_s depends on the total energy and the
potential slope U'. For collisions of thermal electrons (at 300°K) with ground
state atmospheric ions this time will be of order 10^{-15} sec. Typical autoionization rates are of the order of 10^{14} sec^{-1}. Thus we expect the survival
factor to be close to unity, so that the recombination cross section is given
by $\sigma_{cap}(\varepsilon)$. For the rare gases the slope U' is probably smaller, so that the
stabilization time may be larger. Thus we might find survival factors less than
one for these gases

Having calculated the cross section $\sigma(\varepsilon)$, the recombination coefficient α is
obtained by multiplying by the electron velocity and averaging over the
appropriate distribution of incident energies $p(\varepsilon)$.

$$\alpha_v = \int (2\varepsilon/m)^{1/2}\, \sigma_v(\varepsilon)\, p(\varepsilon)\, d\varepsilon. \tag{23}$$

For a Maxwellian distribution of electron velocities at an electron temperature T_e we have

$$\alpha_v(T_e) = (8/\pi mk^3 T_e^3)^{1/2} \int \sigma_v(\varepsilon) \exp(-\varepsilon/kT)\varepsilon \, d\varepsilon. \qquad (24)$$

We have used the subscript v to indicate the vibrational state of the initial ion. The rotational state of the ion is not important, and we will assume that it is in the ground electronic state.

For comparison with experiment we must consider a gas of ions in varying vibrational states. If we assume a Boltzmann distribution of ions appropriate to a temperature T_v, we have

$$\alpha(T_v, T_e) = [\sum_v \alpha_v(T_e) \exp(-E_v/kT_v)]/[\sum_v \exp(-E_v/kT_v)], \qquad (25)$$

in which E_v is the energy of the vibrational level v.

B. Magnitude of the Recombination Rate

A useful formula for estimating the recombination rate for ions in their ground vibrational and electronic state can be gained by making the following four assumptions.

(i) The survival factor is close to unity.

(ii) The incident energy is small, say, much smaller than 1 eV.

(iii) The repulsive potential curve $U_{AB''}(R)$ is sufficiently steep that the capture point R_c is very close to the stabilization point R_s.

(iv) The vibrational wave function for the initial ion can be represented by a harmonic oscillator eigenfunction.

With these assumptions the integral over energy in Eq. (24) can be done, and the recombination coefficient for ground state ions becomes

$$\alpha_0(T_e) = 8.8 \frac{r\Gamma_c}{U'a} \exp\left(-\frac{(R_s - R_0)^2}{a^2}\right)\left(\frac{T_e}{300}\right)^{-0.5} \times 10^{-7} \quad \text{cm}^3 \text{ sec}^{-1},$$

$$(26)$$

when T_e is expressed in °K. Here the amplitude of the initial vibrational motion has been denoted by a, and R_0 is the equilibrium nuclear separation for the ion. For the largest possible recombination rate the crossover R_s must be close to R_0. For the atmospheric gases there are so many electronic states of the neutral molecule that it is likely that several will cross within the range of the ground state vibrations in the ion. However, for H_2 and He_2 it seems that there are no such states. Assuming the exponential to be close to one, the critical factor is the ratio of capture width to the product $(U'a)$. The number of suitable intermediate states and their multiplicities also have considerable influence on the rate.

The molecules in which large recombination rates have been observed are the atmospheric gases and the rare gases other than helium. *Ab initio* calculations of the capture width for these molecules are extremely difficult. However Bardsley (1968) has shown how spectroscopic data can be used to give all the information required for calculating capture cross sections. The potential energy curves AB″ can be found if they have a minimum at large R and so support vibrational states that can be excited by photon absorption. Estimation of the capture width can be accomplished because of the similarity between the wave functions for the scattering of low energy electrons by molecular ions and for the high Rydberg states of the neutral molecule. The matrix element (16) which controls the capture of an electron can be simply related to that which governs the interaction of Rydberg states with non-Rydberg states in the neutral molecule. The perturbations in the molecular spectra caused by such interactions have been observed (Lagerqvist and Miescher, 1958, 1966; Huber, 1964; Jungen, 1966) and analyzed by Felenbok and Lefebvre-Brion (1966). From their results it is possible to estimate the contribution of two states, B $^2\Pi$ and B′ $^2\Delta$, to the recombination coefficient. The relevant parameters deduced from the spectroscopic data are shown in Table I.

TABLE I

THE CONTRIBUTION OF THE STATES B $^2\Pi$ AND B′ $^2\Delta$ TO THE
RECOMBINATION COEFFICIENT FOR NO$^+$

State	r	Γ (eV)	$U'a$ (eV)	$(R_s - R_0)/a$	$\alpha(300°K)$ (cm^3/sec)
B $^2\Pi$	4	0.075	1.36	0.48	1.6×10^{-7}
B′ $^2\Delta$	4	0.038	0.97	0.58	1.0×10^{-7}

In the last column of Table I is given the contribution of the state to the recombination rate as calculated from Eq. (26). The total contribution of 2.6×10^{-7} cm^3 sec^{-1} is to be compared with the experimental value of 4.1×10^{-7} cm^3 sec^{-1}. These calculations show that coefficients of this order can indeed be explained by the model of recombination which we have presented. It is also possible to see how the rate may differ significantly from such values. If there are no states AB″ crossing near the equilibrium separation of the ion, then the recombination rate for ground state ions may be considerably smaller because of the exponential factor in Eq. (26). This is probably the case for H$_2^+$ and He$_2^+$. On the other hand, a larger rate may result from increases in the ratio $\Gamma_c/(U'a)$. This may explain the rates of order 10^{-6} cm^3 sec^{-1} observed in xenon and krypton. If the ionic equilibrium separation is larger than in the atmospheric gases, Γ_c may be increased and U' decreased by factors between 1 and 10. These changes will be offset by a decrease in a, but the overall result may be a much larger recombination coefficient.

Following our general discussion of orders of magnitude we turn to the calculations of the recombination coefficient for H_2^+. Bauer and Wu (1956) calculated the capture width Γ_c for two intermediate states. Unfortunately they used plane waves to represent the incident electron. Both the magnitude of the capture cross section and its dependence on electron temperature are strongly influenced by the long range Coulomb force, and the use of Coulomb wave functions seems essential to reliable calculations. Nevertheless their calculation showed that large capture cross sections were possible.

Coulomb wave functions were used in the calculation of Wilkins (1966). He considered the formation of the state $(2p\sigma_u)(2s\sigma_g)\ ^3\Sigma_u{}^+$. The contribution of this state to the recombination rate at room temperature was estimated to be 3×10^{-8} cm^3/sec. A similar method has been applied by Dubrovsky et al. (1967a, b) for capture into the state $(2p\sigma_u)^2\ ^1\Sigma_g{}^+$. The contribution from this state was estimated to be $4.2 \times 10^{-8}\ (T_e^{-0.5})$ cm^3/sec, which would give a value of about 3×10^{-9} cm^3/sec at 300°K. Unfortunately, as noted in Section II, C, 2, e, to date there appear to be no experimental determinations of $\alpha(H_2^+)$ with which to compare these theoretical predictions.

In all these calculations the potential energy curves used for the intermediate states were calculated by completely neglecting the interelectronic repulsion. It was argued that this led to a consistent application of perturbation theory, since the interelectronic repulsion is responsible for the configuration interaction. Our present understanding of autoionizing states (Bardsley and Mandl, 1968; Taylor, 1969) makes this unnecessary, and better calculations of the potential curves are now feasible (see Section III, D).

Calculations of the recombination rate for several diatomic ions have been reported by Warke (1966) and Chan (1968). We will postpone discussion of their results to the end of Section III, E.

C. Temperature Dependence of the Recombination Rate

The simplifying assumptions listed at the beginning of Section III, B lead to a very simple dependence of the recombination rate on electron temperature; namely,

$$\alpha_v(T_e) = C_v T_e^{-0.5}. \tag{27}$$

Recent experimental evidence in which selective heating of electrons was employed show variations close to this prediction. If the rate is expressed as a power of the electron temperature, the exponent varies between -0.39 (for N_2^+) and -0.67 (for Ar_2^+).

Deviations from the simple law (27) above room temperature can be caused by several factors. The most important is the variation of the capture point R_c with incident energy. This is particularly important for molecules in which there are no favorable curve crossings. For example, in helium where almost all the states of He_2 cross the ionic curve at large R, the recombination cross section is very small for low energy electrons incident on ground state ions. However, electrons with energies of a few electron volts may be able to cause a transition within the Franck–Condon region, so that the recombination rate may increase with increasing temperature. This point can be investigated experimentally either through the microwave heating or the merged beam experiments.

If considerable autoionization of the state AB'' occurs, then this may effect the temperature dependence. As the electron energy is raised, the stabilization time is increased, and so the recombination cross section will be reduced. This may cause a more rapid decrease of the rate with increasing electron temperature.

In NO^+ and O_2^+ some evidence has been found for a stronger dependence of the rate on temperature below 300°K. The presence of dimer ions complicates experiments at low temperatures, but significant deviations from Eq. (27) in this temperature range would suggest the existence of another recombination mechanism, as will be discussed in Section III, E.

The variation of the recombination rate with the ion temperature arises from the dependence of the recombination cross section on the vibrational state of the ion. In several gases it has been found that the effect of heating the electrons selectively differs significantly from that of heating the whole gas. This can be seen most clearly in the case of neon. In afterglow experiments (Frommhold *et al.*, 1968; Philbrick *et al.*, 1969) the electron temperature has been raised to around 10,000°K, and the $T_e^{-0.5}$ law is approximately followed over the whole range. In the shock-tube experiments in which both electrons and ions are heated, a similar behavior is seen below 1000°K, but above this temperature the rate drops off more rapidly with increasing temperature and a $T^{-1.5}$ law fits the results between 1000°K and 3500°K (Cunningham and Hobson, 1969a, b).

Several authors have suggested that the difference between the two experiments is due to a rapid decrease in the recombination cross section with increasing vibrational quantum number of the initial ion. Such a decrease is to be expected if the potential curves for AB'' and AB^+ cross close to the equilibrium separation for AB^+. The shock-tube experiments on neon and argon led O'Malley (1969a, b) to suggest a simple model in which he assumes

$$\alpha_v = 0 \qquad \text{for} \quad v > 0. \tag{28}$$

If we assume the vibrational levels to be those of a harmonic oscillator with spacing $\hbar\omega$, Eq. (28) leads to

$$\alpha(T_v, T_e) = \frac{\alpha_0(T_e)\exp(-E_0/kT_v)}{\sum_v \exp(-E_v/kT_v)}$$

$$= \alpha_0(T_e)[1 - \exp(-\hbar\omega/kT_v)]. \tag{29}$$

In the limit that $kT_v \gg \hbar\omega$, Eq. (29) reduces to

$$\alpha(T_v, T_e) \approx (\hbar\omega/kT_v)\alpha_0(T_e). \tag{30}$$

If we assume a variation of $\alpha_0(T_e)$ as $T_e^{-0.5}$, this shows that the recombination rate for a gas in complete thermal equilibrium should vary as $T^{-0.5}$ at low temperatures $(kT_v \ll \hbar\omega)$ and as $T_e^{-0.5}T_v^{-1.0}$ at high temperatures $(kT_v \gg \hbar\omega)$.

The validity of this model may be questioned in two respects. First, it seems surprising that the recombination of all excited vibrational states should be negligible. Even though one can choose potential curves AB″ for which $\alpha_v < \alpha_0$ for all $v > 0$, one might expect there to be other states which make little contribution to α_0 but which are significant for excited ions. Second, it is doubtful whether the region $kT_v \gg \hbar\omega$ is reached in the laboratory experiments. For the atmospheric gases the condition $kT_v = 2\hbar\omega$ implies an ion temperature over 5000°K. The vibrational spacings for the rare gas ions are not firmly established, but they are almost certainly less than in the atmospheric gases. Thus it is possible that the high temperature region can be reached in the shock-tube experiments with the rare gases.

In order to explore how the recombination might vary with ion vibrational temperature, Bardsley (1970) has reported numerical calculations of the rate for various forms of the potential curves AB″. The model potentials were chosen to represent nitric oxide and neon. For nitric oxide it was found that below 1000°K there should be no significant difference between the results of electron heating and total heating experiments. Nitrogen and oxygen should be similar in this respect, and it is difficult to reconcile this result with the observed differences in the nitrogen experiments of Kasner (1967; gas heating) and Mehr and Biondi (1969; electron heating). Between 1000° and 7000°K Bardsley's calculations were compared with the simple power law

$$\alpha(T_v, T_e) = cT_e^{-\gamma_e}T_v^{-\gamma_v}. \tag{31}$$

It was shown that γ_v can be positive or negative, depending on the position of the crossover of the AB⁺ and AB″ curves. When this crossover point was varied across the Franck–Condon region for the ground state ion, the following ranges of values were found

$$0.2 \leq \gamma_e \leq 0.8, \qquad -0.3 \leq \gamma_v \leq 0.5. \tag{32}$$

This is the range of exponents that might be expected in the atmospheric gases. However, the greatest deviations from the $T_e^{-0.5}$ behavior were found at temperatures above 5000°K, which is outside the range of accurate experimental measurements. For the potential curves derived from spectroscopic data, the calculated recombination coefficients are shown in Fig. 11 and there compared with the experimental results.

For the numerical calculations on neon, the vibrational spacing was chosen as 0.08 eV. This value was calculated by Gilbert and Wahl (see Wahl *et al.*, 1967) and suggested by Cunningham and Hobson (1969a, b). With this value it was not possible to fit both the afterglow and the shock-tube results. To fit both experiments a spacing smaller than 0.050 eV would be necessary. Alternatively, this failure may be an indication that an equilibrium distribution of ionic vibrational states (i.e., $T_v = T_i$) is not appropriate for the shock-tube experiment.

D. Calculation of Potential Energy Curves

The most important factor governing the behavior of the recombination coefficient is the position of the potential energy curves for the intermediate states AB″. The calculation of these curves is difficult, and their formal definition presents a problem. To illustrate this latter point consider that repulsive curves AB″ must cross the infinite number of Rydberg states with potential curves below the ionic curve. Now no two adiabatic curves can cross if they belong to states of the same symmetry. Thus in order to retain the physical reality of the states AB″ we must discover some definition of potential curves for which the no-crossing rule does not apply. Work on this problem is continuing, with considerable progress being made by O'Malley (1969c, d) and Smith (1969).

Very similar problems arise from the effects of autoionization at small nuclear separation. The calculation of autoionizing states of molecules has been given considerable attention in recent years, and several methods are now available (Bardsley and Mandl, 1968; Taylor, 1969).

O'Malley (1969c, d) has studied hydrogen and helium. For H_2 he finds that the lowest non-Rydberg state crosses ground state H_2^+ at about 3 a.u, which is well outside the Franck–Condon region. This result suggests that the recombination rate for H_2^+ should be very small near 300°K, but should increase at higher temperatures. For helium O'Malley has studied the $^3\Pi_u(1\sigma_g)(1\sigma_u)^2(1\pi_u)$ state and again finds the crossing to be at large separations. Until further states have been examined [in particular, $^3\Sigma_g^+(1\sigma_g)(1\sigma_u)^2(2\sigma_g)$], no conclusion can be drawn, although there is considerable experimental evidence that there are no favorable crossings in He_2.

In larger molecules the number of curves that must be calculated may be very large. For example, Harris and Michels (1968) have determined the potential curves in oxygen for the 62 states which dissociate into two oxygen atoms in the lowest configuration. Unfortunately comparable calculations on O_2^+ are not available and so it is not yet possible to predict with confidence which states are important in recombination. It seems likely that several states can participate in the reaction. One of the most interesting results of such calculations is the determination of the final states of the oxygen atoms produced in the recombination process. Most of the curves which appear to have a favorable crossing dissociate into $O(^3P) + O(^1D)$, but there is one state which would produce $O(^1D) + O(^1S)$. By studying the subsequent radiation following each recombination event, Zipf (1970) has measured the direct population of the three oxygen states to be $0.1(^1S)$, $0.9(^1D)$, and $1.0(^3P)$. These values appear to be consistent with the potential curves described above.

An alternative method for determining potential curves is from spectroscopic data (Bardsley, 1968). The application of this technique to NO was discussed in Section III, B, but its use seems to be limited to a few special cases.

E. THE ROLE OF RYDBERG STATES IN RECOMBINATION

Several early experimental studies of the temperature dependence of the recombination rate indicated temperature variation close to $T^{-1.5}$. This led Bardsley (1968) to search for an alternative mode of recombination. One possibility is a three-stage process which was named indirect recombination,

$$AB^+ + e \rightarrow AB' \rightarrow AB'' \rightarrow A + B. \tag{33}$$

In the first transition, a Rydberg state AB' is formed in an excited vibrational level. This transition involves the transfer of energy from the incident electron directly to the vibrational motion of the nuclei. This initial capture step is followed by the predissociation of the state AB'. This predissociation can be considered as a two-stage process. The step $AB' \rightarrow AB''$ arises from configuration interaction between the Rydberg state AB' and the non-Rydberg state AB''. If the potential curve associated with AB'' is repulsive the nuclei will be forced apart and the recombination completed by dissociation. Typical potential curves for this process are shown in Fig. 13.

The most important feature of this process is that the initial capture step can occur only if the energy of the system coincides with the energy of a vibrational level of a Rydberg state. The vibrational spacings in diatomic atmospheric ions are of order 0.2 eV, and the energy spread in a thermal gas at room temperature is close to 0.025 eV. Thus for a given Rydberg state, one vibrational level at most will be important. Theoretical considerations

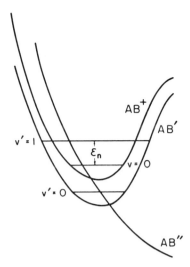

FIG. 13. Hypothetical potential energy curves for the molecular ion AB$^+$, a highlying, stable Rydberg state AB$'$, and an unstable molecule curve AB$''$ involved in the *indirect* dissociative recombination process. The $v = 0$ vibrational level of the molecular ion AB$^+$ and the $v' = 0$ and 1 vibrational levels of the Rydberg state AB$'$ are also shown.

show that the electron capture proceeds most easily when only one vibrational quantum is gained by the nuclear motion. This restricts our attention to those Rydberg states whose binding energies are just smaller than the vibrational spacing. This means principal quantum numbers around 7 or 8. Let ε_n denote the excess energy of the first vibrational level ($v' = 1$) of such a state above the energy of the initial ion ($v = 0$). It can be shown (Bardsley, 1968) that the variation of the recombination rate with electron temperature for ions in the ground vibrational state is

$$\alpha_0'(T_e) = T_e^{-1.5} \sum_n c_n \exp(-\varepsilon_n/kT_e).$$

Thus if there is only one significant Rydberg state and this has $\varepsilon_n < kT$, the temperature variation may appear as $T^{-1.5}$.

If this indirect process is to be important, two conditions must be satisfied. The initial capture step must have a large cross section, and predissociation of the Rydberg state must compete effectively against autoionization of the state. These two requirements are in conflict to a certain extent, because a large capture cross section implies a rapid rate of autoionization. Bardsley (1968) showed that this means that at high temperatures (above, say, 1000°K) the indirect recombination process must always be less important for diatomic molecules than the direct process. At lower temperatures the indirect process should never be completely dominant, but it may have significant effects on the temperature dependence of the recombination rate below 500°K.

The importance of the indirect process at low temperatures depends on the magnitude of the capture cross section and on whether or not there are Rydberg states whose first vibrational level lies just above the energy of the ion. Berry and Nielsen (1970) have calculated the autoionization rates of some Rydberg states in hydrogen. For the $8p\sigma$ and $8p\pi$ states of H_2 they find rates of order 10^{12} sec^{-1}. The electron capture cross section can be calculated from these rates, and it is found that they are 10 times too small for indirect recombination to be important. It seems quite reasonable, however, that the capture rate for s states might be considerably larger than for p states and that the capture rate might be greater for the atmospheric gases than for hydrogen. Thus the indirect process cannot yet be ruled out.

There are two gases, NO^+ and O_2^+, for which there is some experimental evidence that the rate varies more rapidly with temperature below 400°K than above (Weller and Biondi, 1968; Kasner and Biondi, 1968)—see Figs. 10 and 11. However, experiments at low temperatures are very difficult because of the formation of polyatomic ions and this evidence cannot be regarded as conclusive proof of the occurrence of indirect recombination.

The conflict discussed above between large capture cross section and efficient stabilization probably does not occur for polyatomic ions. For molecules with several vibrational modes it is possible to have a large electron capture cross section with a slow rate of decay through autoionization (see e.g., Bardsley and Mandl, 1968, Section 12.2). Thus indirect recombination may be more important in polyatomic molecules. It would be interesting to learn whether the temperature variation of the recombination coefficient for polyatomic ions is significantly different than that of the diatomic ions. In general the recombination rate for polyatomics is an order of magnitude greater than that of diatomics.

Berry and Nielsen (1970) have discussed a third mode of recombination in which the electron is captured into a Rydberg state at an energy above its dissociation limit. Through most Rydberg states this process is likely only for ions in high vibrational states. For ions in their ground state, vibrational capture must take place into Rydberg states with very small principal quantum number. Such a transition involves the transfer of a considerable amount of energy from electronic to vibrational motion, and this would seem to be highly improbable.

Calculations of the rate of dissociative recombination in several gases have been reported by Warke (1966) and Chan (1968). These authors use a one-active-electron approximation. We understand this to mean that capture is due to direct transfer of energy to the nuclear motion through the breakdown of the Born–Oppenheimer approximation, so that excitation of a target electron is not involved. This approximation means that their model is equivalent to that of Berry and Nielsen and that a low-lying Rydberg state is formed. Unfortunately in calculating the capture cross section Warke uses a classical approximation that we believe to be unjustifiable and that has been criticized by Chan (1968). In trying to correct Warke's treatment, Chan assumes that the potential curve for the intermediate state crosses that of the molecular ion. This we believe to be inconsistent with the one-active-electron approximation, since the potential curve of a Rydberg state does not cross the ionization limit of the series to which it belongs. Further doubt on the validity of the calculation of Chan for the case of oxygen arises from his prediction that the dissociating atoms should both be in the ground state, which is in conflict with the experimental data of Zipf (1970).

IV. Recombination in the Upper Atmosphere

An article on dissociative recombination would be incomplete without some discussion of recombination in the ionosphere. Throughout much of the upper atmosphere, dissociative recombination is the dominant recombination mode, and many authors have used this fact to make estimates of the recombination coefficient from observed ionospheric properties. The recombination rate is only one of the many reaction rates which are important in determining the constitution of the ionosphere, however, and the deduction of any one rate from atmospheric data must be regarded as speculative.

Both the chemical composition and the temperature of the ionosphere vary appreciably with changes in time and altitude. Above 250 km the ions are mostly atomic (O^+, H^+, N^+, and He^+). In the E and F regions, at heights between 90 and 250 km, the predominant molecular ions are O_2^+, NO^+, and N_2^+. In the lower E and D regions between 80 and 100 km, atomic ions such as Si^+, Ca^+, Mg^+, and Na^+ appear, and they may react with ozone to form molecules such as NaO^+ and MgO^+ (Donahue, 1966a). In the D region significant concentrations of hydronium ions and hydrated forms (e.g., H_3O^+, $H_3O^+ \cdot H_2O$) are found. The temperature of the lower regions of the ionosphere is in the range 150°–400°K, but it increases up to about 2000°K in the upper F region.

Much of the discussion of ionospheric recombination is in terms of an effective recombination coefficient, as defined in the Introduction. This is necessary because of our uncertain knowledge of the state and identity of the positive ions and because of the considerable concentration of negative ions in the D region. At heights between 50 and 70 km, the negative ion concentration exceeds that of free electrons, so that ion–ion recombination becomes important.

A. Ionic Densities under Quiescent, Equilibrium Conditions

If one knows both the rate of production and the concentration of ions under steady-state conditions, then one can deduce the recombination co-efficient by equating the rate of loss of ions to the production rate. One can either analyze each ion separately or consider only the total ion concentration in terms of the effective recombination rate. Donahue (1966b) has studied the equilibrium concentrations of electrons and the ions N^+, O^+, N_2^+, NO^+, and O_2^+ at heights between 130 and 220 km. The concentrations at 130 km are found to be consistent with a set of values of the rates for seven ion–molecule reactions and with rates for dissociative recombination of NO^+ and of O_2^+ equal to 5.2×10^{-7} cm^3/sec and 1.2×10^{-7} cm^3/sec, respectively. The ions N^+, O^+, and N_2^+ are shown to decay mainly by charge transfer (with subsequent recombination of either NO^+ or O_2^+). The variation of ion density with altitude gives information about the temperature dependence of the reaction rates. Donahue suggests that there is a rapid variation of the recombination rates for NO^+ ($\sim T^{-1.4}$) and for O_2^+ ($\sim T^{-1.0}$) and a slow variation for N_2^+($\sim T^{-0.2}$). While these conclusions must be regarded as tentative, since the temperature dependence of the ion–molecule reaction rates has not been established, they are in reasonable harmony with the dependences deduced from the laboratory studies (see Figs. 9–11).

Donahue (1966a) has made a similar analysis of NO^+ in the D region. The measured concentrations of NO and NO^+ are shown to be consistent with ion production from the Lyman-α flux appropriate to a solar minimum and a recombination coefficient of order 10^{-6} cm^3/sec at 160°K.

Rocket measurements of charge concentrations and radiation flux have been used by Bourdeau et al. (1966) to derive an effective recombination coefficient of 1.8×10^{-7} cm^3/sec between 95 and 115 km, and of 2×10^{-8} cm^3/sec between 83 and 93 km. The latter value is very small compared with both Donahue's values and laboratory measurements. Larsson and Houston (1969) have analyzed data obtained at a height of 65 km, and deduce an effective coefficient of 6×10^{-7} cm^3/sec. At this altitude the presence of negative ions must be taken into account.

B. Diurnal Variations

Since most of the ionization in the upper atmosphere is caused by solar radiation, one would expect the ionization production to be symmetrical about noon and to decrease suddenly at sunset. If lifetimes against recombina-tion were very short, these properties should also be found in the ionization concentrations. However since electron–ion recombination is not instan-taneous, the ionization density is asymmetrical in time about noon and a finite time elapses after sunset before the density reaches its night-time value.

The temporal asymmetry and the accompanying displacement of the time of maximum ionic density from noon were first examined by Appleton (1937) and have been discussed more recently by Appleton and Lyon (1955), Mitra (1964), and Polyakov and Shchukina (1966). The effective recombination coefficient derived by Mitra for the E region is of order 10^{-8} cm^3/sec.

Ionization changes near sunset have been determined by Hirao et al. (1969) from charge density profiles measured during both the ascent and descent of a rocket fired from Uchiura Bay, Japan just before sunset. They estimate the effective recombination rate between 120 and 180 km to be between 1×10^{-7} and 3×10^{-7} cm^3/sec, with the rate decreasing rapidly above 170 km. However the analysis is difficult, since ionization cutoff at sunset is not instantaneous and because of horizontal diffusion of ions from the day-side to the night-side ionosphere.

C. Solar Flares and Eclipses

Solar flare activity leads to an increase in the ionization density in the upper atmosphere. This lowers the effective reflection height of low frequency radio waves (Volland, 1964; Chilton et al., 1965; Nestorov, 1965; Nestorov et al., 1966), and increases the atmospheric absorption of cosmic radio noise (Whitten and Popoff, 1964a, b). Measurements of the increase in ionization have led to estimates of the effective recombination coefficient which increase with decreasing altitude from around 5×10^{-7} cm^3/sec at 80 km to of order 10^{-3} cm^3/sec near 50 km. Although the results near 50 km are only approximate, they indicate a very large ratio of negative ions to free electrons.

In an eclipse the solar radiation incident upon the atmosphere is suddenly reduced, and this leads to a reduction in the electron and ion densities. The consequent increase in the reflection height of vlf radio waves has been observed by Crary and Schneible (1965) who deduced an effective recombination coefficient of 2×10^{-5} cm^3/sec at around 80 km. Lerfald et al. (1965) studied the absorption of cosmic radio noise at 10 and 14 MHz through this eclipse and found that the absorption was reduced to 40% of its normal value. They deduced an effective recombination coefficient of 6×10^{-7} cm^3/sec for the region near 70 km. In the same eclipse (July 20, 1963) measurements of electron density were obtained from four rockets fired from Fort Churchill, Canada. In the analysis of their data Smith et al. (1965) looked for a time lag between the maximum phase of the eclipse and the minimum electron density. The time lag was found to vary between 0 and 3 minutes with height. In the D region Smith et al. calculated an effective recombination coefficient decreasing from 3.5×10^{-6} cm^3/sec at 80 km to 1.5×10^{-6} near 90 km.

For the E and F_1 regions they deduced lower limits to the coefficient of 1×10^{-7} cm^3/sec and 4.6×10^{-7} cm^3/sec, respectively.

During an eclipse the ionization is not reduced to zero, since radiation from the solar corona still reaches the atmosphere. In an analysis of the 1966 eclipse Kane (1969) allows for continuing ionization from X rays. He finds that the calculated effective recombination coefficient in the D region does not decrease monotonically with height. Kane argues from this that during the eclipse there must be additional ionization as well as that from X rays.

D. Aurora and Nuclear Detonations

The ionospheric equilibrium may also be destroyed by other atmospheric disturbances, for example, in aurora and after nuclear explosions. In 1931 Kaplan suggested that the oxygen green line prominent in auroral light is due to the formation of $O(^1S)$ in the dissociative recombination of O_2^+ ions. This suggestion has been confirmed by the analysis of rocket experiments by Donahue et al. (1968) and the laboratory measurements of Zipf (1970). By comparing measurements of magnetic disturbances, auroral luminosity, and cosmic noise absorption, Gustaffson (1964) has estimated the effective recombination coefficient around 80 km to be 2×10^{-6} cm^3/sec. Further analysis of this comparison has been given by Pudovkin (1966).

The additional ionization caused by nuclear explosions has also been studied by rocket experiments (Whitten et al., 1965, 1966; see also Hamlin and Lowen, 1966) and by electromagnetic (EM) absorption analysis (LeLevier, 1964; Kozlov and Raizer, 1966). The results indicate a dissociative recombination coefficient in the D region of around 6×10^{-7} cm^3/sec. This leads to an effective recombination coefficient which increases from 10^{-6} cm^3/sec at 80 km to of the order of 10^{-3} cm^3/sec at 50 km.

E. The Atmospheres of Venus and Mars

Theoretical models have been constructed for the ionospheres of Venus (McElroy, 1968, 1969; McElroy and Strobel, 1969; Shimizu, 1969) and Mars (McElroy, 1967, 1969; Aiken, 1968; Cloutier et al., 1969). In an early model of the Martian ionosphere, McElroy (1967) required an unusually large value of $\alpha(CO_2^+)$, of order 10^{-5} cm^3/sec, in order not to exceed the measured ionospheric density in the face of the implied ionization rates of CO_2. The subsequent laboratory determination that $\alpha(CO_2^+)$ was only 4×10^{-7} cm^3/sec has led to a reexamination of this model, and Cloutier et al. (1969) have shown that, as a result of distortion of the ionosphere by the solar wind, the value of $\alpha(CO_2^+)$ in the Martian upper atmosphere may be as small as indicated by the laboratory value. In the case of Venus, a similar analysis by McElroy (1969) indicates excellent agreement between the model and the observed ionospheric electron density profiles when the laboratory value of $\alpha(CO_2^+)$ is used.

F. SUMMARY

The atmospheric evidence on recombination is obtained in one of two ways. The recombination rate can be estimated by equating the production and loss rates for ions (electrons) either in steady-state conditions or at the maximum or minimum charge concentrations in a disturbed atmosphere. Alternatively one can look for a time lag between the peak of some disturbance and the consequent maximum or minimum in the charge concentrations. In both methods it is necessary to know the rate of production of ionization and the charge concentration. Measurement of the production rate is very difficult and is probably responsible for most of the errors in the atmospheric analyses. Nevertheless a qualitative picture of recombination in the atmosphere is emerging.

The rate of dissociative recombination in the atmosphere has been shown to be of order 10^{-7} to 10^{-6} cm^3/sec. The higher values are found in the lower ionosphere and are due to the lower temperature and possibly to the presence of polyatomic hydrated ions. The atomic ions are neutralized through a charge transfer reaction with subsequent recombination of a molecular ion. The rate governing step is nearly always the charge transfer reaction. This leads to a significant reduction in the effective recombination coefficient above 200 km, where atomic ions become predominant. In the lowest region of the ionosphere the recombination is enhanced by the considerable negative ion concentration, and effective recombination rates around 10^{-3} cm^3/sec have been estimated at ~ 50 km.

V. Conclusions

The determination of dissociative recombination rates has provided a formidable challenge to both experimenters and theorists. The differences in the measured rates for some molecules have shown the need for great care in the design of experiments. The most important points are that mass spectrometers should be used for ion identification and that the decay of ionization should be observed over as large an electron (ion) concentration range as possible. (However, a large range should not be obtained by starting with such a large electron density that 3-body recombination loss is significant.) In addition, the mode of ion generation should be chosen with a view to creating the molecular ions in the desired (specifiable) states.

Measurements of the rate of dissociative recombination at room temperature have been reported for many of the most interesting ions. The best values are included in Table II. Since the quality of these results varies significantly we have given error limits only where we are able to determine them with some confidence. The reader should refer to Section II and the original papers to assess the reliability of these values.

TABLE II

MEASURED VALUES OF THE DISSOCIATIVE RECOMBINATION COEFFICIENT α^a

Ion	α ($\times 10^{-7}$ cm^3/sec)	Notes	References
He_2^+	<0.1	—	See Section II
Ne_2^+	(1.8 ± 0.1)	b	Kasner (1968), Philbrick et al. (1969), Biondi and Brown (1949a)
Ar_2^+	(7.5 ± 1)	—	Oskam and Mittelstadt (1963); Mehr and Biondi (1968)
Kr_2^+	(12 ± 1)	—	Oskam and Mittelstadt (1963)
Xe_2^+	$(14 \pm 1$	—	Oskam and Mittelstadt (1963)
N_2^+	(2.5 ± 0.3)	b	Kasner (1967); Kasner and Biondi (1965); Mehr and Biondi (1969)
$(N_2 \cdot N_2^+)$	≈ 20	b	Kasner and Biondi (1965)
O_2^+	(2.1 ± 0.2)	b	Kasner and Biondi (1968); Mehr and Biondi (1969)
$(O_2 \cdot O_2^+)$	≈ 23	b, c	Kasner and Biondi (1968)
NO^+	4.1	b	Gunton and Shaw (1965); Weller and Biondi (1968)
$(NO \cdot NO^+)$	≈ 17	b	Weller and Biondi (1968)
CO_2^+	(3.8 ± 0.5)	b	Weller and Biondi (1967)
H_3O^+	≈ 13	b	Biondi and Leu (1970)
H_3O^+	≈ 2	b, d	Green and Sugden (1963)
$H_3O^+ \cdot H_2O$	≈ 27	b	Biondi and Leu (1970)

^{a} Under conditions such that $T_e = T_i = T_{gas} \simeq 300°K$, except where otherwise noted.
^{b} Identified the ions undergoing recombination by a mass spectrometer.
^{c} Measured at 205°K.
^{d} Measured at 2100°K.

In *ab initio* calculations of the recombination rate comparatively little has been achieved, and there is not yet a complete, reliable calculation that can be compared with experiment. With present computational facilities it should be possible to complete the calculations for H_2^+ and He_2^+. For large molecules it is doubtful whether the time and expense necessary for accurate calculations can be justified. Nevertheless theory has contributed to the discussion of the temperature dependence of the recombination rate. At temperatures such that the thermal energy is less than the vibrational spacing, the variation of the rate with temperature should be close to $T_e^{-0.5}$, provided that there are favorable curve crossings. Deviations at low temperatures may be due to indirect recombination, although the importance of this mechanism has yet to be confirmed. At higher temperatures deviations can be caused by the dependence of the recombination cross section on the vibrational state of the ion, and by other effects discussed in Section III, C.

The importance of the initial vibrational motion of the ion leads one to question whether there is an equilibrium distribution of vibrational states in the laboratory plasmas, or whether the distribution depends on the ionization mechanism and mode of ion heating. Further study of vibrational relaxation in ionized gases would be of great value.

Interest in the ionosphere has provided much of the motivation for the measurement of the recombination coefficients for N_2^+, NO^+, O_2^+, and CO_2^+. The values of these coefficients near 300°K are fairly well established, but there is still some uncertainty about their values at higher ion temperatures. We would hope that in the future more of the analyses of ionospheric data will use values of the recombination coefficients obtained in laboratory studies rather than treat the recombination rates as parameters to be adjusted as needed.

Other molecular ions of interest in astrophysics include H_3O^+, H_3O^+. H_2O, MgO^+, and NaO^+ in the atmosphere of the earth, H_3^+ in the atmospheres of the outer planets and in the interstellar gas, and CO^+ in comet tails. Thus there is ample scope for further measurement. But perhaps the greatest challenge for the experimenter is an accurate determination of the dissociative recombination coefficient of H_2^+, since it is this ion which offers the best possibility for a quantitative *ab initio* calculation to which the measurement can be compared.

REFERENCES

Aikin, A. C. (1968). *Icarus* **9**, 487.
Albritton, D. L., Miller, T. M., Martin, D. W., and McDaniel, E. W. (1968). *Phys. Rev.* **171**, 94.
Appleton, E. V. (1937). *Proc. Roy. Soc.* (*London*) **A162**, 451.
Appleton, E. V., and Lyon, A. J. (1955). *J. Atmospheric. Terrest. Phys.* **21**, 73.
Bardsley, J. N. (1968). *J. Phys. B.* **1**, 349, 365.
Bardsley, J. N. (1970). To be published.
Bardsley, J. N., and Mandl, F. (1968). *Rept. Progr. Phys.* **32**, 471.
Bates, D. R. (1950). *Phys. Rev.* **77**, 718; **78**, 492.
Bates, D. R., and Massey, H. S. W. (1946). *Proc. Roy. Soc.* (*London*) **A187**, 261.
Bates, D. R., and Massey, H. S. W. (1947). *Proc. Roy. Soc.* (*London*) **A192**, 1.
Bates, D. R., Buckingham, R. A., Massey, H. S. W., and Unwin, J. J. (1939). *Proc. Roy. Soc.* (*London*) **A170**, 322.

Bates, D. R., Kingston, A. E., and McWhirter, R. W. P. (1962). *Proc. Roy. Soc. (London)* **A267**, 297; **A270**, 155.

Bauer, E., and Wu, T. Y. (1956). *Can. J. Phys.* **34**, 1436.

Berlande, J., Cheret, M., Deloche, R., Gonfalone, A, and Manus, C. (1970). *Phys. Rev.* **A1**, 887.

Berry, R. S., and Nielsen, S. E. (1970). *Phys. Rev*, **A1**, 383, 395.

Biondi, M. A. (1951a). *Rev. Sci. Instr.* **22**, 500.

Biondi, M. A. (1951b). *Phys. Rev.* **83**, 1078.

Biondi, M. A. (1956). *Rev. Sci. Instr.* **27**, 36.

Biondi, M. A. (1963). *Phys. Rev.* **129**, 1181.

Biondi, M. A., and Brown, S. C. (1949a). *Phys. Rev.* **75**, 1700.

Biondi, M. A., and Brown, S. C. (1949b). *Phys. Rev.* **76**, 1697.

Biondi, M. A., and Leu, M. T. (1970), private communication.

Bourdeau, R. E., Aikin, A. C., and Donley, J. L. (1966). *J. Geophys. Res.* **71**, 727

Chan, F. T. (1968). *J. Chem. Phys.* **49**, 2533.

Chapman, S. (1931). *Proc. Roy. Soc. (London)* **A132**, 353.

Chen, C. J. (1969). *Phys. Rev.* **177**, 245.

Chen, C. L., Leiby, C. C., and Goldstein, L. (1961). *Phys. Rev.* **121**, 1391.

Chilton, C. J., Conner, J. P., and Steele, F. K. (1965). *IEEE Trans.* **53**, 2018.

Cloutier, P. A., McElroy, M. B., and Michel, F. C. (1969). *J. Geophys. Res.* **74**, 6215.

Collins, C. B., and Hurt, W. B., (1969). *Phys. Rev.* **179**, 203.

Collins, C. B., and Robertson, W. W. (1964). *J. Chem. Phys.* **40**, 701, 2202, 2208.

Collins, C. B., and Robertson, W. W. (1965). *J. Chem. Phys.* **43**, 4188.

Connor, T. R., and Biondi, M. A. (1965). *Phys. Rev.* **140**, A778.

Crary, J. H., and Schneible, D. E. (1965). *Natl. Bur. Std. (U.S.) J. Res.* **69D**, 947.

Cunningham, A. J., and Hobson, R. M. (1969a). *Abstr. Papers, 6th Intern. Conf. Phys. Electron. At. Collisions, Cambridge, Mass.* (unpublished) p. 1038ff.

Cunningham, A. J., and Hobson, R. M. (1969b). *Phys. Rev.* **185**, 98.

Donahue, T. M. (1966a). *J. Geophys. Res.* **71**, 2237.

Donahue, T. M. (1966b). *Planetary Space Sci.* **14**, 33.

Donahue, T. M., Parkinson, T., Zipf, E. C., Doering, J. P. Fastie, W. G., and Miller, R. E. (1968). *Planetary Space Sci.* **16**, 737.

Dubrovsky, G. V., and Ob'edkov, V. D. (1967a). *Soviet Astron. AJ.* **11**, 305.

Dubrovsky, G, V., Ob'edkov, V. D., and Janev, R. J. (1967b). *Abstr. 5th Intern. Conf. Phys. Electrons At. Collisions, Leningrad* p. 342.

Edwin, R. P., and Turner, R. (1969). *J. Chem. Phys.* **50**, 4388.

Felenbok, P., and Lefebvre-Brion, H. (1966). *Can. J. Phys.* **44**, 1677.

Ferguson, E. E., Fehsenfeld, F. C., and Schmeltekopf, A. L. (1965). *Phys. Rev.* **138**, A381.

Fox, J. N., and Hobson, R. M. (1966). *Phys. Rev. Letters* **17**, 161.

Frommhold, L., and Biondi, M. A. (1968). *Ann. Phys. (N.Y.)* **48**, 407.

Frommhold, L., and Biondi, M. A. (1969). *Phys. Rev.* **185**, 244.

Frommhold, L., Biondi, M. A., and Mehr, F. J. (1968). *Phys. Rev.* **165**, 44.

Gray, E. P., and Kerr, D. E. (1962). *Ann. Phys. (N.Y.)* **17**, 276.

Green, J. A., and Sugden, T. M. (1963). *Proc. 9th Symp. Combustion (The Combustion Inst., Pittsburgh)* p. 607.

Gunton, R. C., and Shaw, T. M. (1965). *Phys. Rev.* **140**, A756.

Gustaffson, G. (1964). *Planetary Space Sci.* **12**, 195.

Hackam, R. (1965). *Planetary Space Sci.* **13**, 667.

Hagen, G. (1968). *Air Force Cambridge Res. Lab., Bedford, Mass., Rept. AFCRL-68-0649*, unpublished.

Hamlin, D. E., and Lowen, R. W. (1966). *J. Geophys. Res.* **71**, 5195.

Harris, F. E., and Michels, H. H. (1968). *J. Chem. Phys.* **48**, 4946.
Hess, W. (1965). *Z. Naturforsch.* **20a**, 451.
Hirao, K., Oya, H., Tohmatsu, T., and Ogawa, T. (1969). *Space Res.* **9**, 292.
Holt, R. B., Richardson, J. M., Howland, B., and McClure, B. T. (1950). *Phys. Rev.* **77**, 239.
Huber, M. (1964). *Helv. Phys. Acta* **37**, 329.
Jacquinot, P., and Dufour, C. (1948). *J. Rech. Centre Natl. Rech. Sci. Lab. Bellevue (Paris)* **6**, 91.
Johnson, R. A., McClure, B. T., and Holt, R. B. (1950). *Phys. Rev.* **80**, 376.
Jungen, Ch. (1966). *Can. J. Phys.* **44**, 3197.
Kane, J. A. (1969). *Planetary Space Sci.* **17**, 609.
Kaplan, J. (1931). *Phys. Rev.* **38**, 1048.
Kasner, W. H. (1967). *Phys. Rev.* **164**, 194.
Kasner, W. H. (1968). *Phys. Rev.* **167**, 148.
Kasner, W. H., and Biondi, M. A. (1965). *Phys. Rev.* **137**, A317.
Kasner, W. H., and Biondi, M. A. (1968). *Phys. Rev.* **174**, 139.
Kozlov. S. I., and Raizer, Yu. P. (1966). *Cosmic Res.* **4**, 509.
Lagerqvist, A., and Miescher, E. (1958). *Helv. Phys. Acta* **31**, 221.
Lagerqvist, A., and Miescher, E. (1966). *J. Chem. Phys.* **44**, 1525.
Larsson, L. E., and Houston, R. E. (1969). *J. Geophys. Res.* **74**, 2402.
LeLevier, R. E. (1964). *J. Geophys. Res.* **69**, 481.
Lennon, J. J., and Sexton, M. C. (1959). *J. Electron. Control* **7**, 123.
Lerfald, G. M., Hargreaves, J. K., and Watts, J. M. (1965). *Natl. Bur. Std. (U.S.) J. Res.* **69D**, 939.
Lin, S. C., and Tease, J. D. (1963). *Phys. Fluids* **6**, 355.
McElroy, M. B. (1967). *Astrophys. J.* **150**, 1125.
McElroy, M. B. (1968). *J. Geophys. Res.* **73**, 1513.
McElroy, M. B. (1969). *J. Geophys. Res.* **74**, 29.
McElroy, M. B., and Strobel, D. F. (1969). *J. Geophys. Res.* **74**, 1118.
McLaren, T. I., and Hobson, R. M. (1968). *Phys. Fluids* **11**, 2162.
Massey, H. S. W. (1937). *Proc. Roy. Soc. (London)* **A163**, 542.
Mehr, F. J., and Biondi, M. A. (1968). *Phys. Rev.* **176**, 322.
Mehr, F. J., and Biondi, M. A. (1969). *Phys. Rev.* **181**, 264.
Mentzoni, M. H. (1965). *J. Appl. Phys.* **36**, 57.
Mitra, A. P. (1964). *J. Geophys. Res.* **69**, 4067.
Mott-Smith, H. M., and Langmuir, I. (1926). *Phys. Rev.* **28**, 727.
Mulliken, R. S. (1964). *Phys. Rev.* **136**, A962.
Nestorov, G. (1965). *Geomagnetism Aeronomiya* **5**, 690.
Nestorov, G., Krivsky, L., and Fatkullin, N. (1966). *J. Atmospheric Terrest. Phys.* **28**, 121.
Niles, F. E., and Robertson, W. W. (1964). *J. Chem. Phys.* **40**, 2909.
O'Malley, T. F., (1969a). *Abstr. Papers, 6th Intern. Conf. Phys. Electron. At. Collisions, Cambridge, Mass.* (unpublished) p. 1034 ff.
O'Malley, T. F. (1969b). To be published.
O'Malley, T. F. (1969c). *Abstr. Papers, 6th Intern. Conf. Phys. Electron. At. Collisions, Cambridge Mass.* (unpublished) p. 997 ff.
O'Malley, T. F., (1969d). *Phys. Rev.* **185**, 101.
Oskam, H. J. (1958). *Philips Res. Rept.* **13**, 335.
Oskam, H. J., and Mittelstadt, V. R. (1963). *Phys. Rev.* **132**, 1445.
Persson, K. B., and Brown, S. C. (1955). *Phys. Rev.* **100**, 729.
Phelps, A. V., and Brown, S. C. (1952). *Phys. Rev.* **86**, 102.
Philbrick, J., Mehr, F. J., and Biondi, M. A. (1969). *Phys. Rev.* **181**, 271.

Polyakov, V. M., and Shchukina, T. B. (1966). *Geomagnetism Aeronomiya* **6**, 664.
Pudovkin, M. I. (1966). *Geomagnetism Aeronomiya* **6**, 677.
Richardson, J. M. (1952). *Phys. Rev.* **88**, 895.
Richardson, J. M., and Holt, R. B. (1951). *Phys. Rev.* **81**, 153.
Rogers, W. A., and Biondi, M. A. (1964). *Phys. Rev.* **134**, A1215.
Saporoschenko, M. (1965a). *J. Chem. Phys.* **42**, 2760.
Saporoschenko, M. (1965b). *Phys. Rev.* **139**, A349.
Sayers, J. (1956). *J. Atmospheric Terrest. Phys. Special Suppl.* **6**, 212 ff.
Shimizu, M. (1969). *Icarus* **10**, 11.
Smith, D., and Goddall, C. V. (1968). *Planetary Space Sci.* **16**, 1177.
Smith, F. T. (1969). *Phys. Rev.* **179**, 111.
Smith, L. G., Accardo, C. A., Weeks, L. H., and McKinnon, P. J. (1965). *J. Atmospheric Terrest. Phys.* **27**, 803.
Stein, R. P., Schiebe, M., Syverson, M. W., Shaw, T. M., and Gunton, R. C. (1964). *Phys. Fluids* **7**, 1641.
Taylor, H. S. (1969). To be published.
Theard, L. P. (1969). *Abstr. Papers, 6th Intern. Conf. Phys. Electron. At. Collisions, Cambridge, Mass.*, (unpublished), p. 1042 ff.
Thomas, R. D., Hackam, R., and Lennon, J. J. (1966). *Proc. 7th Intern. Conf. Phenom. Ionized Gases (Belgrade)* **I**, 27.
Varnerin, L. J. (1951). *Phys. Rev.* **84**, 563.
Volland, H. (1964). *J. Atmospheric Terrest. Phys.* **26**, 695.
Wahl, A. C., Bertoncini, P. J., Das, G., and Gilbert, T. L. (1967). *Intern. J. Quant, Chem. Symp.* **1**, 123.
Warke, C. S. (1966). *Phys. Rev.* **144**, 120.
Weller, C. S., and Biondi, M. A. (1967). *Phys. Rev. Letters* **19**, 59.
Weller, C. S., and Biondi, M. A. (1968). *Phys. Rev.* **172**, 198.
Whitten, R. C. and Popoff, I. B. (1964a). *J. Atm. Sci.* **21**, 117.
Whitten, R. C., and Popoff, I. B. (1964b). *Discussions Faraday Soc.* **37**, 185.
Whitten, R. C., Popoff, I. B., Edmonds, R. S., and Berning, W. W. (1965). *J. Geophys. Res.* **70**, 1737.
Whitten, R. C., Popoff, I. B., and Edmonds, R. S. (1966). *J. Geophys. Res.* **71**, 5199.
Wilkins, R. L. (1966). *J. Chem. Phys.* **44**, 1884.
Yamane, M. (1968). *J. Chem. Phys.* **49**, 4624.
Young, R. A., and St. John, G. (1966). *Phys. Rev.* **152**, 25.
Zipf, E. C. (1967). *Bull. Am. Phys. Soc.* **12**, 225.
Zipf, E. C. (1970). To be published.

ANALYSIS OF THE VELOCITY FIELD IN PLASMAS FROM THE DOPPLER BROADENING OF SPECTRAL EMISSION LINES

A. S. KAUFMAN*

Bar-Ilan University, Ramat-Gan, Israel
and
The Hebrew University of Jerusalem, Israel

* Permanent address: Department of Physics, The Hebrew University of Jerusalem, Israel.

I. Introduction

Understanding the plasma state of matter entails extensive measurement of the plasma by a wide variety of methods. Those methods dependent on the measurement of radiant energy enjoy distinct advantages. In general, they do not perturb the plasma and are supported by a wealth of astrophysical experience. A prominent and well-tried method, the Doppler broadening of spectral lines, can be investigated in emission and absorption, and by scattering. In the case of emission, a fair-sized literature dealing with the investigation of hot, diffuse plasmas in the laboratory and of the solar atmosphere has accumulated. Doppler broadening likewise features prominently in the formation of absorption lines in stellar atmospheres (for example, Jefferies, 1968, pp. 86–88 and Chapter 9). Scattering made its debut with the investigation of the ionosphere by radar pulses (Bowles, 1964, pp. 155–168 for a review), and this was followed by an extensive series of measurements of the scattering of laser light in laboratory plasmas (see Kunze, 1968, Chapter 9; Evans and Katzenstein, 1969 for reviews). This account is concerned solely with the case of emission.

An emission line broadened by the Doppler effect represents the envelope of the unresolved displacements of the line caused by the motion of the radiating atoms or ions in the direction of observation. The motion may not be purely thermal but can, in general, be combined with mass motion which in itself can be of two forms, systematic and random. From the point of view of nonequilibrium thermodynamics, systematic mass motion represents the most organized and ordered of these three types of motion. Random mass motion takes up a position in between; it displays facets of both organized and thermal character.

In principle, Doppler broadening provides a simple and fairly accurate means of measuring the temperature of an inaccessible radiating medium. Specifically, when the temperature of a hot plasma is to be measured, the contribution to the broadening by mass motion may be considerable in view of the ability of a plasma to sustain collective forms of motion. Thonemann (1961, p. 257) stated that one should always suspect that mass motion is present rather than absent. Likewise, a purely thermal velocity field appears to be the exception rather than the rule in stellar atmospheres (Gebbie and Thomas, 1968, p. 271). Hence, doubt can be cast on the validity of temperature measurement inferred from Doppler widths. Experimental methods are required to distinguish between thermal motion and mass motion, and to separate the two. Although the prime object in measuring the Doppler broadening of spectral lines radiated by laboratory plasmas is the measurement of temperature, the opportunity also exists of investigating mass motion in general and random mass motion in particular by means of an unperturbing probe. "Plasma turbulence" is a subject still in its infancy, yet it occupies a central position in plasma dynamics. The satisfactory analysis of the velocity field from Doppler profiles should help in providing a detailed picture of plasma dynamics both in the laboratory and in space, for example, in stellar atmospheres (Thomas, 1961) and in the stochastic acceleration of charged particles including "turbulent" heating (Tsytovich, 1966). Investigations of this kind can also be viewed within the wider framework of the investigation of phenomena displaying order within a disordered system (see, for example, Wigner, 1967, p. 41).

The history of the subject under review would appear to date back to Rayleigh (1902, p. 261) who, in a footnote to his celebrated paper on the limit to light interference caused by Doppler broadening, wrote: "It is here assumed that we are dealing with a gas in approximate temperature equilibrium. The case of luminosity under electric discharge may require further consideration." Many years later, at the beginning of the research on hot laboratory-plasmas, it was indeed realized that mass motion should be taken into account in the interpretation of Doppler broadening of spectral lines (Post, 1956, p. 361). At first, progress in making this assessment was slow (Harding et al., 1958, p. 368). Later experimental attempts were more successful, but to date, most of the published work on the subject is inadequate because there are insufficient clear-cut experimental data from which to draw firm conclusions. In short, the problem of separating the two types of motion has not yet been satisfactorily solved. Likewise in astrophysics, the difficulty of interpreting Doppler-broadened lines is as great (for example, Jefferies, 1962, p. 704; Gebbie and Thomas, 1968, for the general background to the problem). Neither has the subject for laboratory conditions been satisfactorily reviewed. Earlier reviews are brief and incomplete (Thonemann, 1961, pp. 66–69; Rusanov, 1962, pp. 138–141; Cooper, 1966, pp. 105–106). The account by Zaidel' et al. (1961a, pp. 99–100, 112–114) is rather dated.

About a decade ago I was engaged in a program of measuring temperatures in the apparatus SCEPTRE (Payne and Kaufman, 1959). Ion temperatures were inferred from the Doppler broadening of ion lines but, as already indicated, the contribution of mass motion to the line widths was obscure. On returning recently to the subject of the measurement of very high temperatures and enquiring what had been accomplished in the interim, it quickly became evident that the interpretation of Doppler-broadened spectral lines still presented a challenge; no coherent picture had yet emerged. The challenge was accepted and this paper is part of the result in attempting to prepare a critical and systematic account of the analysis of the velocity field in plasmas. It combines review and original contributions, a feature in conformity with the editorial policy of this serial publication.

Emphasis is placed on laboratory plasmas, and wherever possible the existence of local thermodynamic equilibrium is *not* assumed. The stage is set in Section II with a discussion of the meaning of kinetic temperature and mass motion in a plasma. A criterion is then established how to distinguish between thermal motion and mass motion in a measurement of line broadening. In Section III the subject matter begins to assume a more specific character: there the experimental methods of separating thermal motion from random mass motion (named *commotion* rather than *plasma turbulence*) are classified. Section IV, which constitutes more than half of this account, is devoted to the method of separation called the *method of different ionic species*. Finally, the short conclusion is followed by three appendixes.

The source of radiant energy is taken to be a hot, optically thin plasma so that effects of absorption can be neglected. This assumption was justified experimentally for the energy radiated by highly ionized atoms in not-too-dense plasmas (Zaidel' *et al.*, 1961b, pp. 1348–1350; Sawyer *et al.*, 1963, p. 1895; Heroux, 1964, p. 133; Boland *et al.*, 1968, p. 60). Theoretical calculations of radiative transfer were reviewed by Cooper (1966, pp. 90–96) and Richter (1968, pp. 60–62). Sufficient has already been written in the literature about the importance of other forms of line broadening in relation to Doppler broadening (for example, Griem, 1962, pp. 617–618 for natural broadening; Margenau, 1960 and Griem, 1962, p. 618 for Stark effect; Cooper, 1966, pp. 111–112, Neumann, 1967, pp. 289–295, and Bötticher, 1968, pp. 650–663 for Zeeman effect; Okazaki and Andō, 1968 for fine structure). The assumption is made here that they can be neglected with respect to Doppler broadening, or that they, including instrumental broadening, have already been separated from it (see Wiese, 1965, pp. 301–304; Armstrong, 1967, pp. 71–76; Lochte-Holtgreven and Richter, 1968, pp. 325–328 for reviews).

SI units and the unit electron volt (eV), which is allowed in conjunction with SI, are used throughout. The symbols for physical quantities conform as far as possible to those recommended in 1965 by the International Union of Pure and Applied Physics [Document U.I.P. 11 (S.U.N. 65–3)]. In addition, in accordance with resolution number 3 of the thirteenth General Conference of Weights and Measures (for example, Barrell, 1968), the degree kelvin is replaced by kelvin with the symbol K. To assist the reader who may use this review piecemeal, a list of symbols is provided at the end.

A note must also be added about nomenclature. The subject matter is connected with frequency distributions in spectroscopy, kinetic theory, and random processes. The language of probability theory is basic to all three and is thus preferentially used in the theory of unfolding a spectral-line profile (Section IV, B). Nomenclature in radiometry conforms to the consensus published by Meyer-Arendt (1968). For example, the radiance at a given point and in a given direction is defined as the radiant power crossing unit surface normal to that direction, per unit solid angle.

Every effort has been made to give references in the published literature and not reports. To be referred to a report which is not readily available is most annoying. In Sections I and II no claim to provide a comprehensive list of references is made. Those references quoted have been selected with an eye to the maximum retrieval of information.

II. Thermal Motion and Mass Motion

A. PLASMA TEMPERATURE

The plasma temperature is defined in terms of the average kinetic energy of the charged particles contained within a volume less than that of either the Debye–Hückel sphere (see Appendix A) for a plasma in thermodynamic equilibrium, or the equivalent screening cloud of charge for a plasma far from equilibrium. The advantages of choosing this temperature, the kinetic temperature, are discussed and reference is made to experimental verification of its existence. The remainder of the section is devoted to thermal fluctuations in number density of the charged particles. The conclusion is that even for a stable plasma not in equilibrium, spatial correlation of the ionic thermal motion exists on a scale roughly defined by the Debye–Hückel radius of the equilibrium case. The assumption is made that this conclusion is also valid for an unstable plasma.

A hot plasma may be far removed from thermodynamic equilibrium. In such a situation it is still possible to define a plasma temperature, the so-called kinetic temperature, which is applicable whether or not the gas is in equilibrium (Chapman and Cowling, 1952, p. 37). In the case of the ions, T, their kinetic temperature, or their temperature for short, is given by

$$\tfrac{1}{2}m\langle(u^{th})^2\rangle = \tfrac{3}{2}kT \tag{1}$$

where m is the ionic mass, u^{th} the translational speed exclusive to each ion, and k a constant of proportionality. A similar relation exists for the electrons. For anisotropic conditions the temperature in a given direction l is defined by

$$\tfrac{1}{2}m\langle(u_l^{th})^2\rangle = \tfrac{1}{2}kT_l. \tag{2}$$

Strictly the term temperature is meaningful when the velocity distribution of the ions is given by Maxwell's law or is close to it. Otherwise, one can only speak of an effective temperature defined by the average kinetic energy of the ions. The coincidence of the kinetic temperature scale and the thermodynamic temperature scale to which all temperature measurements should be ultimately referred (for example, Stimson, 1962, p. 59) is obtained by setting k to the value of the Boltzmann constant (for example, Guggenheim, 1960, p. 18), namely the internationally accepted value of 1.38054×10^{-23} J K^{-1} (1.38054×10^{-16} erg K^{-1}) (Cohen and DuMond, 1965, p. 590).

Apart from its general applicability, the kinetic temperature carries with it two other advantages. Although a plasma does not behave as a perfect gas, the kinetic temperature scale requires no correction to bring it in line with the thermodynamic scale, provided k is selected as the Boltzmann constant. This is so because the relation (1) holds for a perfect gas. Secondly, the numerous methods of measuring either the velocity of the charged particles (for example, Doppler broadening) or the square of their speed (for example, electrostatic-energy analysis) furnish in principle a temperature determination for each method.

Experimental verification of the existence of the electron kinetic temperature was demonstrated by measuring the hydrostatic pressure exerted by plasma electrons (Alexeff and Neidigh, 1962). This pressure was found to be closely equal to $n_e k T_e$, where n_e is the number density of the electrons and T_e, their kinetic temperature. The latter was measured using ion acoustic waves.

To complete this section, random thermal fluctuations in number density of the charged particles are considered. For a plasma in either full or local thermodynamic equilibrium, the fluctuations in charge density are manifest on a microscopic scale corresponding to the size of the Debye–Hückel sphere (for example, Sitenko, 1967, pp. 67–68). Moreover, a similar result holds for the independent fluctuations in ionic number density [Sitenko, 1967, p. 70; the spatial correlation function is determined by integration of Eq. (5.29) as on p. 55]. However, in the general case of a plasma far from thermodynamic equilibrium, the Debye–Hückel radius cannot be defined exactly. Some form of screening of the electric field in the vicinity of a charged particle nevertheless exists, although it is no longer symmetrical nor of permanent shape as in the equilibrium case (for example, Balescu, 1960; and Longmire, 1963, p. 171). In the case of a spatially homogeneous system of electrons not in equilibrium, in the presence of a neutralizing background of ions, the effective screening distance is several times greater than the Debye–Hückel radius (Bernstein and Ahearne, 1968, p. 21). A similar conclusion was drawn by Joyce and Montgomery (1967) for the particular case of counter-moving electron streams in a uniform, immobile background of positive charge. Likewise, the fluctuations in ionic number density of a plasma not in equilibrium can be inferred to exist on a scale of similar size: screening is a function of the kinetic energy and number density of the charged particles, and is not a function of their mass and charge sign. Thus far, thermal fluctuations have been considered for a stable plasma in the absence of a magnetic field. When a magnetic field is present, the domain of the charge-density fluctuations in the equilibrium case remains unaltered (Akhiezer *et al.*, 1967, p. 118). It may be inferred that a similar result should hold for both the fluctuations in charge density and in ionic number density for a plasma not in equilibrium—the magnetic field does not alter the kinetic energy of the charged particles.

The conclusion to be drawn is that in the general case of a stable, spatially homogeneous plasma not in equilibrium, the Debye–Hückel radius can be regarded as a rough measure of the extent of screening, and that thermal fluctuations in number density are discernible within a volume somewhat larger than the Debye–Hückel sphere of the equilibrium case. Since the number density and fluid velocity are related through the equation of continuity, the same conclusion is relevant to thermal velocity fluctuations. Of direct concern here is the form of screening in the very general case where a spatially inhomogeneous plasma far from equilibrium exhibits random mass motion. It is assumed that the above conclusion remains essentially unchanged for this case.

Finally, it should be carefully noted that the speed u^{th} in Eq. (1) is taken to refer to the velocity distribution of a group of ions on the scale of the screening distance. The reason for this is discussed in Section II, C. Such a microscopic velocity distribution is meaningful provided there are many particles in the Debye–Hückel sphere or in the equivalent screening cloud of charge. For the conditions considered here, this is certainly the case (see Appendix A).

B. SYSTEMATIC AND RANDOM MASS MOTION

In this section the minimum length scale and time scale of mass motion as typified by electrostatic plasma waves are discussed. This is followed by reference to rotational motion, and systematic and random mass motion are then defined. The term *commotion* is coined for the random mass motion detected by the Doppler shift or broadening of spectral lines. Commotion is discussed in relation to the prevalent use of the terms *plasma turbulence* and *microturbulence*.

Macroscopic forms of plasma motion in which charged particles move together with identical velocities will be called plasma mass motion, or mass motion for short. This definition requires amplification in two respects. Firstly, the word macroscopic has been inserted to indicate that thermal velocity fluctuations are excluded from the description of mass motion (see Section II, A). Secondly, mass motion does not imply that in a given volume element and during a given short time interval *all* the charged particles present are moving with the same mass velocity. For example, in the case of a plasma situated in an electromagnetic field, the mass velocity can be a function of the charge-to-mass ratio of the ions (for example, Thonemann, 1961, p. 68). One can thus visualize different sets of ions located in the same region and in relative motion. In some circumstances the number of sets of ions with the same charge and mass but of different velocity may be sufficient to warrant the use of a distribution function of mass velocities: a set of ions moving in unison, named here an element, can be regarded as a quasi-particle of kinetic theory.

In order to differentiate between the *microscopic* velocity distribution function pertaining to thermal motion only (Section II, A) and the function for mass velocities, the latter is named the *macroscopic* velocity distribution function. The mass motion can be translational, oscillatory, or rotational.

Oscillatory motion in the form of electrostatic plasma waves, the typical collective pattern in a plasma, deserves special mention. The modes of oscillation, which include ion acoustic and electrostatic ion–cyclotron waves, are longitudinal in that the electric field is parallel, or is approximately parallel, to the wave vector (for example, Stix, 1962, p. 223). A simple physical picture indicates that the charged particles cannot respond in phase if their average velocity is greater than the phase velocity of the waves (for example, Balescu, 1963, p. 85). Hence, wave motion of the ions is damped out for wavelengths less than a critical wavelength λ_0 given approximately by

$$\lambda_0 = \langle (u_l^{th})^2 \rangle^{1/2}/v_p, \tag{3}$$

where v_p is the plasma frequency of the ions (see Kadomtsev, 1968 for a recent review of Landau damping). In the case of thermodynamic equilibrium, $\lambda_0/2\pi$ can be recognized as the Debye–Hückel radius (for example, Balescu, 1963, p. 85). For a plasma not in equilibrium, $\lambda_0/2\pi$ is a measure of the effective average screening distance. Experimental verification of the spatial damping of ion acoustic waves was first demonstrated by Wong *et al.* (1964). Doubts regarding the veracity of this experiment were dispelled by later work (Alexeff *et al.*, 1967). Mass motion of the ions exists on a minimum time scale given approximately by the reciprocal of the ion plasma frequency. This result also holds for the case where the waves propagate in a direction parallel to a magnetic field (Stix, 1962, p. 217). For wave propagation in a direction perpendicular to the field, the minimum time scale remains unchanged provided the plasma frequency is much greater than the cyclotron frequency (for example, Akhiezer *et al.*, 1967, p. 59). This condition is well fulfilled for the type of plasma considered here where the relevant limiting conditions are helium ions of number density 10^{19} m^{-3} (10^{13} cm^{-3}) contained in a magnetic field of strength $(4\pi)^{-1}$ 10^8 A m^{-1} (10^5 Oe). Spectroscopic measurement of hot plasmas suffers from lack of radiant energy should the number density be less than about 10^{19} m^{-3} (for example, Hirschberg and Palladino, 1962, p. 49; Afrosimov and Petrov, 1968, p. 1467). Equation (3) remains relevant for the same limiting conditions; for experimental confirmation, see Hirose *et al.* (1969).

Rotational motion can be resolved into two components (Kovásznay, 1960, p. 817). One component in which ions and electrons rotate together with identical velocities can be called fluid motion. Note that Kovásznay restricted the use of the term mass motion to this component of motion and so used it in a much narrower sense than here. With the second component, the ions and electrons rotate in opposite directions such that the net momentum is zero. Electric current flows and this component of motion may be named an electric eddy.

Two main types of mass motion can be distinguished—systematic and random. Systematic mass motion, or systematic motion for short, is a regular movement of the charged particles. In the literature, the names gross motion, ordered motion, and directed motion have also been used. Systematic motion can be defined by considering a volume element of plasma, considerably larger than the screening cloud of charge and yet small enough to be approximated by a mathematical point. In this way the effect of thermal (Brownian) motion which sets the limit to the accuracy of measurement is smoothed out. Let the plasma be produced in cycles. The velocity of systematic motion at such a given volume element and at a given time from some arbitrary start is the same for repeated observation from cycle to cycle: the error of measurement is assumed negligibly small. In reality, the velocity is measured over a finite time interval and so the time of measurement is an average over this interval.

The same type of measurement is repeated when random mass motion is present instead of systematic motion. In this case, the velocity varies from cycle to cycle in an apparently indeterminable manner and no regularity is discernible, although the macroscopic physical conditions remain unchanged. Should an observer be capable of examining spectroscopically the light emitted by the ions contained within the volume element, random Doppler shifts of the spectral line would be detected from cycle to cycle. If the time taken to record the line is long enough, the shifts would be blurred with the result that the line would appear broader than in the case where the mass motion is absent. All random mass motion revealed by either shift or broadening will here be called *commotion* and the speed of commotion in the line of sight can be aptly termed the *commoving speed*.

There is need to justify the coining of the term commotion instead of using the term plasma turbulence. This can best be done by asking what is meant by plasma turbulence. It has been described as the state of a plasma in which the motion on a macroscopic scale is extremely disordered (Artsimovich, 1964, p. 205). A more explicit definition by Stott (1968, p. 676) is worthy of being quoted in full.

The term "turbulent plasma" will be used here simply to refer to a plasma which supports density fluctuations of a scale length longer than the Debye length yet small compared with the overall dimensions of the plasma and with a time scale which is much shorter than the plasma lifetime.

In following a more analytical approach by Kadomtsev (1965, pp. 1, 31), plasma turbulence may be defined as the state of a plasma in which random motions mix nonlinearly, so that the observed power spectral density indicates the presence of a continuous range of frequencies or at least a large number of frequencies (see Fig. 5). The general impression from all this is that the word turbulence is used indiscriminately to refer to observations which cannot be explained by a streamline theory (Moffatt, 1962, p. 396; Artsimovich, 1964, p. 205). A similar criticism of the understanding of turbulence in stellar atmospheres was voiced by Clauser (1961). One would therefore be well advised not to use the term turbulence to cover all aspects of fluctuation phenomena (other than thermal) in plasmas, since it has at the same time a fairly precise connotation in the motion of electrically nonconducting fluids (for example, Lumley and Panofsky, 1964, pp. 3–5). Moreover, Kovásznay (1960, p. 817) ascribed the term plasma turbulence only to the random *rotational* motion in a plasma. This is reasonable and logical because one of the features of fluid turbulence is that the motion is rotational. Kovásznay's nomenclature is accepted here and hence the suggested use of the synonym commotion for random mass motion detected spectroscopically. Note that commotion is more or less equivalent to plasma turbulence as defined by Stott (1968).

In astrophysics, the term *microturbulence* is frequently used " to designate non-thermal motions of such randomness and scale that their spectroscopic effect is indistinguishable from that of thermal motion " (Billings, 1964, p. 710). Commotion and microturbulence are not exactly equivalent, although they both derive from spectroscopic observation. Microturbulence applies to the motion both of atoms and of ions, whereas here commotion is restricted by definition solely to ionic motion.

C. Distinction between Thermal Motion and Mass Motion

Thermal motion produces line broadening while mass motion results in a line shift. The limitation to this distinguishing criterion in a practical measurement is discussed. Great care must be exercised in deducing temperatures from the results of the measurement of macroscopic plasma parameters.

In an elegant presentation, Sitenko (1967) has shown the unity of fluctuation phenomena in plasmas. Either the Debye–Hückel radius or the effective screening distance serves as a natural length scale to divide the fluctuations into two distinct types, although the transition between them is continuous and not abrupt. Within the Debye–Hückel sphere or the equivalent screening cloud of charge, the fluctuations are predominantly thermal (Section II, A); without, plasma waves dominate (Section II, B).

In the light of this division of fluctuations into separate domains, the ideal conditions conducive to a spectroscopic measurement of the ionic velocity field can be described. Consider a volume element of plasma several times greater than the screening cloud of charge (Section II, B). At any instant of time, thermal motion, characterized by the lack of spatial correlation between the radiating ions contained in this volume element, produces line broadening: the Doppler shifts due to thermal velocity fluctuations are unresolved. Mass motion, on the other hand, by virtue of the correlation between radiating ions leads to a line shift. This basic difference then can serve as the criterion of separating the two types of motion in an experimental arrangement. In practice, this criterion is not perfect in that it is usually limited by the available spatial resolution. For a plasma at a temperature of 10^6 K and of electron number density 10^{20} m^{-3} (10^{14} cm^{-3}), the volume of the Debye–Hückel sphere is about 1 fm^3 (10^{-9} cm^3). In a spectroscopic observation of the plasma, the smallest radiating volume which can be resolved is very much larger than this; see the example considered in Section IV, F, 2. When commotion is present, this volume at any instant of time may contain a number of commoving elements of varying sizes, each element containing a large number of ions moving in unison. Thus the individual line shifts, caused by commotion in the line of sight, may merge to produce broadening, in addition to that resulting from thermal motion. On the other hand, this limitation may not be serious. The condition that the additional broadening can be neglected is given in Appendix B.

A similar criterion for separation on the time scale of measurement can be defined. Here the dividing time scale is fixed by the ion plasma frequency (Section II, B). As for the case of spatial correlation, the criterion for temporal correlation is limited by the available temporal resolution. For a number density of 10^{20} helium ions per cubic meter (10^{14} cm^{-3}), the reciprocal of the ion plasma frequency is about 1 ns, which is less than the resolution available in practice. As for spatial resolution, the seriousness of this limitation may be groundless.

To emphasize the problem of distinguishing between the two types of motion, consider a macroscopic measurement in which the hydrostatic pressure p of the ions is measured. Without detailed examination, the ion temperature T cannot be given by the relation

$$p = nkT, \tag{4}$$

where n is the number density of the ions. Mass motion of the plasma may be so fine-grained that it contributes to this pressure where the velocity distribution of the ions on a macroscopic scale may be close to that of Maxwell (Thonemann, 1961, p. 67). The temperature, on the other hand, is defined by the velocity distribution on the microscopic scale of the screening cloud of charge (Section II, A). The measurement in question is valueless as a means of temperature determination unless the components of pressure due to correlated and noncorrelated motion can be resolved.

The distinction brought out here is of practical importance. In a fusion reactor, the rate of binary encounters of the ions leading to thermonuclear reactions is a function of the temperature given by the microscopic velocity distribution (Thonemann, 1961, p. 67).

III. Classification of the Methods of Separating Thermal Motion from Commotion (Random Mass Motion)

Consider a plasma in which systematic motion is absent. The much more complicated case where thermal and systematic motions exist together with commotion is not included in this paper.

Ideal separation of thermal motion from commotion means that the microscopic and macroscopic velocity distribution functions of the ions can be determined. The former yields the ion temperature and the latter the mean square commoving speed. To achieve this in a particular arrangement may be intrinsically impossible. One may only be able to assign values to the temperature and commoving speed without having precise information about the form of the velocity distribution function. Or it may only be possible to detect commotion without separation being feasible. If it can be shown unambiguously that commotion is absent, the temperature is defined by the measured Doppler width.

The experimental methods of separation can be classified into two main classes according to whether the commotion is examined by its average (or macroscopic) properties or statistically. The former class comprises but a single method, the method of different ionic species, where the commoving elements are unresolved and produce line broadening in addition to that given thermally. See Section II, C for the analogous case of pressure measurement. The basis of the second class, numbering three methods, rests on the difference between thermal motion and mass motion already discussed in Section II, C—thermal motion produces line broadening while commotion is manifest as a line shift and can be examined statistically. The three methods differ according to the number of information channels available. The method of line splitting (for example, Berezin *et al.*, 1963), a two-channel arrangement, is inherently poor in producing information and is only capable of determining the relative thermal and commoving speeds. Although this is not separation in the full sense accepted above, the method is nevertheless included here. The minimum number of information channels required to effect separation is three, hence the triple-slit method (Hearn *et al.*, 1962), but this is only possible at the expense of assuming the line profile. The line-profile method (for example, Payne and Kaufman, 1959, p. 37; Zaidel' *et al.*, 1961a, pp. 113–114), the third method, employs a multichannel arrangement. In this case both the line profile and the instantaneous shift of the line are actually recorded; no assumption need be made about the line profile.

IV. Separation of Thermal Motion from Commotion—
The Method of Different Ionic Species

A. INTRODUCTION AND PRINCIPLE OF SEPARATION

The criterion of separation established in Section II, C is inapplicable when the commoving elements are unresolved, and commotion, in addition to thermal motion, produces line broadening. Instead, separation is based on the well-known concept of unfolding a spectral-line profile into its constituent broadening mechanisms. The theory of unfolding is presented subsequently in Section IV, B. The peculiar feature that both broadening mechanisms are of common origin renders the problem of unfolding rather intractable. Nevertheless, an attempt to unfold can be made and the practical method considered is based on the measurement of broadening of spectral lines from different ionic species. All ions bearing the same electrical charge and having the same mass are said to belong to the same ionic species. An historical note (Section IV, C) prepares the way for a logical exposition of the method of different ionic species. Essential to the method is the concept of a common ion temperature in the direction of observation (Section IV, D). The section following, Section IV, E, consists of a theoretical guide to the relationship between the commoving speed and ionic properties. In practice, application of the theory of unfolding imposes restrictions on the measurement of broadening (Section IV, F), while averaging procedure is discussed in Section IV, G. The various schemes of investigation based on the method are outlined in Section IV, H and illustrated there by a number of experimental results. Finally, in Section IV, I the method is evaluated.

B. THEORY OF UNFOLDING

In this section the measured line profile is unfolded by expressing it in terms of the separate probability density functions of the Doppler shifts for thermal motion and commotion. An alternative approach to separation by means of the second moments of these functions is of more limited scope in the information it can yield, yet it is more useful in practice. The latter is chosen as the basis of the various schemes of investigation (Section IV, H). Both approaches are illustrated by examples of practical importance. Comment is made on the assumption that the thermal motion and commotion are statistically independent.

Consider a spatially homogeneous volume of plasma of unit size containing n_j excited ions in the j level. For an optically thin plasma, the radiance L (Section I) of the spectral line due to the transition from the j level to a lower level is proportional to n_j. Of the number n_j, consider those ions with combined thermal and commoving velocities in the range between $\mathbf{u}$ and $\mathbf{u} + \delta\mathbf{u}$. Let the corresponding range of Doppler shifts in the direction of observation l be from $\Delta\lambda$ to $\Delta\lambda + \delta(\Delta\lambda)$. Then, provided the range is small,

$$L(\Delta\lambda)\, \delta(\Delta\lambda) \propto f(\mathbf{u})\, \delta\mathbf{u} \tag{5}$$

where $L(\Delta\lambda)$ is the spectral radiance, or radiance per unit wavelength interval, corresponding to the shift $\Delta\lambda$ and $f(\mathbf{u})$ the velocity distribution function of the n_j ions. The probability of an ion having any specified thermal or commoving speed in the direction l is independent of the components of velocity in other directions, supposing that perfect randomness of direction prevails. Moreover, assume that the thermal motion and commotion are statistically independent. Then the probability density function of the combined motion can be expressed as the sum of separate (or unfolded) probability density functions of thermal motion and commotion (for example, Chandrasekhar, 1943, p. 83, for discrete functions, as is the case here for an assembly of individual ions). Equation (5) can thus be expressed as

$$L(\Delta\lambda)\, \delta(\Delta\lambda) \propto \sum W^{th}(u_l^{th}) W^c(u_l - u_l^{th})\, \delta u_l \tag{6}$$

where u_l is the velocity component of the combined motion in the direction l, and $W^{th}(u_l^{th})$ and $W^c(u_l^c)$ are the respective probability density functions of thermal and commoving speeds in the same direction. For example, $W^{th}(u_l^{th})\delta u_l^{th}$ represents the probability of values of the thermal velocity component u_l^{th} occurring in the range between u_l^{th} and $(u_l^{th} + \delta u_l^{th})$. The summation is over all possible pairs of the discrete values of u_l^{th} and $(u_l - u_l^{th})$. In this section, the two probability density functions are regarded as statistically steady, hence the explicit omission of the time variable t. For ion speeds much less than the speed of light c, as is the case even at a temperature of 10^8 K, Doppler's principle to a first-order approximation is given by

$$\Delta\lambda/\lambda = \pm(u_l/c) \tag{7}$$

where λ is the undisplaced wavelength. On introducing this equation in relation (6) and after appropriate normalization,

$$L(\Delta\lambda) = \sum W_l^{th}(\Delta\lambda') W_l^c(\Delta\lambda - \Delta\lambda') \tag{8}$$

where $W_l^{th}(\Delta\lambda)$ and $W_l^c(\Delta\lambda)$ are the probability density functions of the Doppler shifts for thermal motion and commotion, respectively, in the direction l. Note that this result is similar to that for the folding of two line profiles where a convolution integral of continuous functions is used (for example, Griem, 1964, p. 101). Measurement of the line profile and knowledge of one of the probability density functions permit the determination of the other.

Now consider separation into thermal and commoving components by means of the second moments of the probability density functions and line profile. The averaging procedure (Section IV, G) is such that the mean speeds of thermal motion and commotion in the direction of observation can be taken as zero; the "center-of-mass" of the spectral line is unshifted. Hence, according to the theorem about the mean square deviation of the sum of two probability density functions (Chandrasekhar, 1943, p. 83, for discrete functions; Hughes and MacDonald, 1961, p. 78, for continuous functions), and by combining this theorem with Eq. (2),

$$\langle(u_l)^2\rangle = (kT_l/m) + \langle(u_l^c)^2\rangle \tag{9}$$

where $\langle(u_l^c)^2\rangle$ is the mean square commoving speed in the direction l. Since the radiance is proportional to the number n_j of excited ions and the speed u_l is related to the shift $\Delta\lambda$ through Doppler's principle (Eq. 7), the line profile reproduces the probability density function of the combined speeds, provided the spectral radiance scale is suitably normalized. The root-mean-square value of half the Doppler linewidth is therefore given by

$$(\langle\Delta\lambda^2\rangle)^{1/2}/\lambda = \pm(1/c)[(kT_l/m) + \langle(u_l^c)^2\rangle]^{1/2} \tag{10}$$

where $(\Delta\lambda)^2$ is abbreviated to $\Delta\lambda^2$. Separation of the two types of motion is reduced to the problem of separation of the two terms on the right-hand side of Eq. (10). In the derivation of this equation one supposes that the second moments are finite. The second moment of the Lorentzian function is infinite (for example, Kittel, 1958, p. 131), but this is a function unlikely to be encountered in the present circumstances. Thus the result given in Eq. (10) is quite general. It is applicable to a plasma not in thermodynamic equilibrium where the microscopic velocity distribution function of the ions is not Maxwellian, and to the case where the probability density function of the commotion is not Gaussian.

Two particular cases of Eqs. (8) and (10) are now considered. In the first, instead of using Eq. (8) itself, the form for continuous functions

$$L(\Delta\lambda) = \int_{-\infty}^{+\infty} W_l^{th}(\Delta\lambda')W_l^c(\Delta\lambda - \Delta\lambda')\,d(\Delta\lambda') \tag{11}$$

is employed. The functions $W_l^{th}(\Delta\lambda)$ and $W_l^c(\Delta\lambda)$ can thus no longer be regarded as discrete. If the commotion is both statistically steady and spatially homogeneous, the function $W_l^c(\Delta\lambda)$ is anticipated to be Gaussian about the undisplaced wavelength for a large number of random and independent Doppler shifts. Thus

$$W_i^c(\Delta\lambda) = [1/\sigma(2\pi)^{1/2}] \exp[-(\Delta\lambda^2/2\sigma^2)] \tag{12}$$

where

$$\sigma \equiv (\lambda/c)\langle(u_i^c)^2\rangle^{1/2}. \tag{13}$$

In this case, the continuous function $W_i^{th}(\Delta\lambda)$ is given by Jones and Misell (1967, p. 1481) to be

$$W_i^{th}(\Delta\lambda) = L(\Delta\lambda) - \sigma^2 L^{II}(\Delta\lambda) + (\sigma^4/2!)L^{IV}(\Delta\lambda) - (\sigma^6/3!)L^{VI}(\Delta\lambda) + \cdots \tag{14}$$

where Roman numeral superscripts denote derivatives, for example, the second derivative is denoted by II. Solving for the microscopic velocity distribution function of the ions is thus possible, provided the mean square commoving speed is known, as for example, through solution of Eq. (10). Or an equation similar to Eq. (14) can be used to find the function $W_i^c(\Delta\lambda)$ when the microscopic velocity distribution function of the ions is Maxwellian and $W_i^c(\Delta\lambda)$ is skew.

The second case is the very special one where in addition to the function $W_i^c(\Delta\lambda)$ being Gaussian, the microscopic velocity distribution function of the ions is Maxwellian, that is

$$W_i^{th}(\Delta\lambda) = (mc^2/2\pi kT_i\lambda^2)^{1/2} \exp[-(mc^2/2kT_i\lambda^2) \cdot \Delta\lambda^2]. \tag{15}$$

The spectral radiance $L(\Delta\lambda)$ is also Gaussian where the half-radiance width $\Delta\lambda_{1/2}$ of the line is related to the root-mean-square value by

$$\Delta\lambda_{1/2} = (2 \ln 2)^{1/2}(\langle\Delta\lambda^2\rangle)^{1/2}. \tag{16}$$

Note that $\Delta\lambda_{1/2}$ is *half* the width between the points where the spectral radiance is half its maximum value. The result of combining Eqs. (16) and (10) is that the half-radiance width is given by

$$\Delta\lambda_{1/2}/\lambda = [(2 \ln 2)^{1/2}/c][(kT_i/m) + \langle(u_i^c)^2\rangle]^{1/2} \tag{17}$$

(for example, Unsöld, 1955, pp. 286–287). An important corollary to this case is that if the line profile is measured and proves to be Gaussian, the individual probability density functions $W_i^{th}(\Delta\lambda)$ and $W_i^c(\Delta\lambda)$ must also be Gaussian (Hughes and MacDonald, 1961, pp. 79–80).

The simple separation of thermal motion from commotion as expressed by Eqs. (8) and (10) is only possible by assuming their statistical independence. Can this assumption be justified? Consider the case of random ion acoustic waves. The obvious connection between the two types of motion is in wave damping (Section II, B), which is strong only for circular wave numbers comparable to the reciprocal of the screening distance (for example, Spitzer,

1962, pp. 82–86). These wave numbers are expected to be associated with little energy (see Appendix B) and so the contribution of the statistically dependent commoving elements to the term $\langle (u_l^c)^2 \rangle$ in Eq. (10) is small. However, this is not the whole story. In nonlinear damping, two waves combine to produce a virtual wave of phase velocity comparable to the thermal velocity of an ion, the wave energy being transmitted to the ion (for example, Kadomtsev, 1965, p. 69; Sagdeev and Galeev, 1969, p. 2 and Chapter III). Since these waves may contain considerable energy, the validity of the assumption under discussion is questionable. Further consideration is beyond the scope of the present account.

C. Historical Note

The method of different ionic species is a development of that for neutral-atom radiators. In the latter, the distinguishing features of the velocity field are that the root-mean-square thermal speed of the radiator varies inversely as the square root of its mass, while it is supposed that radiators of different masses simultaneously present in the same volume element all possess, as in fluid turbulence, the same mass velocity (for example, Bell and Meltzer, 1959, p. 39). Thus, use of Eq. (10) for two elements of different atomic mass should suffice to yield the common temperature T_l and the mean square speed of the mass motion. In the absence of a common temperature, the method of separation fails.

This method was applied fairly successfully by Bell and Meltzer (1959) to the measurement of solar temperatures using absorption lines. Nevertheless the method is seriously limited in scope because the temperature and mass velocity may be a function of depth in the solar atmosphere, while the radiance distribution in depth may differ from element to element (for example, Bell and Meltzer, 1959, p. 41; Jefferies, 1962, pp. 704–705). On the other hand, Hirayama (1964, pp. 108–110 and 1969) obtained very consistent results of the two–dimensional distribution of temperature and mass velocity in solar quiescent prominences. Emission lines of the Balmer series in hydrogen were paired with those of the spectra of Mg I, Fe I, and Sr II in Eq. (10).

When the same method was considered for lines of multi-ionized atoms in hot laboratory-plasmas, difficulties were encountered. One complicating factor was that ions of the same element in different stages of ionization possessed different Doppler widths when they were scaled to the same wavelength: the width increased with increasing ionic charge. While the interpretation of this effect for most of the plasmas considered was ambiguous because temporal and spatial resolution were lacking (see Zaidel' et al., 1962, p. 265), the results from one of the stellarator assemblies (Hirschberg and Palladino, 1962) indicated that, contrary to neutral atoms, the different ionic species were not characterized by a common mass velocity (further details are presented in Section IV, H, 1). Hence, the simple method of analyzing the velocity field as proposed for neutral atoms cannot be generally applicable to spectral lines radiated by ions.

A similar problem is known from spectral observation of spicules, the fine jets in the solar chromosphere (for a review, see Beckers, 1968, especially pp. 386–389, 397). The widths of singly ionized calcium lines are large compared with those of the lines of hydrogen and neutral helium. Two of the explanations offered are similar to those referred to above. The first supposes that the lines of the three elements cannot all originate in the same region. They are radiated, however, simultaneously from the spicule (Pasachoff *et al.*, 1968, p. 143). The second explanation is that the calcium ions and neutral atoms do not possess the same random mass velocity. The argument advanced against this possibility (Beckers, 1968, p. 389) is not convincing. Spicular rotation has also been proposed to explain the large width of the K line (393.368 nm [3933.68 Å]) of Ca II. Much work requires to be done before this astrophysical problem can be solved.

D. COMMON TEMPERATURE

An essential feature of the method of different ionic species is that the various species should be bound by a common temperature in the direction of observation. This condition necessitates discussion of the time of energy equipartition between the ionic species. The subject is difficult and no foolproof argument can be submitted to prove a common temperature, should it exist; nevertheless, the form of the line profiles acts as a pointer in this respect.

The ionic species can relax to a common temperature by processes which are regarded as separable (Section II, C), namely, binary encounters between ions with impact parameter less than the screening distance and collective effects usually known as wave–particle interactions.

For binary encounters, Spitzer (1962, p. 135) calculated the time taken for the equipartition of energy between two ionic species. He assumed that a magnetic field was absent and that the initial microscopic velocity distribution functions of the species were Maxwellian. Fairly satisfactory agreement between Spitzer's theory and experiment was reached by Bottoms and Eisner (1966) for the equipartition of energy between electrons and ions in hydrogen and helium plasmas. Spitzer's theory can be improved upon in various ways, for example, by taking into account the effects of a strong magnetic field where necessary (for example, Silin, 1963) and microscopic velocity distribution functions which are initially not Maxwellian (for example, Lehner, 1967).

The effect of wave–particle interactions was calculated for the case of energy equipartition between electrons and ions in a stable plasma (Ramazashvili *et al.*, 1963; Wu *et al.*, 1965; Lehner and Pohl, 1968). There do not appear to be parallel calculations for the case of equipartition between ionic species. For a plasma in a state of random mass motion, the situation is very complex. However, the general pattern appears to be clear—wave energy can be fed to particles by wave damping while the reverse process in which particle energy is transmitted to waves can also occur (for example, Jukes, 1968, p. 333). Physically, both of these effects can be looked upon as scattering of particles by waves so that the collision frequency can be effectively increased (for example, Petviashvili *et al.*, 1965; Weinstock, 1968).

One could conclude that, in the absence of a strong magnetic field, Spitzer's equipartition time can be regarded as an approximate upper limit to the net effect of both binary encounters and wave–particle interactions. However, there may be a source of energy driving the ions away from relaxation, for example, an externally applied electric field or the stochastic field of electrostatic plasma waves themselves. Thus it is possible to envisage the plasma even settling down to an approximately steady "turbulent" state in which the various ionic species possess their own individual temperatures and no common temperature exists (see, for example, Sawyer *et al.*, 1963, p. 1897).

A criterion is required to indicate a common temperature if it exists. Consider the following argument, although it cannot be rigorously justified. The line profiles for the different ionic species are examined and found to be Gaussian or nearly so (for example, the "turbulent-heating" experiment of Kovan and Spektor, 1968, p. 750); methods of checking this accurately are presented by Griffin *et al.* (1962, pp. 656–659). A Gaussian profile produced by both thermal motion and commotion shows that the probability density function of thermal speeds in the direction of observation is also Gaussian (Section IV, B). One can thus conclude that any source driving the ions away from self-relaxation is not powerful enough to prevent its establishment in that direction. Under the conditions prevailing in the schemes of investigation (Section IV, H), the masses of the ionic species are not expected to differ by more than an order of magnitude while the charges may differ by a factor five at the most (Appendix C). Hence, it seems reasonable to assume that the same source will not prevent the establishment of a common temperature, especially if the line profiles from *all* the ionic species are Gaussian or nearly so.

In the design of an experiment based on the method of different ionic species, a common temperature is supposed a priori to exist. Examination of the line profiles as outlined above may show whether this assumption is justified. A necessary condition in design is that the observation time of the line be much greater than the equipartition time between ionic species. The equipartition time given by Spitzer (1962, p. 135) appears to be reasonably safe and easily calculable for this purpose. Should the driving source be insufficiently strong to prevent the establishment of a common temperature but adequate to prolong the time of equipartition beyond the time fixed to observe the line, a distinct departure from a Gaussian profile may be evident.

E. Mean Square Commoving Speed: Theoretical Guidance

In the light of the evidence presented in Section IV, C, one cannot proceed with the application of Eq. (10), unless the mean square commoving speed can be related to the properties of the ionic species. To obtain an exact relation it is necessary to derive a complete theory of the motion of ions in random electromagnetic fields. To date, such a theory does not exist. Moreover, there is the hope, as expressed in the Introduction, that investigation of Doppler-broadened lines may not only shed light on the magnitude of the ion temperature but may also reveal features of any random mass motion present in the plasma. Over-emphasis of theory could possibly defeat this purpose through the construction of a closed system of logic in which the model has been defined at the start to have just the properties that one is going to deduce (Ziman, 1965, p. 1189). Nevertheless, some theoretical guidance is necessary and this is the theme of the section. The main assumptions are

(a) the resolution of the random mass motion into the two statistically independent forms, random electrostatic plasma waves and plasma turbulence (rotational motion), and

(b) the adoption of simple polynomial forms for the mean square commoving speeds of the ions. Polynomial representation is regarded as the best under the circumstances of generality, and paucity of experimental information; only a few values of the discrete variables, the ionic charge ze and mass m, are available according to the number of ionic species in use.

A fully ionized gas in which random mass motion is prominent can be regarded as a dynamic system under the effect of constraints and the reaction of the system to these constraints. The constraints, which are both of external and internal origin (arising from the plasma itself) and of random character, include electric and magnetic fields and the effects of number density and temperature gradients. The reaction of the plasma expresses itself, for example, in the electrostatic interactions among the particles and collective effects, such as wave damping. Writing down an explicit formulation of the history of the plasma, as, for example, in a generalized Lagrangian coordinate system (Orszag, 1969) is of little value in this discussion. Instead, breakdown of the various constraints and reactions into basic parameters shows that, in general, the mean square speed of random mass motion $\langle (u_l^{mm})^2 \rangle$ in the direction l can be expressed as

$$\langle (u_l^{mm})^2 \rangle = f(ze, m, ze/m, n, T_l, \mathbf{E}, \mathbf{B}, \omega, \mathbf{k}), \tag{18}$$

where $\mathbf{E}$ is the electric field strength, $\mathbf{B}$ the magnetic flux density, and ω and $\mathbf{k}$ are the angular frequency and circular wave vector of the random variables, respectively.

Some comment on Eq. (18) is desirable. The variable ze/m appears because it is related to the acceleration of a charged particle in an electromagnetic field. One of the reasons for the inclusion of the ionic charge ze independently of the ratio ze/m is that particle interactions are electrostatic; in the Hamiltonian describing the energy of the system, the mass m appears in the kinetic energy term in additive form, while the electrostatic energy term involves the charge ze as a product (for example, Balescu, 1963, p. 1). Likewise, the appearance of m by itself is dictated, for example, by transport coefficients and by any effects involving the microscopic velocity distribution, such as wave–particle interaction. A fluid velocity is not specially included, it being supposed that interpenetrating streams of particles derive their motion from existing fields.

The essence of the method of different ionic species is, firstly, to suppose that one can identify the theoretical mean square speed $\langle (u_l^{mm})^2 \rangle$ with the mean square commoving speed $\langle (u_l^{c})^2 \rangle$ of Eq. (10). Secondly, it is assumed that $\langle (u_l^{mm})^2 \rangle$ can be expressed solely as a function of the ionic parameters ze and m, the other parameters in Eq. (18) being regarded constant for all species used. Can this latter assumption be justified? A common temperature T_l is fundamental to the method of separation (Section IV, D). It is for this reason that T_l rather than the root-mean-square thermal speed [Eq. (2)] appears in Eq. (18), although the latter is more basic. Localization and simultaneity of the species in the plasma are a necessity (Section IV, F). It would appear at first glance that if these two conditions are satisfied, the electromagnetic field characterized by $\mathbf{E}$, $\mathbf{B}$, ω, and $\mathbf{k}$ is the same for the various species. However, an ion describes a path in the field that varies according to the values of ze and m (Jones and Wilson, 1962, p. 890). Thus effects of inhomogeneity of the field must be judged in relation to the limited practical availability of localization. That the ionic number density n is the same depends on the type of species used. If they are formed from different elements, the known initial number density of neutral gas is inadequate to fix the ionic number density, one reason being that the atoms are distributed among several stages of ionization. It may be possible to tie n to a common value by measurement of spectral-line radiance (for example, McWhirter, 1965, p. 209, Eq. 16). In the instance where the species are formed from the same element in different stages of ionization, there is no possibility of a common value for n. Another possible factor, albeit a small one, is that the spectral range of frequencies and wave numbers actually varies from ion to ion. The largest permissible frequency is given approximately by the ion plasma frequency (Section II, B) which is proportional to $ze/m^{1/2}$. The largest wave number shows the same dependence [Eq. (3)]. A similar argument applies for ion–cyclotron damping (for example, Spitzer, 1962, pp. 86–88), except that the dependence is ze/m.

Equation (18) is too general to be of practical value. A more specific form can be written by considering the forms of mass motion described in Section II, B, namely, random electrostatic plasma waves and random rotational motion. Ionic motion due to transverse electromagnetic waves (radiation field) is ignored. For a helium plasma at a temperature as high as 10^8 K, the root-mean-square speed of the ions is much smaller than the speed of light.

In the case of random electrostatic waves, one generally expects to find the factor $(ze/m)^2$ in the expression for the mean square speed $\langle (u_i^e)^2 \rangle$ of the ions (for example, Stix, 1962, p. 65, Problems 3 and 4). Thus, for use in the subsequent schemes of investigation, assume the polynomial form

$$\langle (u_i^e)^2 \rangle = g(z/m)^2 z^\alpha m^\beta \tag{19}$$

where the coefficient g and the indices α and β are considered constant for the various ionic species, notwithstanding the qualifying remarks made above. Other than simplicity, the form of Eq. (19) cannot be justified except on the grounds that it is a generalization of the results obtained from a simple model of electrostatic waves: for a noninteracting particle in a "monochromatic" electric field, the factor $z^\alpha m^\beta$ in Eq. (19) is unity (Hirschberg and Palladino, 1962, p. 50). Other models (for example, Jones and Wilson, 1962, p. 889; Tchen, 1969, p. 238) introduce the charge ze and mass m in an increasingly more complex way depending on the degree of approximation. The indices α and β are introduced in an attempt to correct the characteristic factor $(ze/m)^2$ for the effects of the numerous constraints and reactions which are dependent on ze and m.

Now consider random rotational motion called plasma turbulence by Kovásznay (1960, p. 817), who also coined the terms kinetic turbulence and magnetic turbulence to describe the two components of this motion (Section II, B). Kinetic turbulence, in which ions and electrons move in unison, resembles the fluid turbulence in electrically nonconducting fluids. Kinetic turbulence is clearly independent of ze and m (Section IV, C). In magnetic turbulence, a random motion of electric eddies, the net momentum is zero by definition (Section II, B),

$$\sum nm\mathbf{u}^m + n_e m_e \mathbf{u}_e^m = 0 \tag{20}$$

where $\mathbf{u}^m$ and $\mathbf{u}_e^m$ are the respective mass velocities of the ions and electrons and m_e is the electronic mass; hence, ions of different mass move with different mass velocity. For plasma turbulence as with random electrostatic waves, assume polynomial representation of the mean square speed $\langle (u_i{}^p)^2 \rangle$ of the ions, thus,

$$\langle (u_i{}^p)^2 \rangle = hz^\gamma m^\delta \qquad (21)$$

where the coefficient h and the indices γ and δ are constants for the different species. The factor z^γ is inserted to account for departures from charge equilibrium. If δ in Eq. (21) is equal to -1 and there is no charge dependence, plasma turbulence as expressed by the second moment of the probability density function is indistinguishable from thermal motion.

One returns now to Eq. (18) and assumes that the ionic motions in random electrostatic waves and plasma turbulence are statistically independent. Thus, according to the theorem about mean square deviations (Section IV, B),

$$\langle (u_i{}^c)^2 \rangle = \langle (u_i{}^{mm})^2 \rangle = \langle (u_i{}^e)^2 \rangle + \langle (u_i{}^p)^2 \rangle. \qquad (22)$$

Kovásznay (1960, p. 817) argued intuitively that for a plasma confined by a magnetic field and not bounded by physical boundaries, there should be little interaction between the two forms of motion. However, it would appear that this topic requires careful examination in a given situation. On combining Eq. (22) with Eqs. (19) and (21),

$$\langle (u_i{}^c)^2 \rangle = g(z/m)^2 z^\alpha m^\beta + hz^\gamma m^\delta \qquad (23)$$

which is the desired result.

In conclusion, there appears to be a fairly clear-cut division between random electrostatic waves and plasma turbulence, due to the prominence of the factor $(ze/m)^2$ in the former. While the random mass motion is reduced analytically to irrotational and rotational forms, in which the former is considered oscillatory only, one cannot exclude the possibility of velocity fluctuations due to small-scale translational motion. However, Eq. (23), in its polynomial representation of the variables z and m, should be general enough to include all possibilities.

F. CONDITIONS OF MEASUREMENT

Equation (10) in Section IV, B is derived for conditions where the ion temperature T_i and the mean square commoving speed $\langle (u_i{}^c)^2 \rangle$ are single valued. In general, both T_i and $\langle (u_i{}^c)^2 \rangle$ can be functions of both space and time for a given ionic species. Thus, for proper use of Eq. (10), the line profile must be formed from radiant energy emanating from ions present simultaneously (the condition of *simultaneity*) in the same volume element (the condition of *localization*). If the commotion is spatially homogeneous and statistically steady, these conditions apply only to T_i. Moreover, T_i and $\langle (u_i{}^c)^2 \rangle$ can only be single valued provided all the ions of a given species experience, on the average, the same physical processes during the plasma cycle until the measurement of broadening is complete. This condition is given the name *same plasma history*.

Simultaneity and localization of the radiating ions of the different ionic species are also a necessity, allusion to which is made in the historical note (Section IV, C). On its face value, the condition of simultaneity means observing the spectral lines of the various species during the same time interval, which is kept as short as possible. However, this concept of simultaneity must be modified to accommodate the opposing condition—that the observation time of the line be much greater than the equipartition time between ionic species (Section IV, D). It is in this modified sense that the condition of simultaneity is to be understood here.

1. Simultaneity

In general, temporal resolution of the spectral lines is a necessity to satisfy the condition of simultaneity. Here there is no need to dwell on the methods of time-resolved spectroscopy, since up-to-date and adequate reviews are available [for example, Griem, 1964, pp. 229–232, 234, 236; Turner, 1965, pp. 327–332, 336; Neumann, 1967, pp. 31–45; Lochte-Holtgreven and Richter, 1968, pp. 330–342 (see also pp. 368–369 of the same volume); Pimentel, 1968; and solely for the Fabry–Perot etalon, Greig and Cooper, 1968].

A calculated equipartition time for binary encounters (Spitzer, 1962, p. 135) of about 5 μs is typical for some of the plasmas considered here. A considerably longer observation time of, say, 50 μs is well within the capabilities of available instrumentation, even in the vacuum ultraviolet region of the spectrum. Temporal resolution in the X-ray region is also feasible (Bogen, 1968, p. 461).

2. Localization

Spatial resolution of the emitted radiant energy is necessary to satisfy the condition of localization. A stigmatic spectrometer (for example, Griem, 1964, p. 229) can be used for this purpose. For high accuracy, two alternative arrangements consisting of a lens and a pair of aperture stops are available (Webb, 1968, pp. 5444–5448, 5466–5469). In the vacuum ultraviolet region of the spectrum where lenses cannot be used, care has to be taken to choose a spectrometer with little or no astigmatism (for example, Lincke, 1968, pp. 360–368; Namioka, 1962). In the X-ray region, a pinhole arrangement in conjunction with a crystal spectrometer can be employed to supply spatial resolution (Bogen, 1968, pp. 461–462). All these methods are not discussed here in any detail. Instead, special reference is made to observation in the end-on and side-on (or transverse) directions.

Consider first end-on observation where the line of sight is coincident with, or parallel to, the axis of symmetry of the plasma vessel, such as a cylinder or race track (Fig. 1). The resolved volume element can be a cylinder parallel to the axis and of cross section given by the scanning aperture (Webb, 1968, p. 5444). If the temperature and commotion are uniform in the end-on direction, the problem of spatial resolution is simplified and localization is determined by the scanning aperture. This condition can be checked to a certain extent by side-on observation at intervals along the length of the vessel: an unchanged line profile should be a reasonable indicator of uniformity in the end-on direction. The minimum size of scanning aperture is limited by diffraction (for example, Webb, 1968, pp. 5445–5446), which is only of theoretical interest for most hot plasmas because of their size. It is perhaps only in a plasma-focus experiment where typically the radius and length of the hot-plasma region are 1 mm and 5 mm, respectively (for example, Mather, 1965, p. 369), that diffraction could be significant. For example, at a wavelength of 100 nm (1000 Å), the approximate useful limit of a lens, the theoretical resolution limit in Webb's arrangements is 20 μm: the corresponding resolvable volume in the plasma focus is 5 pm^3 (5×10^{-6} cm^3). Of more practical value in defining the size of scanning aperture is the availability of radiant energy. A further point meriting consideration is end effects. For example, in race-track geometry, correction of the results may be necessary to account for the effect of the optical path traversing the curved portion of the vessel (Fig. 1).

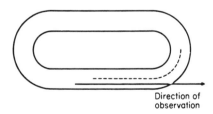

Direction of
observation

FIG. 1. End-on observation with race-track geometry.

Side-on observation (Fig. 2) of a plasma contained in a vessel in the form of a cylinder, torus, or race track is more complicated. The spatial distribution of line width across a section perpendicular to the axis of symmetry can be measured with a scanning accuracy comparable to that for end-on observation. However, in this case, because the measured distribution lacks circular symmetry, the line profile represents the integrated effect of the radiance and broadening distributions along a chord of the cross section. Inversion of the

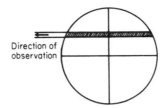

FIG. 2. Side-on observation with the cross section of the plasma vessel shown.

measured distribution to a radial distribution can be performed by a method based on the Abel integral equation in a first-order approximation and assuming a Gaussian line profile (Bohn *et al.*, 1967). Naturally the assumption of a Gaussian profile limits the generality of inversion but this may not be a drawback. The method of different ionic species is itself limited in scope because of the need for a common temperature, a possible criterion for which is a Gaussian line profile (Section IV, D). Numerical methods of data reduction using the Abel integral equation (for example, Griem, 1964, pp. 176–178; Lochte-Holtgreven, 1968, pp. 184–186, 203–209) were reviewed and compared by Cremers and Birkebak (1966) and by Bielski *et al.* (1968).

While end-on observation obviates the need of inversion, side-on observation is generally necessary. The former cannot be undertaken with toroidal geometry, and one of the ways of detecting systematic motion is to observe in two different directions (for example, Eberhagen *et al.*, 1965). Moreover, provided inversion can be satisfactorily performed, side-on observation provides proper localization, whereas end-on observation assumes uniformity of temperature and commotion along the line of sight.

3. Same Plasma History

This condition implies that in the case where the ions are formed from gas initially present in the plasma vessel, injection of further gas (either neutral or ionized) into the observed plasma region during the plasma cycle cannot be tolerated (Jones and Wilson, 1962, p. 890). Obviously, the injected gas would not experience the same physical processes, for example heating, as the gas initially present.

Injection of gas originating in the walls of the plasma vessel is known to occur (Dimock, 1960; Burton and Wilson, 1961, Sections 4 and 5; and Kunze and Gabriel, 1968, Section V), but little has apparently been done to observe and understand the phenomenon in detail. Nevertheless, certain guiding principles on how to eliminate or reduce this form of injection can be discerned. With this object in mind, chemically inert gases such as argon and neon are much preferred to other elements such as carbon, nitrogen, oxygen, and silicon (Burton and Wilson, 1961, p. 1426; Kunze and Gabriel, 1968, p. 1220). The gas concentration (Kunze and Gabriel, 1968, p. 1219) and number of plasma cycles appear to be important factors; even with argon, effects of injection were eventually perceptible after many discharges in the apparatus ZETA (Burton and Wilson, 1961, p. 1429). The wall material upon which the gas is adsorbed may be another factor, carefully cleaned stainless steel being possibly superior to fused silica in reducing injection. Otherwise, it is difficult to reconcile the apparently conflicting results obtained with methane (Dimock, 1960; Kunze and Gabriel, 1968). The chemical form of the element injected is also significant (for methane versus carbon dioxide, see Dimock, 1960; for molecular oxygen deliberately introduced versus oxygen presumably released from walls, see Kunze and Gabriel, 1968). Finally, injection can be reduced when the strength of the magnetic field containing the plasma is increased (Burton and Wilson, 1961, p. 1426; Heroux, 1964, pp. 130–131). To summarize, it appears that the rare gases are particularly immune to the injection process; immunity may exist for other elements depending on circumstances.

The problem of gas injection can presumably be averted in highly transient plasmas (I am grateful to Dr. J. R. Greig of the University of Maryland for making this point). The condition for this is that the time from the commencement of the plasma cycle (including preliminary heating phases) to completing the observation of the spectral line should be less than the time taken for the injected gas to enter the region of the radiating ions and to be ionized to the requisite degree of ionization.

G. Averaging Procedure

That commotion results in line broadening means that the individual line shifts associated with the commoving elements are averaged. The ensemble average over a large number (strictly, an infinite number) of plasma cycles cannot be achieved in practice. Any averaging procedure must include an average of the commotion over the resolved volume of plasma and observation time of the line (Sections IV, F, 2 and IV, F, 1). The averaging possibilities are (a) Combined time and space average. Observation of the line takes place for a single plasma cycle. (b) Combined ensemble, time, and space average. Observation takes place for a large number of plasma cycles.

Consider the merits of the two averaging procedures for end-on observation. In the former, the largest commoving elements may be of a size greater than the scanning aperture and so their effect would not be properly averaged. On the other hand, any attempt to incorporate the effect of these elements by the use of inferior spatial resolution increases the chance that radial variations in temperature and commotion are introduced. In view of this difficulty, the latter procedure is expected to give results closer to the ensemble average than the former. A similar argument applies to the time average.

The ion temperature is defined by the average kinetic energy of a group of ions contained within the screening cloud of charge (Section II, A). In practice, the measured ion temperature is an average over a group of radiating ions contained within the spectroscopically resolved volume during the total observation time, whether this be during one plasma cycle or over a number of cycles.

H. Schemes of Investigation

Three schemes of investigation are considered, depending on whether the ionic charge, mass, or both charge and mass, are varied.

A note about nomenclature as used in this section. The "apparent temperature" is defined as that temperature deduced from the Doppler width assuming thermal motion alone. Use of the "apparent temperature" permits a direct association between information deposited here and that provided in the literature. The quotation marks are a reminder that, in general, an "apparent temperature" is *not* to be equated to a kinetic temperature.

1. Varying the Ionic Charge

This scheme of investigation employs ionic species formed from the same element in different stages of ionization. The results of Hirschberg and Palladino (1962) indicate the feasibility of employing it in practice. However, measurements made on many other plasmas defy spectroscopic interpretation because little or no attention was paid to the conditions of measurement, especially localization (Section IV, F, 2).

a. Procedure. The polynomial form of the mean square commoving speed of the ions, as expressed in Eq. (23), is inserted in Eq. (10) to give

$$(\langle \Delta \lambda^2 \rangle)^{1/2}/\lambda = \pm(1/c)[(kT_i/m) + g_1 z^{(2+\alpha)} + h_1 z^\gamma]^{1/2}. \tag{24}$$

The constants g_1 and h_1, which incorporate the common factors involving the mass m, replace the constants g and h of Eq. (23). To determine the five unknowns, namely, T_i, g_1, α, h_1, and γ, five ionic species are required. Appendix C is devoted to the problem of their selection. If this number of species is not available, one way of reducing the number of unknowns is to forego the dependence of the plasma turbulence term on charge ze. The thermal and plasma turbulence terms are now indistinguishable and can be replaced by a single constant H. The unknowns H, g_1, and α can be determined provided the line widths of at least three species are known.

b. B-65 Stellarator. Hirschberg and Palladino (1962) performed measurements of Doppler broadening with oxygen and nitrogen ions in various stages of ionization in a stellarator assembly, the B-65 with divertor, at Princeton. The most significant result is the increase in Doppler width with ionic charge.

Considerable care was exercized to satisfy the conditions of localization and simultaneity (Section IV, F). Observation was end-on and was thus in a direction parallel to the applied magnetic field. The light accepted was restricted to a beam of cross section 100 mm^2. Within the accuracy of measurement, the Doppler widths remained unchanged as the aperture of the beam was varied from 50 mm^2 to 400 mm^2 and when the beam was displaced 10 mm to one side. It was apparently assumed that the ion temperature did not vary along the length of the light path, although this is not stated. No measurements were possible in the side-on direction. Line widths were evaluated at time intervals of 100 μs during the plasma cycle with an observation time for each measurement estimated as between 20 μs and 50 μs (Hirschberg, private communication). No obvious attention was paid to the history of each ionic species in the plasma (Section IV, F, 3). It may be inferred that the various species originated in the walls of the plasma vessel (see Burnett *et al.*, 1958) and were thus injected into the plasma, but little more can be deduced. The method of averaging used is not obvious.

Direct information on the existence of a common temperature is lacking. The form of the line profiles is not given, nor is it certain that the observation time was greater than the equipartition time for binary encounters because the number densities of the ionic species are not known. Nevertheless, for the most highly ionized species O V and O IV, this condition was probably satisfied.

Hirschberg and Palladino (1962) analyzed their results in a simple manner by taking the index α in Eq. (24) as zero and neglecting the term of plasma turbulence, thus,

$$\langle\Delta\lambda^2\rangle = (\lambda/c)^2(g_1 z^2 + H). \qquad (25)$$

This expression was applied graphically to the measured data for the ionic species O II, O III, O IV, and O V (Fig. 3). The fact that their fitted straight line cuts the ordinate at about 15 eV was interpreted to mean that a common temperature of this value existed in the plasma. In contrast, note that the "apparent temperature" of the O V species is as high as 47 eV. They tentatively concluded that random electrostatic waves parallel to the magnetic field were responsible for the dependence of Doppler width on ionic charge.

The following comments on their results are pertinent. The ion temperature was deduced assuming the absence of plasma turbulence. Secondly, the straight-line fit in Fig. 3 was obtained for results which appear to represent a kind of time average over the plasma cycle, although this is not explicitly stated. The

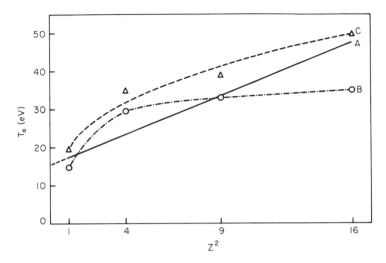

FIG. 3. The "apparent temperature" T_a as a function of z^2 for oxygen ions with the B-65 stellarator. *A*, straight line fitted by Hirschberg and Palladino (1962); *B*, their data for 0.4 ms; *C*, their data for 0.6 ms.

individual time-resolved results do not fit a straight line mainly because of the low values of the Doppler widths for the O II species (see Fig. 3). This is perhaps significant. A satisfactory solution of Eq. (24) should incorporate the case where the charge ze is unity, the commotion terms then being insensitive to the value of the index. No results could be evaluated for side-on observation. Despite these shortcomings, the general impression gained is that the method is workable, at least in a semiquantitative manner, and that the results lend credence to the existence of random electrostatic waves which produce line broadening in addition to thermal broadening.

c. Other Plasmas. In a variety of plasmas, for example, high-current toroidal pinched discharges (Kaufman *et al.*, 1960, pp. 19–20; Zaidel' *et al.*, 1961b; Platz and Hirschberg, 1965), a carbon arc (McNally and Skidmore, 1963, pp. 703–704), a vacuum spark (Edlén and Svensson, 1965, pp. 440–441) and a plasma injector (Gryziński *et al.*, 1968), an increase in Doppler width with ionic charge for ions of the same element was also observed. An examination of the relevant papers reveals that there is no clear-cut *spectroscopic* evidence to explain this effect in terms of random electrostatic waves. Either the measurements were performed without spatial resolution (Section IV, F, 2) or no information about this was given. An alternative or additional explanation is that lines of ions bearing different charges correspond to different regions and times of the plasma (for example, Kaufman *et al.*, 1960, p. 20); see in addition the historical note (Section IV, C).

2. Varying the Ionic Mass

a. *Procedure.* In this scheme, the species are formed from atoms of different elements in the same stage of ionization. Thus, the mass m is variable while the charge ze is constant. Equation (10), by combining it with Eq. (23), is in the appropriate polynomial form,

$$(\langle \Delta \lambda^2 \rangle)^{1/2}/\lambda = \pm(1/c)[(kT_i/m) + g_2 m^{(\beta - 2)} + h_2 m^{\delta}]^{1/2} \qquad (26)$$

where g_2 and h_2, by incorporating factors common to the charge ze, replace the constants g and h of Eq. (23). Solution for the unknowns T_i, g_2, β, h_2, and δ calls for the measurement of line widths for five elements. If this number of ionic species is not available because, for example, the conditions of measurement cannot be met (see Appendix C), one possibility of reducing the number of unknowns is to assume that the commotion terms can be represented by a single polynomial, thus

$$(\langle \Delta \lambda^2 \rangle)^{1/2}/\lambda = \pm(1/c)[(kT_i/m) + qm^{\phi}]^{1/2} \qquad (27)$$

where the coefficient q and index ϕ are constants for the available species.

b. *Results.* As far as is known, this scheme has not been tried. An attempt can be made to analyze some of the data provided by Hirschberg and Palladino (1962, Tables I and II). Nothing consistent can be gleaned from that pertaining to the species N III, O III, and Si III. On the other hand, the results for singly ionized species at times 0.4 ms and 0.6 ms during the plasma cycle are reasonably consistent (Table I): they appear to indicate the absence

TABLE I

"APPARENT TEMPERATURES" (eV) IN THE B-65 STELLARATOR

Ionic species	Time (ms)			
	0.4	0.5	0.6	0.7
He II	14	15	18	21
N II	16.6	12.3	20.2	13.9
O II	14.8	17.1	19.5	20.5
Cl II	16.5	12.6	16.5	13.6

of any significant commotion, the singularity of form m^{-1} excepted. An "ion temperature" of about 15 eV at 0.4 ms to 18 eV at 0.6 ms is consistent with the findings from varying the ionic charge (Section IV, H, 1).

3. Varying the Ionic Charge and Mass

a. Procedure. The ionic species are chosen with both charge and mass as variables. Equation (10), by combining it with Eq. (23), reads as

$$(\langle \Delta \lambda^2 \rangle)^{1/2}/\lambda = \pm(1/c)[(kT_i/m) + g(z/m)^2 z^\alpha m^\beta + hz^\gamma m^\delta]^{1/2}, \qquad (28)$$

so there are seven unknowns.

b. Results. The results obtained with this scheme of investigation may be conveniently divided into two categories, those in which the "apparent temperature" was found to be a function of ionic charge and mass, and those in which it was sensibly independent of them. The results with the apparatuses ZETA and SCYLLA I are in the former category.

(1) ZETA. Jones and Wilson (1962, pp. 891–892), in their analysis of results of measurements performed on the apparatus ZETA, assumed that the various ionic species were bound by a common temperature and common commoving velocity. Thus, instead of using Eq. (28) they resorted to the basic Eq. (10)

$$(\langle \Delta \lambda^2 \rangle)^{1/2}/\lambda = \pm(1/c)[(kT_i/m) + \langle (u_i^c)^2 \rangle]^{1/2}, \qquad (10)$$

where the mean square commoving speed $\langle (u_i^c)^2 \rangle$ is taken as constant. The temperature T_i and $\langle (u_i^c)^2 \rangle$ were evaluated three times, for three ionic species of neon, argon and krypton (with different charges) taken two at a time in Eq. (10). The values obtained were consistent with each other. However, Zaidel' *et al.* (1962) claimed that the form

$$(\langle \Delta \lambda^2 \rangle)^{1/2}/\lambda = \pm(1/c)[g(z/m)]^{1/2}, \qquad (29)$$

where the temperature term in Eq. (28) is regarded as very small compared with the commotion term, fitted the results equally as well. Another possibility is to suppose, like Jones and Wilson (1962), that electrostatic waves play no significant part and that there is no dependence on charge in the plasma turbulence term. One can then attempt to fit their data to the form

$$(\langle \Delta \lambda^2 \rangle)^{1/2}/\lambda = \pm(1/c)[(kT_i/m) + hm^\delta]^{1/2} \qquad (30)$$

and solve once for T_i, h, and δ. Note that Eqs. (30) and (27) are equivalent.

Of greater significance than their analysis is the painstaking care with which Jones and Wilson (1962) undertook the attempt to fulfill the conditions of measurement (Section IV, F), more so than in any other investigation employing the method of different ionic species. The conditions of simultaneity and same plasma history were properly satisfied. No attempt was made to satisfy the condition of localization; on the other hand, there was good reason to suppose that the spatial distribution of light emission was similar for the species used. As regards averaging procedure (Section IV, G), it can be inferred that the line profiles were the result of averaging over a number of discharges (Jones and Wilson, 1962, p. 890).

No direct reference is made to the form of the line profiles. On the other hand, the observation time of a line of about 100 μs or more (Burton and Wilson, 1961, p. 1427) was much longer than the equipartition time of about 10 μs calculated for binary encounters, so that a common temperature in the direction of observation may have existed (Section IV, D).

(2) *SCYLLA I.* Doppler widths were measured with the θ-pinch apparatus SCYLLA I for the ionic species C V, O VII, O VIII, and Ne IX (Sawyer *et al.*, 1963). Examination of their investigation shows that the conditions of measurement (Section IV, F) were only partially satisfied. Observation was available only in the end-on direction. The "apparent temperature" as a function of ionic mass is shown in Fig. 4.

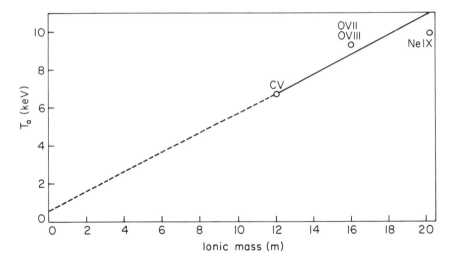

FIG. 4. The "apparent temperature" T_a as a function of ionic mass with SCYLLA I.

In their analysis of the results it was assumed that, as in the case of ZETA, Eq. (10), in which the mean square commoving speed is constant, applied. This was justified on the grounds that the straight-line plot of Fig. 4 fits the experimental data reasonably well, whereas there is no obvious linear relationship involving $(z/m)^2$. The intercept on the ordinate corresponds to an ion temperature of about 600 eV but the uncertainty is large. This value is different from that quoted by the authors since only *optical* measurements are considered here; moreover, their plot of the data is incorrect. For the limited data available, their analysis seems reasonable. However, the existence of a common temperature is in doubt. The calculated equipartition times for binary encounters, although less than the observation time of the lines, are comparable to it. The true line profiles after separation of instrumental broadening could not be deduced; they were just assumed to be Gaussian (see also Sawyer *et al.*, 1962, pp. 473–474). Furthermore, since side-on optical measurements were not performed, it was impossible to prove that the common mass velocity given by the slope of the line in Fig. 4 was not associated with systematic motion in the end-on direction.

(*3*) "*Apparent Temperature*" *Independent of* z *and* m. In a number of cases, the "apparent temperatures" for different ionic species were sensibly identical (for example, Gibbons and Mackin, 1962, p. 1779; Koloshnikov, 1964, p. 27; and Hirschberg, 1964). The claim to the existence of a quiescent plasma without random mass motion, although reasonably justified in some cases, should be treated with caution. In the first instance, the conditions of measurement were not properly satisfied or insufficient information was made available on this point. Secondly, the selection of, say, only two ionic species (for example, He II and C III by Koloshnikov, 1964) seems inadequate in the face of the possible combinations of commotion in Eq. (28) that could fortuitously resemble thermal motion. On the other hand, Hirschberg (1964), although employing two species, namely, He II and O V, showed more convincingly the absence of appreciable commotion in the C stellarator at Princeton. This was partly achieved by altering the operating conditions and remeasuring the broadening. On the latter occasion, the "apparent temperature" was larger, indicative of appreciable commotion. Note that the captions to Figs. 5 and 6 of his paper should be interchanged.

I. APPRAISAL OF THE METHOD

This account of the method of different ionic species concludes with an appraisal.

The method of different ionic species is indirect in that commotion is not observed by individual line shifts but by additional broadening. Consequently, only the average properties of commotion can be determined; it cannot be examined statistically (c.f. Kadomtsev, 1965, p. 123). On the other hand, this disadvantage is somewhat balanced by the fact that broadening due to commotion is the product of *all* commoving elements, even of those beyond the limits of spatial and temporal resolution (see Section II, C).

One of the basic weaknesses of the method lies in finding a suitable formula for the mean square commoving speed in terms of ionic properties. The assumptions of earlier work (for example, Jones and Wilson, 1962) or the polynomial representation in this account are open to criticism. Moreover, evaluation of the ion temperature depends on the mean square commoving speed not possessing a singularity of the same form as thermal motion, namely, m^{-1}. The method is also severely limited in scope on account of the severe restrictions placed on the selection of ionic species (Appendix C).

The need for the various ionic species to be bound by a common temperature in the line of sight limits further the generality of the method. Each species can be at its own individual temperature, in which case the method is useless for temperature evaluation. However, in the case where the contribution to the line width from thermal motion is very small compared with that from commotion, the method can still be used to investigate the average properties of commotion. At the opposite limit, a quiescent plasma in which the fluctuations are purely thermal is defined by the temperature, evaluated directly from the Doppler width, being independent of variations in both ionic charge and mass over wide limits.

When the method of different ionic species was first tried, one of the chief difficulties was the lack of suitable classified spectral lines with the requisite degree of ionization (Kaufman *et al.*, 1960, p. 17; Eberhagen *et al.*, 1965, p. 1376). Since then, the spectra of multi-ionized atoms have been extensively investigated (see Moore, 1968 and 1969, for bibliography) and today this problem carries less weight than formerly, unless one is limited to the visible and near ultraviolet regions of the spectrum.

In plasmas where random mass motion is suspected, it seems advizable to conduct two sets of investigations, one in which the ionic species differ in charge, and the other in mass. Random electrostatic waves are more easily revealed by the former. The results of the two should be consistent before they can be accepted as trustworthy. The scheme of investigation in which both the ionic charge and mass are variables should, in general, be avoided. Apart from the problem of selecting the number of species required, it appears impracticable to try to fit seven parameters to experimental data of the type available here—errors of measurement may be comparable in effect to that due to a correcting parameter, say, z^γ in Eq. (28). On the other hand, this scheme is less restrictive to selecting species that exist simultaneously in the same volume element. Thus it can serve the most useful purpose of indicating a quiescent plasma where there is no problem of interpreting the data.

In the light of these remarks one may ask why in those investigations where commotion evidently contributed to the Doppler broadening, the ion temperature was not unambiguously determined. One or more of the following explanations can be offered in a particular case. (1) Hot laboratory-plasmas are far too complicated to permit the formulation of the mean square commoving speed in terms of simple models of commotion. (2) Systematic motion may have contributed to the line broadening. (3) The conditions of measurement may not have been satisfied. (4) A common temperature may not have existed.

In conclusion, it appears that an unambiguous solution of separating the two types of motion by the method of different ionic species can only be expected in special cases rather than generally. In view of the magnitude of the separation problem, the semiempirical approach adopted here seems reasonable. Nevertheless, the method should be tried only if there is a fair chance of success after careful examination of all the factors involved, especially of the conditions of measurement.

V. Conclusion

A hot plasma does not easily reveal its secrets. Persistent and methodical effort by various methods of measurement is required, and Doppler's principle is no exception in this respect. It is therefore hoped, in view of current interest in plasma fluctuations in general and in their investigation by spectroscopic methods in particular (Alexeff *et al.*, 1969; Griem and Kunze, 1969; Kunze and Griem, 1968), that this account of the spectroscopic analysis of the velocity field may be of value towards setting forth the relative merits of different methods of investigation. Moreover, some of the problems exposed here may stimulate ideas for future work. For example, random mass motion in a plasma under controlled conditions can be established by forcing neutral gas at high speed through the positive column of an arc or glow discharge (for example, Granatstein, 1967). The Doppler broadening of atom and ion spectral lines in this arrangement can be investigated with and without the fluid turbulence. An investigation of this type could perhaps be of value for the interpretation of line broadening in solar spectra and for a better understanding of hot plasmas far from local thermodynamic equilibrium.

The length of this paper precluded adding detailed sections on the remaining methods of separating thermal motion from commotion and on separation of systematic motion. It is to be hoped that this omission can be rectified in the not too distant future.

ACKNOWLEDGMENTS

Many have assisted me in the preparation of this work and to all of them I extend my thanks and gratitude. Professor M. Jammer arranged the association with Bar-Ilan University for the academic year 1967–1968. Parts of earlier drafts bore the scrutiny of Professor G. Bekefi, Dr. S. Goldsmith, and Dr. V. L. Granatstein. Discussion with Dr. C. Breton and Dr. P. Platz, and with a group of final-year undergraduate students at a seminar of the Hebrew University on the subject of this paper, helped me to clarify a number of points, the students perhaps not realizing my indebtedness to them. My wife Josephine, who is not a physicist, read most of the manuscript; her feel for grammar and good style has left its impression on the final text. And her display of patience while this text was written is aptly described by Professor E. H. S. Burhop in his tribute to Professor Sir Harrie Massey (*Advan. At. Mol. Phys.* **4**, 11, 1968). A few short quotations are reprinted with permission from the Cambridge University Press, The Institute of Physics and the Physical Society, and the University of Chicago Press (D. E. Billings, *Astrophys. J.* **139**, 710, 1964). Mrs. R. Vainstein worked hard to produce accurately typed versions of both draft and final manuscripts. This research has been sponsored in part by the Office of Naval Research, Department of the Navy, United States of America.

Appendix A. Electrostatic Screening in a Plasma—The Case of Thermodynamic Equilibrium

Consider the case where the plasma is in full or local thermodynamic equilibrium. The first major advance in the investigation of the screening of the electric field in an assembly of electrically charged particles was due to Debye and Hückel in their celebrated theory of electrolytes (for example, Landau and Lifshitz, 1958, pp. 229–232). Notable is the appearance of the so called Debye–Hückel radius λ_D given by

$$\lambda_D{}^2 = \varepsilon_0 \, kT/(e^2 \sum_a n_a z_a{}^2), \tag{A1}$$

where e is the electronic charge, n_a the number density of particles bearing charge $z_a e$, ε_0 the permittivity of vacuum, and T the equilibrium temperature. At a distance of several Debye-Hückel radii, the electrical potential in the vicinity of a charged particle is effectively screened by particles of opposite sign.

Later calculations attempted to improve upon the original Debye–Hückel theory. For example, in the "ring approximation" of the statistical–mechanical theory of plasmas, the Debye–Hückel potential emerges in a rigorous derivation of the equilibrium case (for example, Balescu, 1963, Chapter 12). The ring approximation is characterized by the dimensionless parameter $e^2 n_e^{1/3}/4\pi\varepsilon_0 kT$ being much smaller than unity (Balescu, 1963, p. 59). At a temperature of 10^6 K and for a number density as high as 10^{24} m^{-3} (10^{18} cm^{-3}), the value of this parameter is only about 10^{-3}. Hence for the plasma conditions considered here, the original Debye–Hückel theory is perfectly adequate for the equilibrium situation.

The total number of charged particles N in the screening cloud of charge, the Debye–Hückel sphere, is given by

$$N = \tfrac{4}{3}\pi\lambda_D{}^3 \sum_a n_a. \tag{A2}$$

For fully ionized hydrogen at a temperature of 10^6 K and of number density 10^{22} m^{-3} (10^{16} cm^{-3}), the Debye–Hückel sphere contains 10^4 charged particles.

Appendix B. Contribution of Unresolved Commoving Elements to the Line Width

The energy spectrum function of commotion $E(k, t)$ is expected to be a decreasing function of the circular wave number k for large values of k (Fig. 5). For example, with electrostatic drift waves, one of the forms of micro-instabilities, $E(k, t)$ was experimentally found to be close to the form k^{-3} (Chen, 1965). Let k_1 be the wave number of the smallest commoving element which can just be resolved as a line shift, and k_0 the wave number corresponding to the reciprocal of the screening distance (Section II, B). The mean square speed $\langle (u_l^c)_{k_1, k_0}^2 \rangle$ in the direction of observation of those ions contained in the unresolved commoving elements is approximately given by

$$\langle (u_l^c)_{k_1, k_0}^2 \rangle = \int_{k_1}^{k_0} E(k, t) \, dk. \tag{B1}$$

It follows from Eq. (10) that the contribution of unresolved commoving elements to the Doppler width is very small in comparison with thermal broadening, provided

$$\left[\int_{k_1}^{k_0} E(k, t) \, dk \right] / (kT_l/m) \ll 1. \tag{B2}$$

Whether or not this condition can be realized depends on the spatial resolution available, and on the plasma conditions such as the relative thermal and average commoving energies and the form of $E(k, t)$.

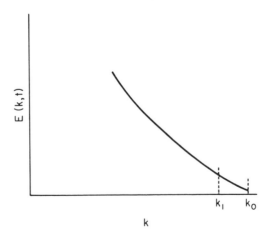

FIG. 5. Energy spectrum function $E(k, t)$ versus the circular wave number k (schematic only).

The corresponding topic as a function of frequency can be treated in an analogous manner.

Appendix C. Selection of Ionic Species

The schemes of investigation varying either the ionic charge or the ionic mass are analyzed using five ionic species (Sections IV, H, 1 and 2). For given excitation conditions, the feasibility of selecting this number of species depends on satisfying the conditions of measurement (Section IV, F). Although this is a stringent criterion, experimental results are presented to show that selection of five species is not an impossible task. It should be much easier to select three species. In any event, for a given set of species, the conditions of measurement can only be satisfied over a limited interval of time during the plasma cycle and for a limited region of the plasma.

Ionic Species by Charge Difference

Consider first the scheme in which the ionic charge is varied. A convenient starting point is to choose one of the rare gases—neon, argon, krypton, or xenon—which are particularly free from injection effects (Section IV, F, 3). Helium is excluded because only one ionic species that produces a spectrum exists. In order to deal with five ionic species, at least the sixth spectrum, for example Ne VI, must be excited. This condition is expected to be met in optically thin plasmas for electron temperatures exceeding about 10^5 K.

A difficult proposition is the question of simultaneity and localization of the selected species. In general, their appearance at different times during the plasma cycle runs counter to the condition of simultaneity. However, their lifetimes may be sufficiently long so that overlap exists during a restricted time interval; likewise for the spatial distribution of the various species. To investigate this problem, one could select a model of ionization and excitation supposedly appropriate to the plasma in question (for example, the time-dependent corona model, McWhirter, 1965, pp. 214–222) and attempt to check theoretically the conditions of simultaneity and localization. This method demands detailed knowledge of the plasma which may be lacking. It appears safer to resort to experiment and perhaps to some principles on which the model is based, rather than to detailed calculation. This approach is now considered with examples from the literature.

In the C stellarator at Princeton, the five species Ne III, Ne IV, Ne V, Ne VI, and Ne VII coexisted in measurable quantity for a short time (Hinnov, 1966, p. 1181). No measurements of spatial resolution were apparently made; nevertheless, there is good reason to suppose that all these species were located simultaneously in a cylindrical shell of plasma (Hinnov, 1966, p. 1187). On the other hand, in the apparatus ZETA, only the four species Ar V, Ar VI, Ar VII, and Ar VIII appeared simultaneously, while the Ar IV species did not (Fawcett *et al.*, 1961, p. 1224). The spatial distribution of these species is not given. An instructive example is provided by a θ-pinch discharge where both spatial and temporal resolution of the spectral lines were available (Elton *et al.*, 1966, p. 191). The ionic species were formed in oxygen and not in a rare gas; the effects of gas injection were not explored. At a time 2 μs from the start of the plasma cycle, the four species O III, O IV, O V, and O VI were excited locally in a region extending over 7 mm in the radial direction. The fifth species O VII was not excited simultaneously, presumably because it is helium-like and so is particularly stable. A helium-like species is not expected to be excited in any appreciable amount until most of the ions of previous stages of ionization are ionized (for example, Williams and Kaufman, 1960, pp. 332–333; Kunze *et al.*, 1968, p. 275; and Bogen, 1968, p. 427). Thus, in the selection of species, a sharp discontinuity in the range of values of ionization potential should be avoided.

Ionic Species by Mass Difference

In the scheme where the mass is varied, a wide spread in ionic masses is desirable. The rare gases satisfy this requirement, as is shown in the accompanying table.

Element:	C	Ne	Ar	Kr	Xe
Mass ratio:	0.59	1	2.0	4.1	6.5

If helium is included, the five species can only be singly ionized. Since few singly ionized xenon ions will be present in a plasma with high electron temperature, the inclusion of helium may have to be reserved for examining conditions at the beginning and end of the plasma cycle. Instead of helium, the fifth element could be carbon, which is satisfactory from the point of view of mass ratio and can be free from injection effects (Section IV, F, 3).

A similar difficulty in satisfying the conditions of simultaneity and localization arises, as it does for ions of different charge. The ionization potential is a determining factor in the appearance of an ion. Although the ionization potentials of argon, krypton, and xenon ions in the same stage of ionization are different but not strikingly so, those of neon are very different (for example, Allen, 1963, pp. 37–38). However, adequate overlap of the appearance of species including neon and in their location may exist. Note that if carbon is chosen, the helium-like C V and hydrogen-like C VI species should be avoided for the reason already given. In the case where the electron temperature is not too high and neon is unsatisfactory, the set of species B III, N III, Ar III, Kr III, and Xe III can be tried provided the nitrogen (Burton and Wilson, 1961, p. 1425) and boron are free from injection effects. However, Kunze and Gabriel (1968, p. 1220) found that boron in the form of boron trifluoride showed severe injection. Diborane (Williams and Kaufman, 1960, p. 331) may be more successful.

Experimental data which can be considered appear to be meager. With the apparatus ZETA, one can infer indirectly that the ionic species Ar VI, Kr VI, and O VI (with low gas injection) would probably have just appeared simultaneously (see Burton and Wilson, 1961, p. 1427, Fig. 5; Jones and Wilson, 1962, Fig. 3). The Ne VI species would not.

NOTE ADDED IN PROOF

E. V. Lifshitz [*Soviet Phys.—JETP (English Transl.)* **29**, 1015–1017 (1969)] in a theoretical analysis of Doppler broadening produced by ions moving in random fields, adds weight to the case for investigating fluctuation phenomena in plasmas by means of Doppler's principle.

LIST OF SYMBOLS

The section in which a symbol is first introduced is listed on the right.

c	Speed of light in vacuum	IV, B
e	Electronic charge	Appendix A
$f(\mathbf{u})$	Velocity distribution function of n_j ions	IV, B
g, g_1, g_2	Constant coefficients in expressions for $\langle (u_l^e)^2 \rangle$	IV, E
h, h_1, h_2	Constant coefficients in expressions for $\langle (u_l^p)^2 \rangle$	IV, E
k	Boltzmann constant	II, A
k	Circular wave number of random variable	IV, E
k_0	Reciprocal of screening distance	Appendix B
k_1	Circular wave number of smallest commoving element that can be resolved	Appendix B
m	Ionic mass	II, A
m_e	Electronic mass	IV, E
n	Number density of ions	II, C
n_a	Number density of particles bearing charge $z_a e$	Appendix A
n_e	Number density of electrons	II, A
n_j	Number density of excited ions in the j level	IV, B
p	Hydrostatic pressure	II, C
q	Constant coefficient in polynomial expression for commotion	IV, H, 2
t	Time	IV, B
u	Speed of ion	II, A
u_e	Speed of electron	IV, E
$\langle (u_l^c)^2 \rangle$	Mean square commoving speed of ions in direction l	IV, B
z_a	Integral number of elementary charges	Appendix A

ze	Ionic charge	IV, E
B	Magnetic flux density	IV, E
E	Electric field strength	IV, E
$E(k, t)$	Energy spectrum function	Appendix B
H	Constant representing thermal and plasma-turbulence terms	IV, H, 1
L	Radiance of spectral line	IV, B
$L(\Delta\lambda)$	Spectral radiance of spectral line	IV, B
N	Number of charged particles in Debye–Hückel sphere	Appendix A
T	Kinetic temperature of ions	II, A
T	Equilibrium temperature	Appendix A
T_a	"Apparent temperature"	IV, H, 1
T_e	Kinetic temperature of electrons	II, A
$W(u)$	Probability density function of speeds	IV, B
$W(\Delta\lambda)$	Probability density function of Doppler shifts	IV, B
α	Index in expression for $\langle (u_l^e)^2 \rangle$	IV, E
β	Index in expression for $\langle (u_l^e)^2 \rangle$	IV, E
γ	Index in expression for $\langle (u_l^p)^2 \rangle$	IV, E
δ	Index in expression for $\langle (u_l^p)^2 \rangle$	IV, E
ε_0	Permittivity of vacuum	Appendix A
λ	Wavelength of undisplaced spectral line	IV, B
λ_D	Debye–Hückel radius	Appendix A
λ_0	Limiting wavelength of plasma waves	II, B
$\Delta\lambda$	Doppler shift	IV, B
$\Delta\lambda^2$	Abbreviated form of $(\Delta\lambda)^2$	IV, B
$\Delta\lambda_{1/2}$	Half-radiance width of spectral line	IV, B
ν_p	Ion plasma frequency	II, B
σ	Identically equal to $(\lambda/c) \langle (u_l^c)^2 \rangle^{1/2}$	IV, B
ϕ	Index in polynomial expression for commotion	IV, H, 2
ω	Angular frequency of random variable	IV, E
Superscripts		
c	(e.g. u^c) Commoving	IV, B
e	(e.g. u^e) Pertaining to random electrostatic waves	IV, E
m	(e.g. u^m) Pertaining to magnetic turbulence	IV, E
mm	(e.g. u^{mm}) Pertaining to random mass motion	IV, E
p	(e.g. u^p) Pertaining to plasma turbulence	IV, E
th	(e.g. u^{th}) Thermal	II, A
Subscripts		
l	(e.g. T_l) Specified direction or direction of observation	II, A

REFERENCES

Afrosimov, V. V., and Petrov, M. P. (1968). *Soviet Phys.—Tech. Phys.* (*English Transl.*) **12**, 1467–1475.

Akhiezer, A. I., Akhiezer, I. A., Polovin, R. V., Sitenko, A. G., and Stepanov, K. N. (1967). "Collective Oscillations in a Plasma." Pergamon Press, Oxford.

Alexeff, I., and Neidigh, R. V. (1962). *Phys. Rev.* **127**, 1–3.

Alexeff, I., Jones, W. D., and Montgomery, D. (1967). *Phys. Rev. Letters* **19**, 422–425.

Alexeff, I., Guest, G. E., McNally, Jr., J. R., Neidigh, R. V., and Scott, F. R. (1969). *Phys. Rev. Letters* **23**, 281–283.

Allen, C. W. (1963). "Astrophysical Quantities," 2nd ed. Oxford Univ. Press (Athlone), London and New York.

Armstrong, B. H. (1967). *J. Quant. Spectry. Radiative Transfer* **7**, 61–88.

Artsimovich, L. A. (1964). "Controlled Thermonuclear Reactions." Oliver & Boyd, Edinburgh and London.

Balescu, R. (1960). *Phys. Fluids* **3**, 52–63.

Balescu, R. (1963). "Statistical Mechanics of Charged Particles." Wiley (Interscience), New York.

Barrell, H. (1968). *Nature* **220**, 651–652.

Beckers, J. M. (1968). *Solar Phys.* **3**, 367 433.

Bell, B., and Meltzer, A. (1959). *Smithsonian Contrib. Astrophys.* **3** (5), 39–46.

Berezin, A. B., Zaidel', A. N., and Malyshev, G. M. (1963). *Soviet Phys.—Tech. Phys.* (*English Transl.*) **8**, 213–216.

Bernstein, I. B., and Ahearne, J. F. (1968). *Ann. Phys.* (*N.Y.*) **49**, 1–38.

Bielski, A., Kaczmarek, W., Kubrycht, J., and Wolnikowski, J. (1968). *Acta Phys. Polon.* **33**, 701–709.

Billings, D. E. (1964). *Astrophys. J.* **139**, 710–722.

Bogen, P. (1968). *In* "Plasma Diagnostics" (W. Lochte-Holtgreven, ed.), pp. 424–477. North-Holland Publ., Amsterdam.

Bohn, W. L., Beth, M.-U., and Nedder, G. (1967). *J. Quant. Spectry. Radiative Transfer* **7**, 661–676.

Boland, B. C., Jones, T. J. L., and McWhirter, R. W. P. (1968). *Calibration Methods Ultraviolet and X-ray Regions Spectrum Symp., Munich, 1968*, **ESRO SP-33**, pp. 59–65. European Space Research Organisation, 92-Neuilly-sur-Seine.

Bötticher, W. (1968). *In* "Plasma Diagnostics" (W. Lochte-Holtgreven, ed.), pp. 617–667. North-Holland Publ., Amsterdam.

Bottoms, P. J., and Eisner, M. (1966). *Phys. Rev. Letters* **17**, 902–903.

Bowles, K. L. (1964). *Advan. Electron. Electron Phys.* **19**, 55–176.

Burnett, C. R., Grove, D. J., Palladino, R. W., Stix, T. H., and Wakefield, K. E. (1958). *Phys. Fluids* **1**, 438–445.

Burton, W. M., and Wilson, R. (1961). *Proc. Phys. Soc.* (*London*) **78**, 1416–1438.

Chandrasekhar, S. (1943). *Rev. Mod. Phys.* **15**, 1–89.

Chapman, S., and Cowling, T. G. (1952). "The Mathematical Theory of Non-uniform Gases," 2nd ed. Cambridge Univ. Press, London and New York.

Chen, F. F. (1965). *Phys. Rev. Letters* **15**, 381–383.

Clauser, F. H. (1961). *Nuovo Cimento Suppl.* **22**, 129–130.

Cohen, E. R., and DuMond, J. W. M. (1965). *Rev. Mod. Phys.* **37**, 537–594.

Cooper, J. (1966). *Rept. Progr. Phys.* **29** (part 1), 35–130.

Cremers, C. J., and Birkebak, R. C. (1966). *Appl. Opt.* **5**, 1057–1064.

Dimock, D. (1960). *Proc. 4th Intern. Conf. Ionization Phenomena Gases, Uppsala, 1959* (N. R. Nilsson, ed.), Vol. II, pp. 1136–1140. North-Holland Publ., Amsterdam.

Eberhagen, A., Bernstein, M. J., and Hermansdorfer, H. (1965). *Z. Naturforsch.* **20a**, 1375–1385.

Edlén, B., and Svensson, L. Å. (1965). *Arkiv Fysik* **28**, 427–446.

Elton, R. C., Hintz, E., and Swartz, M. (1966). *Proc. 7th Intern. Conf. Phenomena Ionized Gases, Beograd, 1965* (B. Perović and D. Tošić, eds.), Vol. III, pp. 190–194. Gradevinska Knjiga Publ. House, Beograd.

Evans, D. E., and Katzenstein, J. (1969). *Rept. Progr. Phys.* **32**, 207–271.

Fawcett, B. C., Jones, B. B., and Wilson, R. (1961). *Proc. Phys. Soc.* (*London*) **78**, 1223–1226.

Gebbie, K. B., and Thomas, R. N. (1968). *Astrophys. J.* **154**, 271–283.

Gibbons, R. A., and Mackin Jr., R. J. (1962). *Proc. 5th Intern. Conf. Ionization Phenomena in Gases, Munich, 1961* (H. Maecker, ed.) Vol. II, pp. 1769–1786. North-Holland Publ. Amsterdam.

Granatstein, V. L. (1967). *Phys. Fluids* 10, 1236–1244.

Greig, J. R., and Cooper, J. (1968). *Appl. Opt.* 7, 2166–2170.

Griem, H. R. (1962). *In* " Temperature " (F. G. Brickwedde, ed.), Vol. 3, Part 1. Reinhold, New York.

Griem, H. R. (1964). "Plasma Spectroscopy." McGraw-Hill, New York.

Griem, H. R., and Kunze, H.-J. (1969). *Phys. Rev. Letters* 23, 1279–1281.

Griffin, P. M., McNally Jr., J. R., and Werner, G. K. (1962). *In* " Temperature " (F. G. Brickwedde, ed.), Vol. 3, Part 1. Reinhold, New York.

Gryziński, M., Nowikowski, J., Sadowski, M., Skladnik-Sadowska, E., and Suckewer, S. (1968). *Plasma Phys.* 10, 450–451.

Guggenheim, E. A. (1960). "Elements of the Kinetic Theory of Gases." Pergamon, Oxford.

Harding, G. N., Dellis, A. N., Gibson, A., Jones, B., Lees, D. J., McWhirter, R. W. P., Ramsden, S. A., and Ward, S. (1958). *Proc. 2nd U.N. Intern. Conf. Peaceful Uses Atomic Energy, Geneva, 1958* 32, 365–378. United Nations, Geneva.

Hearn, A. G., Jones, B. B., and Ramsden, S. A. (1962). *Plasma Phys.* 4, 23–30.

Heroux, L. (1964). *Proc. Phys. Soc. (London)* 83, 121–136.

Hinnov, E. (1966). *J. Opt. Soc. Am.* 56, 1179–1188.

Hirayama, T. (1964). *Publ. Astron. Soc. Japan* 16, 104–134.

Hirayama, T. (1969). *Bull. Am. Astron. Soc.* 1 (3), 279.

Hirose, A., Alexeff, I., Jones, W. D., Krall, N. A., and Montgomery, D. (1969). *Phys. Letters* 29A, 31–32.

Hirschberg, J. G. (1964). *Phys. Fluids* 7, 543–547.

Hirschberg, J. G., and Palladino, R. W. (1962). *Phys. Fluids* 5, 48–51.

Hughes, D. G., and MacDonald, D. K. C. (1961). *Proc. Phys. Soc. (London)* 78, 75–80.

Jefferies, J. T. (1962). *In* " Temperature " (F. G. Brickwedde, ed.), Vol. 3, Part 1. Reinhold, New York.

Jefferies, J. T. (1968). "Spectral Line Formation." Ginn (Blaisdell), Waltham, Massachusetts.

Jones, A. F., and Misell, D. L. (1967). *Brit. J. Appl. Phys.* 18, 1479–1483.

Jones, B. B., and Wilson, R. (1962). *Nucl. Fusion Suppl.*, Part 3, 889–893.

Joyce, G., and Montgomery, D. (1967). *Phys. Fluids* 10, 2017–2020.

Jukes, J. D. (1968). *Rept. Progr. Phys.* 31 (part 1), 305–339.

Kadomtsev, B. B. (1965). "Plasma Turbulence." Academic Press, New York.

Kadomtsev, B. B. (1968). *Soviet Phys.—Usp.* (English Transl.) 11, 328–337.

Kaufman, A. S., Hughes, T. P., and Williams, R. V. (1960). *Proc. Phys. Soc. (London)* 76, 17–24.

Kittel, C. (1958). "Elementary Statistical Physics." Wiley, New York.

Koloshnikov, V. G. (1964). *Soviet Phys.—Tech. Phys. (English Transl.)* 9, 24–28.

Kovan, I. A., and Spektor, A. M. (1968). *Soviet Phys.—JETP (English Transl.)* 26, 747–751.

Kovásznay, L. S. G. (1960). *Rev. Mod. Phys.* 32, 815–822.

Kunze, H.-J. (1968). *In* " Plasma Diagnostics " (W. Lochte-Holtgreven, ed.), pp. 550–616. North-Holland Publ., Amsterdam.

Kunze, H.-J., and Gabriel, A. H. (1968). *Phys. Fluids* 11, 1216–1220.

Kunze, H.-J., and Griem, H. R. (1968). *Phys. Rev. Letters* 21, 1048–1052.

Kunze, H.-J., Gabriel, A. H., and Griem, H. R. (1968). *Phys. Rev.* 165, 267–276.

Landau, L. D., and Lifshitz, E. M. (1958). "Statistical Physics," 2nd ed. Pergamon, Oxford.

Lehner, G. (1967). *Z. Phys.* 206, 284–292.

Lehner, G., and Pohl, F. (1968). *Z. Phys.* **216**, 488–498.
Lincke, R. (1968). *In* "Plasma Diagnostics" (W. Lochte-Holtgreven, ed.), pp. 347–423. North-Holland Publ., Amsterdam.
Lochte-Holtgreven, W. (1968). *In* "Plasma Diagnostics" (W. Lochte-Holtgreven, ed.), pp. 135–213. North-Holland Publ., Amsterdam.
Lochte-Holtgreven, W., and Richter, J. (1968). *In* "Plasma Diagnostics" (W. Lochte-Holtgreven, ed.), pp. 250–346. North-Holland Publ., Amsterdam.
Longmire, C. L. (1963). "Elementary Plasma Physics." Wiley (Interscience), New York.
Lumley, J. L., and Panofsky, H. A. (1964). "The Structure of Atmospheric Turbulence." Wiley (Interscience), New York.
McNally Jr., J. R., and Skidmore, M. R. (1963). *Appl. Opt.* **2**, 699–706.
McWhirter, R. W. P. (1965). *In* "Plasma Diagnostic Techniques" (R. H. Huddlestone and S. L. Leonard, eds.), pp. 201–264. Academic Press, New York.
Margenau, H. (1960). *Proc. 4th Intern. Conf. Ionization Phenomena Gases, Uppsala, 1959* (N. R. Nilsson, ed.), Vol. II, pp. 799–807. North-Holland Publ., Amsterdam.
Mather, J. W. (1965). *Phys. Fluids* **8**, 366–377.
Meyer-Arendt, J. R. (1968). *Appl. Opt.* **7**, 2081–2084.
Moffatt, H. K. (1962). *Proc. Intern. Colloq. Mécanique Turbulence, Marseille, 1961*, pp. 395–404. Centre National de la Recherche Scientifique, Paris.
Moore, C. E. (1968, 1969). "Bibliography on the Analyses of Optical Atomic Spectra," Spec. Publ. 306, Sections 1, 2, and 3. Natl. Bur. Std., Washington, D.C.
Namioka, T. (1962). *J. Quant. Spectry. Radiative Transfer* **2**, 697–704.
Neumann, W. (1967). *In* "Ergebnisse der Plasmaphysik und der Gaselektronik" (R. Rompe and M. Steenbeck, eds.), Vol. 1. Akademie Verlag, Berlin.
Okazaki, K., and Andō, K. (1968). *Japan. J. Appl. Phys.* **7**, 910–915.
Orszag, S. A. (1969). *Proc. Intern. Symp. Turbulence Fluids Plasmas, Brooklyn, 1968*, pp. 17–28. Polytechnic Press, Brooklyn, New York.
Pasachoff, J. M., Noyes, R. W., and Beckers, J. M. (1968). *Solar Phys.* **5**, 131–158.
Payne, R. M., and Kaufman, S. (1959). *Proc. Inst. Elec. Engrs.* (*London*) *Pt. A* **106**, Suppl. 2, 36–42.
Petviashvili, V. I., Ramazashvili, R. R., and Tsintsadze, N. L. (1965). *Nucl. Fusion* **5**, 315–317 (Transl. AEC-tr-6672, pp. 83–89. U.S. Atomic Energy Commission, Oak Ridge, Tennessee).
Pimentel, G. C. (1968). *Appl. Opt.* **7**, 2155–2160.
Platz, P., and Hirschberg, J. G. (1965). *Compt. Rend. Acad. Sci.* **261**, 1207–1210.
Post, R. F. (1956). *Rev. Mod. Phys.* **28**, 338–362.
Ramazashvili, R. R., Rukhadze, A. A., and Silin, V. P. (1963). *Soviet Phys.—JETP* (*English Transl.*) **16**, 939–944.
Rayleigh, Lord (1902). "Scientific Papers," Vol. III. Cambridge Univ. Press, London and New York [reprinted from *Phil. Mag.* **27**, 298–304 (1889)].
Richter, J. (1968). *In* "Plasma Diagnostics" (W. Lochte-Holtgreven, ed.), pp. 1–65. North-Holland Publ., Amsterdam.
Rusanov, V. D. (1962). "Modern Plasma Research Methods," AEC-tr-6348, Dept. of Commerce, Washington (also German transl. by J. Wilhelm and R. Leven under title "Methoden der Plasmadiagnostik," 1965. Akademie Verlag, Berlin).
Sagdeev, R. Z., and Galeev, A. A. (1969). "Nonlinear Plasma Theory," revised and edited by T. M. O'Neil and D. L. Book. Benjamin, New York.
Sawyer, G. A., Jahoda, F. C., Ribe, F. L., and Stratton, T. F. (1962). *J. Quant. Spectry. Radiative Transfer* **2**, 467–475.
Sawyer, G. A., Bearden, A. J., Henins, I., Jahoda, F. C., and Ribe, F. L. (1963). *Phys. Rev.* **131**, 1891–1897.

104   A. S. Kaufman

Silin, V. P. (1963). *Soviet Phys.—JETP* (*English Transl.*) **16**, 1281–1285.
Sitenko, A. G. (1967). "Electromagnetic Fluctuations in Plasma." Academic Press, New York.
Spitzer Jr., L. (1962). "Physics of Fully Ionized Gases," 2nd revised ed. Wiley (Interscience), New York.
Stimson, H. F. (1962). *In* "Temperature" (F. G. Brickwedde, ed.), Vol. 3, Part 1. Reinhold, New York.
Stix, T. H. (1962). "The Theory of Plasma Waves." McGraw-Hill, New York.
Stott, P. E. (1968). *J. Phys. A* (*Proc. Phys. Soc.*) **1**, 675–689.
Tchen, C. M. (1969). *Proc. 2nd Orsay Summer Institute Nonlinear Effects in Plasmas* (G. Kalman and M. Feix, eds.), pp. 233–250. Gordon and Breach, New York.
Thomas, R. N. (1961). *Nuovo Cimento Suppl.* **22**, 497.
Thonemann, P. C. (1961). *In* "Optical Spectrometric Measurements of High Temperatures" (P. J. Dickerman, ed.). Univ. of Chicago Press, Chicago, Illinois.
Tsytovich, V. N. (1966). *Soviet Phys.—Usp.* (*English Transl.*) **9**, 370–404.
Turner, E. B. (1965). *In* "Plasma Diagnostic Techniques" (R. H. Huddlestone and S. L. Leonard, eds.), pp. 319–358. Academic Press, New York.
Unsöld, A. (1955). "Physik der Sternatmosphären," 2nd ed. Springer, Berlin.
Webb, C. E. (1968). *J. Appl. Phys.* **39**, 5441–5470.
Weinstock, J. (1968). *Phys. Rev. Letters* **20**, 1149–1153.
Wiese, W. L. (1965). *In* "Plasma Diagnostic Techniques" (R. H. Huddlestone and S. L. Leonard, eds.), pp. 265–317. Academic Press, New York.
Wigner, E. P. (1967). "Symmetries and Reflections." Indiana Univ. Press, Bloomington, Indiana.
Williams, R. V., and Kaufman, S. (1960). *Proc. Phys. Soc.* (*London*) **75**, 329–336.
Wong, A. Y., Motley, R. W., and D'Angelo, N. (1964). *Phys. Rev.* **133**, A436–A442.
Wu, C.-S., Klevans, E. H., and Primack, J. R. (1965). *Phys. Fluids* **8**, 1126–1133.
Zaidel', A. N., Malyshev, G. M., and Shreider, E. Ya. (1961a). *Soviet Phys.—Tech. Phys.* (*English Transl.*) **6**, 93–119.
Zaidel', A. N., Malyshev, G. M., Shreider, E. Ya., Berezin, A. B., Belyaeva, V. A., Gladushchak, V. I., Skidan, V. V., and Sokolova, L. V. (1961b). *Soviet Phys.—Tech. Phys.* (*English Transl.*) **5**, 1346–1355.
Zaidel', A. N., Konstantinov, O. V., and Malyshev, G. M. (1962). *Soviet Phys.—Tech. Phys.* (*English Transl.*) **7**, 265–266.
Ziman, J. M. (1965). *Nature* **206**, 1187–1192.

THE ROTATIONAL
EXCITATION OF
MOLECULES BY
SLOW ELECTRONS

KAZUO TAKAYANAGI and YUKIKAZU ITIKAWA

Institute of Space and Aeronautical Science,
University of Tokyo,
Komaba, Meguro-ku, Tokyo, Japan

I. Introduction

The excitation of molecular rotation is an important energy-loss mechanism for low-energy electrons in molecular gases. Thus, the process plays an important role in determining the electron-velocity distribution in a gaseous discharge, in electron-drift experiments, and in the ionospheres of the earth and of other planets. It has been shown experimentally that the mean energy of electrons is much lower in molecular gases than in monatomic gases.

When an electron beam is fired into a gas, the far-infrared emissions from rotationally excited molecules will increase. However, the expected intensity increase is too low, especially for nonpolar targets, to be useful for determination of the relevant excitation cross section. Another experimental approach is to measure the energy loss of the scattered electron to identify the inelastic process that has occurred in collision with a target molecule. For homonuclear diatomic molecules, for instance, the most important excitation is of the type $J \rightarrow J + 2$, where J is the rotational quantum number

TABLE I

LINE SEPARATION IN THE ENERGY LOSS
SPECTRUM OF ELECTRONS IN THE
ROTATIONAL EXCITATION OF SOME
DIATOMIC MOLECULES

Target molecule	$2B$ (eV)	$4B$ (eV)
H_2	—	0.029
N_2	—	0.00099
O_2	—	0.00071
CO	0.00048	—
HCl	0.0026	—

of the molecule. Since molecules in a gas are distributed in various rotational states at room temperature, the energy-loss spectrum for scattered electrons will consist of a large number of lines with nearly equal separations, $4B$, where B is the rotational constant [the rotational energy is approximately given by $J(J + 1)B$]. This separation is about 1×10^{-3} eV for N_2, one of the most important target gases. This is the separation which the most advanced experimental technique of the present day can barely resolve. For polar molecules, CO, NO, etc., the transition of primary importance is $J \rightarrow J + 1$, which gives rise to a narrower line splitting $2B$ in the energy-loss spectrum. Some diatomic examples for these separations are shown in Table I. The only exceptionally large rotational constant is that of H_2. For this target molecule, Ehrhardt and Linder (1968) have measured the excitation cross section for $J = 1 \rightarrow 3$ for the incident electron energy above 1 eV (see also Linder, 1969). In the future, the target molecules in a particular rotational state may be produced in a sufficient number by appropriate molecular beam techniques. Then the energy-loss spectrum will consist of a single line or sufficiently separated lines and thus can be observed more easily. However, the energy region where the rotational transitions play the most important role is around and below 1 eV. Here the production of the monoenergetic electron beam itself may present a hard task to experimentalists.

Most of our knowledge of the experimental cross section for rotational excitation comes from the drift-tube and other electron-swarm experiments. Here the observed quantities (drift velocity, rate of lateral diffusion, etc.) depend not only on a particular rotational transition, but also on a large number of other rotational excitation and deexcitation processes, the elastic scattering, and often on the vibrational excitations—sometimes even on the electronic excitation of low-lying levels. As in the beam–gas experiment, the gas molecules are distributed among various rotational levels. It is almost impossible to determine the various cross sections involved without ambiguities. It will be extremely helpful in reducing such ambiguities if theory can predict a general shape of the relevant cross section as a function of the collision energy. Then a small number of parameters in a suggested analytic form of the cross section can be determined empirically. As we see in the following sections, the theory of rotational excitation can predict such analytic forms, at least approximately, for the cross section in some cases, especially at sufficiently low collision energies. In some other cases, however, numerical calculation for each individual example is the only way to estimate the theoretical excitation cross section.

In the present review article, we restrict ourselves to theoretical discussions of the problem. Experimental data will be referred to when comparison with the theoretical result is made.

The term "slow electrons" in the title of this article means the electrons with velocity comparable to or below that of the molecular electrons. The absolute magnitude of the velocity of slow electrons is still fairly large even at the thermal energy at room temperature. (The electron velocities at 0.01 and 0.1 eV are, respectively, 5.9×10^6 and 1.9×10^7 cm/sec.) If the range of the electron–molecule interaction is of the order of 10 Å, the collision duration is of the order of 10^{-14} sec, which is somewhat shorter than the period of molecular vibration and much shorter than the rotational period. Practically, therefore, the molecule does not rotate during an encounter with an electron. This characteristic feature of electron–molecule collision is the basis on which the so-called adiabatic theory is built (see Section IV, A).

Most of the theoretical works discussed in this article have been published within the last few years. In other words, the theory of electron–molecule collision with rotational transitions is a quite new part of physics. We review here those papers published by early 1969, with some additional material added at the last moment to cover some of more recent studies reported at 6th International Conference on Physics of Electronic and Atomic Collisions held on July 28 through August 2, 1969 at the Massachusetts Institute of Technology.

II. General Considerations

A. QUALITATIVE DISCUSSIONS

As we have seen in Section I, the molecule does not rotate appreciably during the encounter with an electron. When an electron passes nearby, each atom composing the molecule will receive a certain amount of momentum, and this momentum will eventually give rise to the excitation of rotation (and translational motion). Since the mean fractional energy loss of an electron in electron–atom collisions is known to be of the order of m_e/M, where m_e is the electron mass and M the atomic mass, the mean fractional energy loss to the rotational excitation may also be expected to be small. In fact, early investigators (Morse, 1953; Carson, 1954) obtained theoretically the mean energy loss of this order of magnitude. They assumed that the electron–molecule interaction is approximately given by the sum of the electron–atom interaction potentials, which are exponentially decaying as functions of distance, and then they applied the Born approximation. The observed energy loss, however, is many orders of magnitude larger than what these treatments predicted (see, e.g., Massey, 1969). The origin of such a large cross section is found in the molecular multipoles produced by deformation of electron cloud in the formation of molecular bonds. Most molecules have either the electric dipole or quadrupole moment, which results in a long-range and orientation-dependent interaction.

Let us first consider the homonuclear diatomic molecule. As we see in a later section, the main contribution to the inelastic cross section comes from jumps satisfying the selection rule

$$\Delta l = 0, \pm 2 \qquad (l = 0 \leftrightarrow 0 \quad \text{forbidden}),$$

where l is the orbital quantum number of the scattered electron. In the energy region of 0.01–0.1 eV, only s ($l = 0$) and p ($l = 1$) waves appreciably interact with the molecular field. Within these restrictions, the only possible channel to the excitation is $l = 1 \leftrightarrow 1$. The classical impact parameter b ($= l/k$, where k is the wave number of the electron) corresponding to the p wave is about 20 Å at 0.01 eV and 6 Å at 0.1 eV. At these distances the undeformed electron–atom interactions are much smaller than the long-range quadrupole interaction that arises from the deformation of the electron cloud. In Fig. 1 is shown the electrostatic interaction between H_2 and an electron on the extrapolation of the molecular axis (calculated by Dr. S. Hara; the polarization interaction has not been included in this figure). The potential can be artificially divided into the quadrupole interaction

$$-(Q/r^3)P_2(\hat{\mathbf{r}} \cdot \hat{\mathbf{s}}),$$

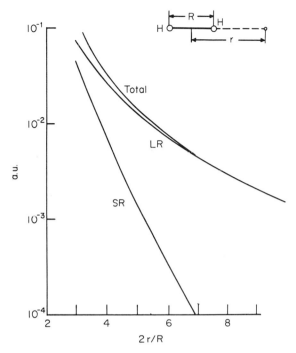

FIG. 1. Electrostatic interaction between H_2 and an electron on the extrapolation of the molecular axis. The potential is divided artificially into the long-range quadrupole interaction (LR) and a short-range part (SR).

and a short-range interaction, where Q is the molecular quadrupole moment, r the distance from the center of the molecule to the electron, $\hat{r}$ and $\hat{s}$ the unit vectors along $\mathbf{r}$ and the molecular axis, respectively, and P is the Legendre polynomial. Here, and in the following, atomic units (a.u.) are used unless otherwise stated. Thus the distance is measured in terms of the Bohr radius a_0, and the energy is expressed in units of $e^2/a_0 = 27.2$ eV. It is seen that in the region where the p wave comes in, the quadrupole interaction is many orders of magnitude larger than the rest part of the electron–molecule interaction. That is, the quadrupole interactions plays an essential part in the rotational excitation of homonuclear diatomic molecules by thermal electrons. The effective cross section obtained is much larger than one would expect from the short-range interaction model.

A similar situation is found for the heteronuclear diatomic molecule. Here, the dipole interaction plays an essential role, and $\Delta l = \pm 1$ is the corresponding selection rule. At energies just above the excitation threshold (usually less than 0.01 eV), the most important transition in the scattered electron is $l = 1 \to 0$. Hence, the effective collision distance is again fairly large and thus the long-range part of the interaction, which can be represented by the point–dipole interaction

$$-(D/r^2)P_1(\hat{\mathbf{r}} \cdot \hat{\mathbf{s}}),$$

where D is the dipole moment of the molecule, determines the excitation cross section. In this case, $\mathbf{r}$ is the position of the electron relative to the gravity center of the molecule. In Fig. 2 is shown the part of the electron–HCl interaction that depends on the orientation as $P_1(\hat{\mathbf{r}} \cdot \hat{\mathbf{s}})$. The charge distribution adopted in obtaining this figure was that of an undeformed Cl atom at the origin and of the electron cloud of the H atom shifted from its original position around the proton toward the Cl atom by a distance which is adjusted to give the known dipole moment of the molecule. It is seen from Fig. 2 that the point–dipole interaction is indeed a good approximation at large distances. As the collision energy becomes large, higher partial waves ($l > 1$) contribute appreciably to the cross section, as we see in the following sections. Therefore, the long-range dipole interaction still predominates. Other interactions do not affect the scattering of electrons in these higher partial waves.

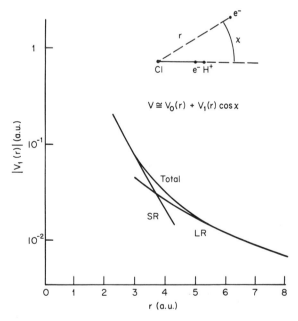

FIG. 2. The part of the electron–HCl interaction, $v_1(r)$, dependent on the orientation as $P_1(\cos \chi)$. As the distance r increases, the potential $v_1(r)$ approaches the long-range point–dipole interaction (*LR*). In short distances it is represented by an exponentially decaying term (*SR*). For details of the calculation of the e–HCl interaction, see Itikawa and Takayanagi (1969).

For polyatomic molecules, the situation is more or less the same except that the contribution of short-range forces becomes important as the size of the molecule increases. Otherwise, the long-range interaction will primarily determine the rotational excitation cross section, which gives the fractional energy loss much larger than m_e/M.

B. Mathematical Formulation

1. Outline of the Procedure. Nature of the Electron–Molecule Interaction

We have to solve the relevant Schrödinger equation for the system of a large number of electrons and nuclei. The standard method of approach is to expand the wave function Ψ of the whole system in terms of the eigenfunctions of the target molecule, thus obtaining a set of coupled equations for the motion of the scattered electron.

The molecular eigenstate is usually described in the Born–Oppenheimer approximation, where the eigenfunction is a product of the electronic eigenfunction for fixed nuclei and functions of molecular vibration and rotation. In this review we assume that the final electronic state of the molecule is the same as the initial state. However, the polarization force, which arises from a temporary and partial excitation of the molecule during the encounter, has an important influence on the scattering of slow electrons—as we know in the study of electron–atom scattering. In the case of molecules, this interaction becomes orientation-dependent, and thus contributes to the rotational transitions directly. Since the polarization potential goes as r^{-4} at large distances, it is a fairly long-range interaction. At low incident energies, this potential is often calculated by fixing the position of the incident electron at various distances from the molecule. The adiabatic polarization potential thus obtained is expected to be fairly reliable, at least at larger separations. At smaller distances the fact that the electron is moving results in a correction term (Callaway and LaBahn, 1968; Kleinman et al., 1968), which Callaway calls "the distortion potential." This correction term is effectively a repulsive interaction and diminishes the effect of attractive polarization interaction.

As the velocity of the incident electron goes up, it becomes more and more difficult for the molecular electrons to adjust themselves adiabatically and thus the velocity dependence of the polarization interaction appears. However, we shall not consider collisions of such high velocities in this article.

The electron-exchange effect is taken into account explicitly by antisymmetrizing the function Ψ with respect to the spatial and spin coordinates of electrons in the system. In the case where the short-range interaction does not appreciably affect the transition under consideration, one may approximately replace the exchange term (with an integral operator) by a local potential function, similar to that suggested by Slater in the calculation of

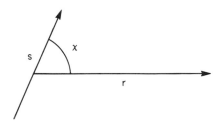

FIG. 3. Coordinate system for electron scattering by a linear molecule. The position vector **r** of the electron is taken relative to the gravity center of the molecule.

atomic structure (Slater, 1960). Hara (1967) has applied this approximate method to electron–H_2 elastic scattering. His effective exchange potential was attractive, corresponding to the fact that the antisymmetrization of Ψ tends to keep electrons with the same spin direction separate (Pauli's exclusion principle) and thus the mutual repulsion due to the Coulomb interaction is reduced.

For the time being, we shall assume that the exchange effect can be replaced by an effective local potential without introducing a large error. In the case of a linear molecule in a $^1\Sigma$ state, the electron–molecule interaction V will be a function of r, s, and $\hat{\mathbf{r}} \cdot \hat{\mathbf{s}} = \cos \chi$ (see Fig. 3). This interaction is thus expanded in terms of the Legendre polynomials as

$$V = \sum_{\nu} v_{\nu}(r, s) P_{\nu}(\hat{\mathbf{r}} \cdot \hat{\mathbf{s}}). \tag{1}$$

The mathematical formulation of our collision problem is then obtained straightforwardly as we see in the next subsection. It is noted that the nuclei where the potential V has poles are not at the origin ($r = 0$) of our coordinate system, so that quite a large number of terms are needed in (1) to represent these poles to a reasonable accuracy (Takayanagi, 1967). It is more convenient to use the two-center coordinate system (see Section IV, A) to represent the potential of a diatomic molecule, but then the mathematical formulation should be different from what we describe below. The so-called adiabatic approximation may be easily combined with the two-center coordinate description of the scattering.

Quite often, the rotational excitation is primarily determined by the long-range interactions. The asymptotic form of the electron–molecule interaction, in the case of the linear neutral molecule, can be written as (see Appendix)

$$V(\mathbf{r}) \xrightarrow{r \to \infty} - (D/r^2) P_1(\hat{\mathbf{r}} \cdot \hat{\mathbf{s}}) - (Q/r^3) P_2(\hat{\mathbf{r}} \cdot \hat{\mathbf{s}}) - \cdots$$
$$- (\alpha/2r^4) - (\alpha'/2r^4) P_2(\hat{\mathbf{r}} \cdot \hat{\mathbf{s}}) - \cdots, \tag{2}$$

where α and α' are the spherical and nonspherical parts of the polarizability of the molecule.

For a nonlinear molecule or a linear molecule in a state other than $^1\Sigma$, the interaction potential cannot be determined by the same number of variables and, hence, its asymptotic form is not so simple as (2). In the case of a symmetric-top molecule, for instance, the potential V depends also on the azimuthal angle ψ around the top axis and is expanded as (see Section III, B, 1, especially Fig. 4)

$$V(r, \chi, \psi) = \sum_{l, m} v_{lm}(r) Y_{lm}(\chi, \psi), \qquad (3)$$

where Y is the spherical harmonic function.

2. Derivation of the Basic Equations and the Formula for the Cross Section

In this section, we assume that the molecule is a linear rigid rotor. The wave function for the whole system can be expanded in terms of the function

$$\mathscr{Y}_{lJ}^{\lambda\mu}(\hat{\mathbf{r}}, \hat{\mathbf{s}}) \equiv \sum_{mM} (lJmM \mid lJ\lambda\mu) Y_{lm}(\hat{\mathbf{r}}) Y_{JM}(\hat{\mathbf{s}}), \qquad (4)$$

which represents the state with well-defined total angular momentum quantum number λ and its z component μ. The Clebsch–Gordan coefficient $(lJmM \mid lJ\lambda\mu)$ has a nonvanishing value only when $m + M = \mu$. The wave function of the system Ψ is now written as

$$\Psi = \sum_{lJ\lambda\mu} (1/r) f_{lJ}^{\lambda\mu}(r) \mathscr{Y}_{lJ}^{\lambda\mu}(\hat{\mathbf{r}}, \hat{\mathbf{s}}). \qquad (5)$$

The radial functions f are determined by the conditions that the function Ψ satisfies the relevant Schrödinger equation and that it has an appropriate asymptotic form. The formulation of a scattering problem along this line was originally developed for nuclear collisions (see, e.g., Blatt and Biedenharn, 1952), and then applied to molecular collisions (Takayanagi, 1954; Takayanagi and Ohno, 1955). The electron–molecule scattering was first treated along this line by Arthurs and Dalgarno (1960).

The Schrödinger equation in the present problem is

$$[H_R - \tfrac{1}{2}\nabla_r^2 + V(\mathbf{r}, \hat{\mathbf{s}}) - E]\Psi = 0, \qquad (6)$$

where the operator H_R is the Hamiltonian for the molecular rotation and the spherical harmonics are its eigenfunctions:

$$[H_R - J(J + 1)B]Y_{JM}(\hat{\mathbf{s}}) = 0.$$

The quantity E in (6) is the total energy of the system determined by the initial condition. By substituting the expanded form of the wave function (5) into (6), and by using the orthonormality of $\mathscr{Y}$, we obtain

$$\left[\frac{d^2}{dr^2} + \left\{k_J{}^2 - \frac{l(l+1)}{r^2}\right\}\right] f_{lJ}^{\lambda\mu} = 2 \sum_{l'} \sum_{J'} \langle lJ | V | l'J'\rangle_\lambda f_{l'J'}^{\lambda\mu}, \tag{7}$$

where

$$k_J{}^2 = 2[E - J(J+1)B],$$

and the matrix element

$$\langle lJ | V | l'J'\rangle_\lambda = \iint d\hat{\mathbf{r}} \, d\hat{\mathbf{s}} \, \mathscr{Y}_{lJ}^{\lambda\mu *}(\hat{\mathbf{r}}, \hat{\mathbf{s}}) V(\mathbf{r}, \hat{\mathbf{s}}) \mathscr{Y}_{l'J'}^{\lambda\mu}(\hat{\mathbf{r}}, \hat{\mathbf{s}}) \tag{8}$$

is known independent of μ. In Eqs. (7), the functions $f_{lJ}^{\lambda\mu}$ with different λ or μ are not coupled to each other. Therefore, the set of equations is divided into many separate sets, each being characterized by definite λ and μ. Since the dependence of the solution on μ comes from an arbitrary constant factor which is determined later by imposing a proper asymptotic form upon Ψ, we can omit the superscript μ while we are solving these sets of equations.

So far we have assumed that the electron exchange can be represented by a local potential. If the wave function Ψ is properly antisymmetrized with respect to the electrons in the system, Eqs. (7) are replaced by a set of integro-differential equations. For detailed discussions, see Henry and Lane (1969).

Equations (7), or more accurately the set of integrodifferential equations, are to be solved with the asymptotic condition

$$f_{lJ}^\lambda(l_0 J_0 | r) \xrightarrow{r \to \infty} \delta_{ll_0} \delta_{JJ_0} \exp\left[-i\left(k_{J_0} r - \frac{l_0}{2}\pi\right)\right]$$

$$- (k_{J_0}/k_J)^{1/2} S^\lambda(lJ; l_0 J_0) \exp\left[i\left(k_J r - \frac{l}{2}\pi\right)\right], \tag{9}$$

where l_0 and J_0 are, respectively, the initial angular momenta of the electron and the molecule. The radial function $f_{lJ}^{\lambda\mu}(r)$ in (5) is given by a linear combination of $f_{lJ}^\lambda(l_0 J_0 | r)$.

The differential cross section is obtained by the formula (Arthurs and Dalgarno, 1960)

$$q(J_0 \to J; \theta) = \sum_{v=0}^\infty A_v(J_0 \to J) P_v(\cos\theta), \tag{10}$$

where θ is the scattering angle and

$$A_v(J_0 \to J) = \frac{1}{4k_{J_0}^2} \frac{1}{2J_0+1} \sum_{\lambda=0}^\infty \sum_{\bar{\lambda}=0}^\infty \sum_{l=|\lambda-J|}^{\lambda+J} \sum_{\bar{l}=|\bar{\lambda}-J|}^{\bar{\lambda}+J} \sum_{l_0=|\lambda-J_0|}^{\lambda+J_0} \sum_{\bar{l}_0=|\bar{\lambda}-J_0|}^{\bar{\lambda}+J_0}$$

$$\times T^\lambda(lJ; l_0 J_0) T^{\bar{\lambda}}(\bar{l}J; \bar{l}_0 J_0)^* (-1)^{J+J_0}$$

$$\times Z(l\lambda\bar{l}\bar{\lambda}; Jv) Z(l_0 \lambda \bar{l}_0 \bar{\lambda}; J_0 v). \tag{11}$$

Here the quantities T and Z are defined by

$$T^\lambda(lJ; l_0 J_0) = \delta_{l l_0} \delta_{J J_0} - S^\lambda(lJ; l_0 J_0), \tag{12}$$

and

$$Z(abcd; ef) = i^{f-a+c}[(2a+1)(2b+1)(2c+1)(2d+1)]^{1/2}$$
$$\times (ac00 \,|\, acf0)W(abcd; ef), \tag{13}$$

where W is the Racah coefficient (Racah, 1942; Biedenharn et al., 1952). By integrating (10) over the scattering direction, we obtain the total cross section for the transition

$$\sigma(J_0 \to J) = 2\pi \int_0^\pi q(J_0 \to J; \theta) \sin \theta \, d\theta$$

$$= 4\pi A_0(J_0 \to J)$$

$$= \frac{1}{2J_0 + 1} \frac{\pi}{k_{J_0}^2} \sum_{\lambda=0}^{\infty} \sum_{l=|\lambda-J|}^{\lambda+J} \sum_{l_0=|\lambda-J_0|}^{\lambda+J_0}$$

$$\times (2\lambda + 1)|\delta_{l l_0} \delta_{J J_0} - S^\lambda(lJ; l_0 J_0)|^2, \tag{14}$$

where the relation $Z(abcd; e0) = \delta_{ac} \delta_{bd}(-1)^{b-c}(2b+1)^{1/2}$ has been used.

Similarly, the momentum-transfer cross section is calculated by

$$\sigma_m(J_0 \to J) = 2\pi \int_0^\pi (1 - \cos \theta)q(J_0 \to J; \theta) \sin \theta \, d\theta$$

$$= 4\pi [A_0(J_0 \to J) - \tfrac{1}{3} A_1(J_0 \to J)]. \tag{15}$$

Crawford et al. (1967) have pointed out that the following definition is more appropriate:

$$\sigma_m(J_0 \to J) = 2\pi \int_0^\pi \left(1 - \frac{k_J}{k_{J_0}} \cos \theta\right) q(J_0 \to J; \theta) \sin \theta \, d\theta. \tag{16}$$

However, the difference between (15) and (16) is negligibly small except for extremely low-energy electrons in hydrogen gas.

3. Resonance Effects

Resonance phenomena are often important in the low-energy electron scattering from atoms and molecules. These are the scattering processes through a compound state of the incident electron and the target. A review article by Bardsley and Mandl (1968) discusses electron–molecule resonant scattering. Two major types of resonance are usually distinguished in the electron–atom collision. One is the electron-excited resonance which corresponds to the temporary electronic excitation of the target with the incident electron captured in its field. The other is due to the temporary capture of the incident electron by the target through the adiabatic interaction. Usually the centrifugal potential barrier prevents the electron captured in the inner region of the molecular field from escaping to the outside. This type of resonance is called potential resonance or shape resonance. All resonance processes depend strongly on the incident electron energy. Each resonance process takes place only within a small energy range, where in the simplest situation the elastic and/or inelastic scattering cross section has a pronounced peak or dip. The resonance region, in typical cases, is very narrow for the electron-excited resonance, while it is somewhat broader for the potential resonance.

Since the lowest electronically excited state is often at several electron volts above the ground state, and since the electron-excited resonance occurs just below the threshold of an excitation process, we need not pay much attention to this type of resonance in the present article where we restrict ourselves to low-energy regions. However, the potential resonance often takes place at low-electron energies. In N_2 and CO, for instance, the vibrational excitation in the energy range 1–3 eV predominantly proceeds through a resonant state. The rotational excitation may also be affected by the presence of these resonant states. Chen (1966) has discussed a resonant excitation of the rotation for electron–N_2 scattering. However, he assumed that the resonance in this case is of the electron-excited type.

In the case of molecular targets, the resonance may take place through the temporary excitation of rotation or vibration. This possibility will be discussed in Section VI.

III. Born Approximation

The Born approximation is derived from the basic Eqs. (7) by assuming that all the matrix elements $\langle lJ|V|l'J'\rangle_\lambda$ are small and by applying the first-order perturbation theory. Or, we may start with the well-known formula for the Born approximation

$$\sigma(\alpha \rightarrow \beta) = \frac{k_\beta}{k_\alpha} \int d\hat{\mathbf{k}}_\beta \left| \frac{1}{2\pi} \int d\mathbf{r} \exp(-i\mathbf{K} \cdot \mathbf{r}) \langle \beta| V |\alpha\rangle \right|^2, \tag{17}$$

where $\mathbf{k}_\alpha$ and $\mathbf{k}_\beta$ are the wave vectors of the incident and outgoing electrons, $\mathbf{K} = \mathbf{k}_\beta - \mathbf{k}_\alpha$, and α, β represent the initial and final states of the target molecule.

A. Linear Molecules in $^1\Sigma$ State

The interaction is given by (1), or

$$V = 4\pi \sum_v [v_v(r)/(2v + 1)] \sum_\gamma Y_{v\gamma}^*(\hat{\mathbf{r}}) Y_{v\gamma}(\hat{\mathbf{s}}). \qquad (18)$$

Its matrix element is

$$\langle JM| V |J_0 M_0\rangle = 4\pi \sum_v \frac{v_v(r)}{2v + 1} \sum_\gamma Y_{v\gamma}^*(\hat{\mathbf{r}})$$

$$\times \int Y_{JM}^*(\hat{\mathbf{s}}) Y_{v\gamma}(\hat{\mathbf{s}}) Y_{J_0 M_0}(\hat{\mathbf{s}})\, d\hat{\mathbf{s}}. \qquad (19)$$

The integral in this formula vanishes unless

$$\gamma = M - M_0, \qquad |J_0 - J| \leqq v \leqq J_0 + J, \qquad v + J_0 + J = \text{even}. \qquad (20)$$

The nonvanishing integral is calculated by the formula (Condon and Shortley, 1935; Racah, 1942)

$$\int Y_{JM}^*(\hat{\mathbf{s}}) Y_{vM - M_0}(\hat{\mathbf{s}}) Y_{J_0 M_0}(\hat{\mathbf{s}})\, d\hat{\mathbf{s}}$$

$$= (-1)^{M_0} \left[\frac{(2J_0 + 1)(2J + 1)}{4\pi(2v + 1)}\right]^{1/2} (JJ_0\, 00\,|\, JJ_0\, v0)$$

$$\times (JJ_0 - MM_0\,|\, JJ_0\, v\, M_0 - M).$$

From conditions (20), it is seen that the term $v = 0$ in interaction (1) gives rise to only the elastic process $J = J_0$ in the Born approximation, while the term $v = 1$ gives the processes $J = J_0 \pm 1$, the term $v = 2$ the processes $J = J_0, J_0 \pm 2$ ($0 \to 0$ forbidden), and so on.

If we take average over M_0 and sum up over M, the cross section becomes, after some straightforward calculations,

$$\sigma(J_0 \to J) = \frac{1}{2J_0 + 1} \sum_{MM_0} \sigma(J_0 M_0 \to JM)$$

$$= \frac{k}{k_0} \sum_{v\gamma} \frac{2J + 1}{(2v + 1)^3} \frac{(JJ_0\, 00\,|\, JJ_0\, v0)^2}{\pi}$$

$$\times \left| \int d\hat{\mathbf{k}} \left| \int d\mathbf{r}\, \exp(-i\mathbf{K} \cdot \mathbf{r}) v_v(r) Y_{v\gamma}^*(\hat{\mathbf{r}}) \right|^2, \qquad (21)$$

where $\mathbf{k}$ and $\mathbf{k}_0$ are the abbreviations for $\mathbf{k}_J$ and $\mathbf{k}_{J_0}$. In order to proceed further, we use the partial-wave expansion of the incident and outgoing plane waves in $\exp(-i\mathbf{K} \cdot \mathbf{r}) = \exp(i\mathbf{k}_0 \cdot \mathbf{r}) \times \exp(i\mathbf{k} \cdot \mathbf{r})^*$. Angular integrals can be performed and the final result is

$$\sigma(J_0 \to J) = \frac{k}{k_0} \sum_{v l_0 l} \frac{(2l_0 + 1)(2l + 1)(2J + 1)}{\pi(2v + 1)^3} (JJ_0 \, 00 \,|\, JJ_0 \, v0)^2$$

$$\times (ll_0 \, 00 \,|\, ll_0 \, v0)^2 \left| \int_0^\infty r^2 \, dr \, v_v(r) j_{l_0}(k_0 r) j_l(kr) \right|^2. \qquad (22)$$

This formula can be used both for excitation ($J > J_0$, $k < k_0$) and for de-excitation ($J < J_0$, $k > k_0$). The relation of detailed balance is satisfied:

$$k_0{}^2(2J_0 + 1)\sigma(J_0 \to J) = k^2(2J + 1)\sigma(J \to J_0).$$

The dependence of the cross section on J_0, J comes from the simple factors

$$(2J + 1) \qquad \text{and} \qquad (JJ_0 \, 00 \,|\, JJ_0 \, v0)^2, \qquad (23)$$

and through k_0 and k. When the target molecule is an ionized species, a more accurate result is obtained by replacing the incident and outgoing plane waves by the appropriate Coulomb wave functions (Coulomb–Born approximation). A modification of formula (22) along this line has been proposed by Boikova and Ob'edkov (1968). The rotational excitation of molecular ions has been also studied by Stabler (1963a) and by Sampson (1965).

When we use the point–multipole interaction (see Appendix), we have

$$v_v(r) = - M_v / r^{v+1}. \qquad (24)$$

Then, the radial integral in (22) can be evaluated by making use of the formula

$$\int_0^\infty r^{-v} J_a(k_1 r) J_b(k_2 r) \, dr \qquad (k_1 > k_2 > 0)$$

$$= \frac{k_2{}^b \Gamma(\alpha)}{2^v k_1^{b+1-v} \Gamma(b+1)\Gamma(1-\beta)} F\left(\alpha, \beta, b+1; \frac{k_2{}^2}{k_1{}^2}\right),$$

where

$$\alpha = \tfrac{1}{2}(-v + a + b + 1), \qquad \beta = \tfrac{1}{2}(-v - a + b + 1),$$

and F is the hypergeometric function. The radial integral converges at the origin in spite of the singularity of r^{-v-1} of interation (24) since the other factors become zero more strongly as the origin is approached. This situation arises because of the factor $(ll_0 \, 00 \,|\, ll_0 v0)$ in (22) which vanishes unless $v \leq l + l_0$. The radial integral in (22) is proportional to k_0^{-2} in the limit of very high incident energy. Therefore, the dipole transition ($v = 1$) has an effective cross section approaching zero at high energy collision, while the quadrupole excitation ($v = 2$) gives a constant cross section for high incident energy. For all transitions arising from $v \geq 3$ the cross section diverges for $k_0 \to \infty$. This is because more and more partial waves come in the neighborhood of the singularity at the origin, and the calculated transition probability per collision for these partial waves increases without limit (usual failure of the first-order perturbation method in the strong coupling problem).

In (22), l_0 and l have physical meaning as the orbital quantum numbers for the incident and the scattered electrons, respectively. If we are not interested in these quantum numbers, formula (21) can be evaluated in a different way. Namely, the exponential function in (21) may be expanded directly as

$$\exp(-i\mathbf{K} \cdot \mathbf{r}) = 4\pi \sum_{lm} (-i)^l j_l(Kr) Y_{lm}(\hat{\mathbf{r}}) Y_{lm}^*(\hat{\mathbf{K}}).$$

Here the quantity l has nothing to do with the electron orbital angular momentum. The integration over $\hat{\mathbf{r}}$ results in

$$\sigma(J_0 \to J) = \frac{k}{k_0} \sum_{v\gamma} \frac{2J+1}{(2v+1)^3} \frac{(JJ_0 00 \,|\, JJ_0 \, v0)^2}{\pi}$$

$$\times \int d\hat{\mathbf{k}} (4\pi)^2 |Y_{v\gamma}(\hat{\mathbf{K}})|^2 \left| \int_0^\infty r^2 \, dr \, v_v(r) j_v(Kr) \right|^2. \qquad (25)$$

The summation over γ makes the integrand independent of $\hat{\mathbf{K}}$, so that the integral operation may be replaced by

$$\int d\hat{\mathbf{k}} \to 2\pi \int_{|k_0-k|}^{k_0+k} K \, dK/k_0 k.$$

In this way, we obtain the following result:

$$\sigma(J_0 \to J) = \frac{8\pi}{k_0^2} \sum_v \frac{2J+1}{(2v+1)^2} (JJ_0 00 \,|\, JJ_0 \, v0)^2$$

$$\times \int_{|k_0-k|}^{k_0+k} K \, dK \left| \int_0^\infty r^2 \, dr \, v_v(r) j_v(Kr) \right|^2. \qquad (26)$$

This formula can be used both for excitation and for deexcitation. Formula (25) before integrating over $d\hat{\mathbf{k}}$ gives us the differential cross section.

For the point–multipole interaction (24), the radial integral in (26) becomes

$$\int_0^\infty r^2 \, dr \, v_v(r) j_v(Kr) = \pi^{1/2} \frac{M_v}{2^v K^{2-v} \Gamma(v + \frac{1}{2})}. \qquad (27)$$

If we have only the $v = 1$ term (dipole interaction), the excitation cross section for the transition $J_0 \to J_0 + 1$ will be

$$\sigma(J_0 \to J_0 + 1) = \frac{8\pi}{3k_0^2} D^2 \frac{J_0 + 1}{2J_0 + 1} \ln \frac{k_0 + k}{k_0 - k}, \qquad (28)$$

where $D = M_1$ is the dipole moment of the molecule (Massey, 1931; Takayanagi, 1966). The differential cross section for the dipole excitation is easily calculated. The result is

$$q(J_0 \rightarrow J_0 + 1; \theta) = \frac{4}{3} \frac{k}{k_0} D^2 \frac{1}{K^2} \frac{J_0 + 1}{2J_0 + 1}, \tag{29}$$

where $K^2 = k_0^2 + k^2 - 2k_0 k \cos \theta$.

Similarly, the differential cross section for the dipole deexcitation is

$$q(J_0 \rightarrow J_0 - 1; \theta) = \frac{4}{3} \frac{k}{k_0} D^2 \frac{1}{K^2} \frac{J_0}{2J_0 + 1}, \tag{30}$$

where k and K are not numerically equal to those in (29).

Unless the incident electron energy is very low, we may neglect the difference between k_0 and k to the first approximation. Then we have $K^2 \cong 2k_0^2 (1 - \cos \theta)$. Thus the momentum-transfer cross section of the point dipole is

$$\sigma_m = \int d\hat{k}(1 - \cos \theta) \sum_J q(J_0 \rightarrow J; \theta)$$

$$= (8\pi/3k_0^2)D^2. \tag{31}$$

This result was first obtained by Altshuler (1957) for a linear molecule with a point dipole, but Crawford (1967b) has shown that this result holds irrespective of the shape of the molecule, provided that (1) we assume that the rotational level spacings are sufficiently small, (2) we adopt the point–dipole interaction, and (3) the Born approximation is applied.

Similarly, the quadrupole excitation due to the term $v = 2$ of (24) is associated with the cross section

$$\sigma(J_0 \rightarrow J_0 + 2) = \frac{8\pi}{15} \frac{k}{k_0} \frac{(J_0 + 1)(J_0 + 2)}{(2J_0 + 1)(2J_0 + 3)} Q^2, \tag{32}$$

where $Q = M_2$ is the quadrupole moment of the molecule (Gerjuoy and Stein, 1955). The differential cross section for the excitation is

$$q(J_0 \rightarrow J_0 + 2; \theta) = \frac{2}{15} Q^2 \frac{k}{k_0} \frac{(J_0 + 1)(J_0 + 2)}{(2J_0 + 1)(2J_0 + 3)}. \tag{33}$$

The corresponding differential cross section for the deexcitation is

$$q(J_0 \rightarrow J_0 - 2; \theta) = \frac{2}{15} Q^2 \frac{k}{k_0} \frac{(J_0 - 1)J_0}{(2J_0 - 1)(2J_0 + 1)}. \tag{34}$$

The sum of these two, together with the elastic part due to the quadrupole interaction

$$q(J_0 \rightarrow J_0; \theta) = \frac{4}{45} Q^2 \frac{J_0(J_0 + 1)}{(2J_0 - 1)(2J_0 + 3)}, \tag{35}$$

with the approximation $k \cong k_0$, gives the total differential cross section due to the point–quadrupole interaction

$$\sum_J q(J_0 \to J; \theta) = (4/45)Q^2, \tag{36}$$

which is independent of J_0 and is spherically symmetric. The momentum-transfer cross section is thus equal to the total (integrated) cross section

$$\sigma_m = (16\pi/45)Q^2. \tag{37}$$

The results, Eqs. (36) and (37), have been obtained by Wijnberg (1966).

The polarization interaction has a component in proportion to $P_2(\hat{\mathbf{r}} \cdot \hat{\mathbf{s}})$. This component has the asymptotic form (see Appendix)

$$-(\alpha'/2r^4)P_2(\hat{\mathbf{r}} \cdot \hat{\mathbf{s}}).$$

The combined effect of this and the point–quadrupole interaction gives the cross section for the excitation $J_0 \to J_0 + 2$

$$\sigma(J_0 \to J_0 + 2) = \frac{8\pi}{15} \frac{k}{k_0} \frac{(J_0 + 1)(J_0 + 2)}{(2J_0 + 1)(2J_0 + 3)}$$

$$\times \left[Q^2 + \frac{\pi}{4} Q\alpha' k_0^{-1} \left(k_0^2 - \frac{\Delta k^2}{4} \right) + \frac{9}{32} \left(\frac{\pi}{4} \alpha' \right)^2 \left(k_0^2 - \frac{\Delta k^2}{2} \right) \right], \tag{38}$$

where $\Delta k^2 - k_0^2 - k^2$ (Dalgarno and Moffett, 1963). When the electron energy is very low, formula (38) becomes practically equal to the simpler formula (32).

B. NONLINEAR MOLECULES

1. Symmetric-Top Molecules

The symmetric-top molecule, such as NH_3, has a permanent dipole in the direction of the symmetry axis of the molecule. The electron–dipole interaction is primarily responsible for the rotational excitation in this type of molecule. The rotational eigenfunction of a symmetric-top molecule is now of the form

$$\psi_{JK_JM} = N(JK_J M)S_{JK_JM}(\Theta)e^{iM\Phi}e^{iK_J\psi}, \tag{39}$$

where N is a normalization constant, Θ and Φ are the angles specifying the direction of the molecular axis in the space (the unit vector in this direction will again be denoted by $\hat{\mathbf{s}}$), and ψ is the angle around the axis (see Fig. 4). The

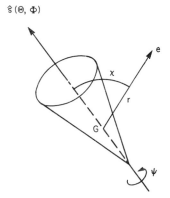

$\hat{s}\,(\Theta,\,\Phi)$

Fɪɢ. 4. Coordinate system for electron scattering by a symmetric-top molecule. The symmetry axis is denoted by $\hat{s}$, whose direction in space is specified by (Θ, Φ). The electron position r is relative to the molecular gravity center G.

function S can be expressed in terms of the hypergeometric function (see, e.g., Wollrab, 1967). If we assume the point–dipole interaction

$$-(D/r^2)P_1(\hat{r} \cdot \hat{s}),$$

which is independent of ψ, we have immediately the selection rule

$$\Delta K_J = 0. \tag{40}$$

The rotational energy level is determined by J and K_J as

$$E = BJ(J+1) + (A - B)K_J^2 \qquad (K_J \leqq J), \tag{41}$$

where $A = 1/2I_A$, $B = 1/2I_B$, (I_A, I_B, I_B) are the three principal moments of inertia, $A - B > 0$ for a prolate symmetric top, and $A - B < 0$ for an oblate symmetric top.

The Born cross section (17) becomes

$$\sigma(J_0 K_{J_0} M_0 \to J K_J M) = \delta_{K_{J_0} K_J} \frac{k}{k_0} \int d\hat{\mathbf{k}}$$

$$\times \left| \frac{D}{2\pi} \int \frac{d\mathbf{r}}{r^2} \exp(-i\mathbf{K} \cdot \mathbf{r}) \langle J K_{J_0} M | \hat{s} \cdot \hat{r} | J_0 K_{J_0} M_0 \rangle \right|^2.$$

If we take the z axis along the vector $\mathbf{K}$, and integrate over $\hat{r}$, it is seen that only the matrix element of $(\hat{s})_z \equiv \hat{s} \cdot \hat{\mathbf{K}}$ survives, so that we have the second selection rule $\Delta M = 0$ in this particular choice of the coordinate system. The relevant matrix element for two possible values of J is (Wollrab, 1967)

$$|\langle J_0 + 1K_{J_0} M_0| \hat{\mathbf{s}} \cdot \hat{\mathbf{K}} | J_0 K_{J_0} M_0 \rangle|^2 = \frac{[(J_0 + 1)^2 - M_0^2][(J_0 + 1)^2 - K_{J_0}^2]}{(J_0 + 1)^2 (2J_0 + 1)(2J_0 + 3)}$$

$$|\langle J_0 - 1K_{J_0} M_0| \hat{\mathbf{s}} \cdot \hat{\mathbf{K}} | J_0 K_{J_0} M_0 \rangle|^2 = \frac{[J_0^2 - M_0^2][J_0^2 - K_{J_0}^2]}{J_0^2 (4J_0^2 - 1)}.$$

When we average these quantities over M_0, we have, respectively,

$$\frac{1}{3} \frac{(J_0 + 1)^2 - K_{J_0}^2}{(J_0 + 1)(2J_0 + 1)} \quad \text{and} \quad \frac{1}{3} \frac{J_0^2 - K_{J_0}^2}{J_0(2J_0 + 1)}.$$

The radial integral (over r) can be done by making use of the formula

$$\int_0^\infty j_1(Kr) \, dr = K^{-1}.$$

The excitation cross section is thus

$$\sigma(J_0 K_{J_0} \rightarrow J_0 + 1K_{J_0}) = \frac{k}{k_0} \int d\hat{\mathbf{k}} \, \frac{4D^2}{K^2} \frac{(J_0 + 1)^2 - K_{J_0}^2}{3(J_0 + 1)(2J_0 + 1)}$$

$$= \frac{8\pi D^2}{3k_0^2} \frac{(J_0 + 1)^2 - K_{J_0}^2}{(J_0 + 1)(2J_0 + 1)} \ln \frac{k_0 + k}{k_0 - k}. \qquad (42)$$

Similarly, the deexcitation cross section is

$$\sigma(J_0 K_{J_0} \rightarrow J_0 - 1K_{J_0}) = \frac{k}{k_0} \int d\hat{\mathbf{k}} \, \frac{4D^2}{K^2} \frac{J_0^2 - K_{J_0}^2}{3J_0(2J_0 + 1)}$$

$$= \frac{8\pi D^2}{3k_0^2} \frac{J_0^2 - K_{J_0}^2}{J_0(2J_0 + 1)} \ln \frac{k + k_0}{k - k_0}. \qquad (43)$$

Formulas (42) and (43) have been obtained by Crawford (1967b).

Generally speaking, the interaction between the incident electron and a symmetric-top molecule may be expanded as in Eq. (3). If the molecule is NH_3 or other symmetric top with symmetry C_{3v}, we have $m = 3n (n = 0, 1, 2, \ldots)$ in that equation. The interaction terms with nonvanishing m may give rise to the excitation of rotation of the three H atoms (in the case of NH_3) around the molecular axis ($\Delta K_J \neq 0$). The lowest order terms of this nature are $(l, m) = (3, \pm 3)$, and the corresponding radial functions $v_{3, \pm 3}(r)$ are asymptotically proportional to r^{-4}.

2. Asymmetric Rotors

The eigenfunction of an asymmetric rotor is usually written as a linear combination of the symmetric-top eigenfunctions (39) in the form

$$\phi_{\tau JM} = \sum_{K_J=-J}^{J} A_{\tau K_J}^{JM} \psi_{JK_JM}. \qquad (44)$$

Here the index τ specifies each of $2J + 1$ discrete energy levels for a given J. Substituting this function into (17), we can get the relevant excitation cross section in the Born approximation. In the case of polar molecules, the matrix element in (17) is directly related to the dipole line strength, as pointed out by Crawford (1967b). With use of the table of the calculated line strength for the asymmetric rotor (Cross *et al.*, 1944), therefore, the Born cross section can be obtained for rotational transitions due to the dipole interaction. When interactions other than the dipole term dominate, the calculation of the matrix element is very difficult; no such calculations have been made as yet.

C. Partial-Wave Analysis—Validity of the Born Approximation

The assumption of the point–dipole or point–quadrupole interaction, combined with the Born approximation, has been used frequently in the past to obtain the rotational excitation cross section. To test the validity of this type of treatment, we may assume a more realistic interaction and adopt a more accurate method (say, close-coupling method) to calculate the cross section. Such accurate calculations will be discussed in a later section (see Section IV, C). Another way to study the validity of the simple treatment is to evaluate the relative importance of various partial cross sections in the treatment. If the higher partial waves play the major role, the point–dipole or point–quadrupole interaction may be adopted since only distant encounters are important. Furthermore, the Born approximation is also valid in that case, since the distortion of the electron wave function is relatively small in distant collisions. It is interesting, therefore, to investigate the relative importance of the partial waves.

1. Dipole Interaction

Among the orientation-dependent interactions, the dipole interaction [the term $v = 1$ in (1)] is of the longest range, so that the distant collisions should be more important here than in other interactions. We may expect that the simple treatment mentioned above should give a fairly accurate cross section. The partial-wave analysis can be done by evaluating individual terms in the cross section (22) for the dipole term $v_1(r) = -D/r^2$. Numerical evaluation of the radial integrals shows (Takayanagi, 1966; Inokuti and Takayanagi, unpublished) that for the incident electron energy up to a few times the threshold of the excitation ΔE, the main contribution comes from the outgoing s wave. Since the s wave can always penetrate into the innermost part of the target, it will be distorted considerably. However, the selection rule $\Delta l = \pm 1$, which arises from the factor $(ll_0 \, 0 \, 0 | ll_0 \, 1 0)$ in (22), requires that the outgoing s wave should result only from the incoming p wave. In the energy range under consideration (usually 1–10 meV), the p wave passes by the target at a considerable distance. Therefore, in the spatial region from where the main contribution to the transition matrix comes, the distortion of the electron wave function is expected to be very small. It is true that the s wave will have a finite phase shift in the transition region. However, because of the long-range nature of the interaction, the transition region is expected to be fairly broad, so that the phase shift will not affect the result drastically. From these considerations, we may conclude that the Born approximation is not bad in the near-threshold region. As the incident energy E_0 increases to ten times ΔE, the contribution from the outgoing s, p, and d waves put together becomes only 50% of the total Born cross section. About 16% comes from the higher partial waves with $l \geq 10$. As the energy E_0 increases further, say to 200 ΔE, the combined contribution of the s, p, and d waves becomes only 26% of the total. This energy corresponds, in typical cases, to the incident energy around 1 eV. At this energy, the distortion of the partial waves with $l \geq 3$ is expected to be negligibly small. Therefore, the Born approximation should be quite good in the whole energy range of practical importance. As we see in Section IV, C, a close-coupling calculation confirms the validity of the Born approximation for the dipole transition.

If the dipole moment of the molecule is very large, the dipole interaction itself may distort the electron wave function and the first-order perturbation theory may not be applicable. A rough estimation indicates that the first-order theory may fail when the dipole moment becomes as large as 1 a.u. (Massey, 1931; Takayanagi, 1966). A close-coupling calculation (Section IV, C) shows, however, that the Born approximation gives still fairly accurate rotational cross sections for $D = 1$ a.u.

2. Quadrupole Interaction

The term $v = 2$ in (1) with the assumption $v_2(r) = -Q/r^3$ was adopted by Gerjuoy and Stein (1955) in their calculation of the rotational cross section for H_2 and N_2. With this model, they could explain the observed behavior of thermal electrons in these gases. Since the quadrupole interaction is of a shorter range than the dipole interaction, the lower partial waves should be comparatively more important than in the polar molecules. However, the selection rule $\Delta l = 0, \pm 2$ ($0 \leftrightarrow 0$ forbidden) arises from the Clebsch–Gordan coefficient in (22). Thus it is expected that only the $l = 1 \leftrightarrow 1$ jump gives the important contribution to the rotational excitation except in the immediate neighborhood of threshold. In the near-threshold region, only the s wave can come out, so that the incident wave should be a d wave ($l = 2$). It is interesting, therefore, to see whether the p wave is distorted considerably by the molecular field in the thermal energy region. If so, the Born approximation cannot be expected to be very accurate.

As the *l*th partial wave approaches the molecular field, the spherical interaction it first meets is the spherical part of the polarization interaction. (The nonspherical interactions should have a smaller effect since they vanish on the average.) The effective scattering potential (including the centrifugal term) is, therefore,

$$V_{eff}(r) = [l(l + 1)/2r^2] - (\alpha/2r^4).$$ (45)

This potential has a maximum value of

$$V_{eff}^m(l) = [l(l + 1)]^2/8\alpha$$ (46)

at

$$r = r_m \equiv [2\alpha/l(l + 1)]^{1/2}.$$ (47)

If the incident electron has an energy greater than this barrier height for $l = 1$, the p wave function will be fairly large in the inner region ($r < r_m$), and the cross section is expected to be quite different from what we have in the Born approximation. For simple molecules, a typical value of the polarizability is $\alpha = 10$ a.u. With this value, it is seen that at the incident energy in the range 0.01–0.1 eV only the s wave is considerably distorted. At 1.0 eV, however, both the s and p waves can come into the inner region, so that these waves should be distorted considerably. Since the p wave is responsible for the rotational excitation under consideration, we may expect a large deviation from the Born result when the incident energy approaches 1 eV. We shall see later (in Section IV, B) that a distorted-wave calculation for N_2 and O_2 shows a steep increase in the calculated cross section when the incident energy becomes comparable to 1 eV, corresponding to the deep penetration of the p wave. On the other hand, in the energy range from thermal (at room temperature) to about 0.1 eV the distortion of the p wave is still fairly small, so that the Born approximation is expected to be valid. This is the reason why the Gerjuoy–Stein treatment was a considerable success. In the case of oxygen, however, the quadrupole moment is so small that the short-range interaction is comparatively more important. In this case, therefore, a small distortion in the electron wave function may have an appreciable effect on the cross section obtained. (The transition region is not sufficiently far outside the strong molecular field.)

For the nonpolar molecules, the nonspherical part of the polarization interaction does contribute to the rotational excitation to an important degree as we see in (38). However the polarization interaction is of a shorter range than the quadrupole interaction, so that the latter is primarily responsible for the excitation in the low collision energies. As the energy increases, the polarization interaction also comes in, but then the distortion effect due to the polarization potential should be also appreciable. Therefore, formula (38), as distinguished from (32), is applicable only within a fairly limited energy region.

3. Higher Multipole Interactions

For the terms $v \geq 3$ in (1), the interaction is of shorter range than in the quadrupole interaction, so that the applicability of the Born approximation is further limited. If the target has neither dipole nor quadrupole moment (as in the case of CH_4), $v = 3$ will be the leading term in (1) responsible for the rotational excitation. At low collision energies, the main contribution comes from $l = 1 \leftrightarrow 2$. Therefore, the transition region is again outside the main part of the molecular field so that the distortion of the wave function is expected to be small. However, because of the strong decay of the interaction as a function of the distance, it is expected that the relevant interaction is fairly small in the region where the p and d waves considerably overlap. In other words, the excitation cross section should be fairly small in the thermal energy region. As the collision energy increases, the short-range interaction becomes more important than the point–octupole interaction ($v = 3$), so that the distortion effect cannot be neglected.

D. Sample Calculations in the Born Approximation

The above arguments show that as far as the dipole and quadrupole interactions are concerned the Born approximation should give a fairly accurate excitation cross section in the thermal energy region. Therefore, the rotational cross section is primarily determined by the molecular constants such as the dipole or quadrupole moment and the rotational constant B. As the collision energy becomes somewhat larger, the polarizability also begins to play a role. Values of the relevant molecular constants for some linear molecules and ammonia are shown in Table II. Using these values, the Born excitation

TABLE II

MOLECULAR CONSTANTS RELEVANT TO ROTATIONAL EXCITATION CROSS SECTION

	$2B^a$		D^b	Q^b	α^c	α'^c
	10^{-4} eV	10^{-5} a.u.	a.u. (ea_0)	a.u. (ea_0^2)	a.u. (a_0^3)	a.u. (a_0^3)
H_2	147.0	54·0	0	$+0.490^d$	5.53	1.41
N_2	4.96	1.824	0	-1.13	11.9	3.13
O_2	3.56	1.310	0	-0.29	10.8	4.95
HCl	25.8	9.52	0.42	$+2.8$	17.5	1.40
CO	4.76	1.752	0.044	-1.9	13.3	2.40
CN	4.69	1.723	0.57^e	?	?	?
NO	4.20	1.546	0.062	-1.3	11.7	3.80
CO_2	0.966	0.355	0	-3.2	17.7	9.45
N_2O	1.037	0.381	0.065	-2.2	20.2	13.3
NH_3	24.7^f	9.06^f	0.578	-0.74	15.0	1.30

[a] Herzberg (1950), unless otherwise stated.
[b] Stogryn and Stogryn (1966), unless otherwise stated.
[c] Bridge and Buckingham (1966).
[d] Kołos and Wolniewicz (1964).
[e] Thomson and Dalby (1968).
[f] Herzberg (1945). Another rotational constant for NH_3 is $A(NH_3) = 2.87 \times 10^{-5}$ a.u.

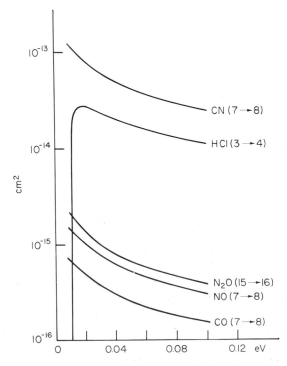

FIG. 5. The Born cross sections for the dipole excitation of rotation in several polar molecules, plotted against the incident electron energy. The type of the rotational transition studied, which is representative at room temperature, is indicated in the parentheses following the name of each molecule.

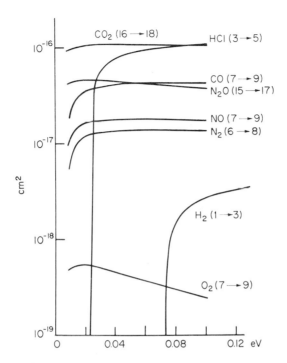

FIG. 6. The Born cross sections for the quadrupole excitation of rotation in several molecules, plotted against the incident electron energy. The type of the rotational transition studied, which is representative at room temperature, is indicated in the parentheses following the name of each molecule.

cross section has been calculated by either (28) or (38). The initial rotational state studied is that of the largest population at room temperature (300°K). The result for linear molecules is shown in Fig. 5 for the dipole interaction and in Fig. 6 for the quadrupole interaction. It is seen that the excitation cross section due to the dipole interaction is many orders of magnitude larger than the cross section due to the quadrupole interaction. In the incident energy region near 0.03 eV, corresponding to the mean thermal energy at room temperature, the excitation cross section for the quadrupole interaction is more or less flat as a function of energy, except for H_2 and HCl where the threshold energy is rather high, while the cross section for the dipole inter- action is generally a decreasing function of energy. For oxygen, the quad- rupole moment is so small that the polarization interaction is more important than for the other molecules. The mutual cancellation of the quadrupole and polarization interactions makes the Born cross section a decreasing function of energy (see, however, Section IV, B). In Fig. 7, the K_J dependence is shown, with use of (42), for the rotational cross section of ammonia due to the dipole interaction.

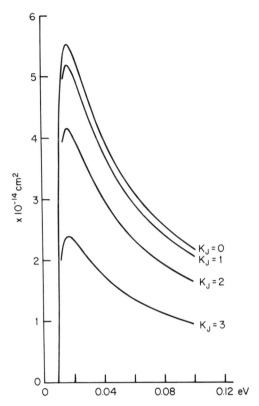

FIG. 7. The Born cross sections for the dipole excitation of rotation, $J = 3 \to 4$, in NH_3, calculated using formula (42) for several values of K_J.

IV. More Accurate Calculations

A. The Adiabatic Approximation

First, we assume that the target molecule is a rigid linear rotor. The wave function of the whole system is expanded as

$$\Psi = \sum_{JM} F_{JM}(\mathbf{r}) Y_{JM}(\hat{\mathbf{s}}). \tag{48}$$

The functions F are determined by requiring that Ψ satisfies the Schrödinger equation for the collision problem under consideration, or equivalently, by solving the following integral equation:

$$F_{JM}(\mathbf{r}) = \delta_{JJ_0}\, \delta_{MM_0} \exp(i\mathbf{k}_0 \cdot \mathbf{r})$$

$$-(1/2\pi) \iint G_J(\mathbf{r}, \mathbf{r}') V(\mathbf{r}', \hat{\mathbf{s}}) \Psi(\mathbf{r}', \hat{\mathbf{s}})\, d\mathbf{r}'\, d\hat{\mathbf{s}}. \tag{49}$$

Here, J_0, M_0 specify the initial rotational state, $\mathbf{k}_0$ is the wave vector of the incident electron, and G_J the Green function defined by the outgoing condition:

$$G_J(\mathbf{r}, \mathbf{r}') = \frac{\exp[ik_J|\mathbf{r}' - \mathbf{r}|]}{|\mathbf{r}' - \mathbf{r}|}. \tag{50}$$

In the inelastic scattering, F_{JM} $(JM \neq J_0 M_0)$ has the asymptotic form

$$F_{JM}(\mathbf{r}) \xrightarrow{r \to \infty} f(J_0 M_0 \to JM; \theta, \phi) \exp[ik_J r]/r \tag{51}$$

with

$$f(J_0 M_0 \to JM; \theta, \phi) = -(1/2\pi) \iint \exp[-i\mathbf{k}_J \cdot \mathbf{r}] Y_{JM}^*(\hat{\mathbf{s}}) V(\mathbf{r}, \hat{\mathbf{s}}) \Psi(\mathbf{r}, \hat{\mathbf{s}}) \, d\mathbf{r} \, d\hat{\mathbf{s}}, \tag{52}$$

where $\mathbf{k}_J$ is the wave vector in the direction (θ, ϕ). Different approximations can be obtained by substituting various approximate functions into Ψ in the right-hand side of (52). The Born approximation can be obtained if we take

$$\Psi = \exp[i\mathbf{k}_0 \cdot \mathbf{r}] Y_{J_0 M_0}(\hat{\mathbf{s}}).$$

A better approximation to Ψ is the so-called adiabatic approximation. As we have seen in Section I, the electron velocity is fairly high even for the thermal electron, so that the molecule does not change its orientation appreciably during the encounter with an electron. Therefore, the following form of the wave function may be suggested as a good approximation:

$$\Psi = F_{\hat{\mathbf{s}}}(\mathbf{r}) Y_{J_0 M_0}(\hat{\mathbf{s}}) \tag{53}$$

The function $F_{\hat{\mathbf{s}}}(\mathbf{r})$ is the wave function of an electron scattered by the molecule with a fixed orientation $\hat{\mathbf{s}}$.

In the case of rotational transitions, a further approximation seems to be quite reasonable; that is, we may assume the difference between k_J and k_0 to be negligible. Except for H_2, and for highly excited states of other molecules, the rotational level spacings are much smaller than the mean thermal energy at room temperature. Therefore, the difference between k_J and k_0 is indeed small in most cases. The outgoing wave vector $\mathbf{k}_J$ is then replaced without serious error by $\mathbf{k}$, which has the same magnitude as k_0 and the direction of (θ, ϕ). In this way, we have, to a good approximation

$$f(J_0 M_0 \to JM; \theta, \phi) = \int Y_{JM}^*(\hat{\mathbf{s}}) f_{\hat{\mathbf{s}}}(\mathbf{k}_0 \to \mathbf{k}) Y_{J_0 M_0}(\hat{\mathbf{s}}) \, d\hat{\mathbf{s}}, \tag{54}$$

where

$$f_{\hat{\mathbf{s}}}(\mathbf{k}_0 \to \mathbf{k}) = -(1/2\pi) \int \exp[-i\mathbf{k} \cdot \mathbf{r}] V(\mathbf{r}, \hat{\mathbf{s}}) F_{\hat{\mathbf{s}}}(\mathbf{r}) \, d\mathbf{r} \tag{55}$$

is nothing but the amplitude of elastic scattering of an electron from the target molecule with fixed $\hat{s}$. The rotational differential cross section is then given by

$$q(J_0 M_0 \to JM; \theta, \phi) = \left| \int Y_{JM}^*(\hat{s}) f_{\hat{s}}(\mathbf{k}_0 \to \mathbf{k}) Y_{J_0 M_0}(\hat{s}) \, d\hat{s} \right|^2. \tag{56}$$

If we sum this up over J, M, the result is

$$\sum_{JM} q(J_0 M_0 \to JM; \theta, \phi) = \int |Y_{J_0 M_0}(\hat{s})|^2 |f_{\hat{s}}(\mathbf{k}_0 \to \mathbf{k})|^2 \, d\hat{s}. \tag{57}$$

This is the differential scattering cross section obtained first for fixed $\hat{s}$, and averaged later over the molecular orientation. If we further take average over M_0, the result is

$$\frac{1}{2J_0 + 1} \sum_{M_0} \sum_{JM} q(J_0 M_0 \to JM; \theta, \phi) = (1/4\pi) \int |f_{\hat{s}}(\mathbf{k}_0 \to \mathbf{k})|^2 \, d\hat{s}. \tag{58}$$

The quantity on the right-hand side, which has been calculated for some simple molecules in many papers, is thus approximately equal to the sum of the elastic and rotational transition cross sections.

It is important to notice that, according to (56), the rotational cross section can be calculated immediately as soon as the scattering amplitude $f_{\hat{s}}$ is obtained. Only a single amplitude $f_{\hat{s}}$ is required to calculate the rotational cross section for all the transitions $(J_0 M_0) \to (JM)$. The accuracy of the resulting cross section thus depends on the accuracy of $f_{\hat{s}}$. It does not matter whether $f_{\hat{s}}$ is calculated in the one-center coordinate system or in the two-center system. Effects of electron exchange and target polarization can be taken into account in obtaining $f_{\hat{s}}$.

When we study the electron scattering from a space-fixed molecular field in the one-center coordinates, the wave function may be expanded as

$$\psi_{\hat{s}}(\mathbf{r}) = \sum_{lm} r^{-1} f_{lm}(r) Y_{lm}(\hat{r}).$$

If the incident direction is chosen as the z direction, the asymptotic form of the radial functions f will be of the form

$$f_{lm}^{l_0}(r) \xrightarrow{r \to \infty} \text{const}\{\delta_{l l_0} \delta_{m 0} \exp[-i(k_0 r - \tfrac{1}{2} l_0 \pi)]$$
$$- S_{\hat{s}}(lm \,|\, l_0 0) \exp[+i(k_0 r - \tfrac{1}{2} l\pi)]\},$$

which defines the S matrix. This matrix should, in principle, be unitary. Here l_0 specifies the initial orbital quantum number of the electron. When we consider the rotational transitions in the adiabatic approximation [Eq. (54)], the S matrix of the problem becomes

$$S_{\hat{s}}(JMlm \mid J_0 M_0 l_0 0) = \int Y_{JM}^{*}(\hat{s}) S_{\hat{s}}(lm \mid l_0 0) Y_{J_0 M_0}(\hat{s}) \, d\hat{s}. \tag{59}$$

It can be shown that this matrix is also unitary, so that the particle flux is conserved in this approximate treatment.

The adiabatic approximation (53) becomes the exact solution of the wave equation (6), if we neglect the operation of H_R on the parameter $\hat{s}$ in $F_{\hat{s}}(\mathbf{r})$. The difference

$$H_R[F_{\hat{s}}(\mathbf{r}) Y_{J_0 M_0}(\hat{s})] - F_{\hat{s}}(\mathbf{r}) H_R \, Y_{J_0 M_0}(\hat{s})$$

may be considered as a small perturbation when an improvement over the adiabatic approximation is desirable.

Formula (56), with the Born amplitude substituted into $f_{\hat{s}}$, has been used by Altshuler (1957) and Wijnberg (1966) to study the rotational transitions in simple molecules. Mittleman et al. (1968) substituted into (56) the exact solution of the scattering problem in the space-fixed point–dipole field and found a considerable deviation from the Born approximation when the dipole moment became larger (above about 0.5 a.u.).

Fisk (1936) calculated the total scattering cross section for a potential function which made the wave equation separable in the two-center spheroidal coordinates

$$\xi = (r_A + r_B)/s, \qquad \eta = (r_A - r_B)/s, \tag{60}$$

and the angle ϕ around the molecular axis, r_A and r_B being the distances from the two nuclei. Adjustable parameters in the potential function were determined by fitting the calculated total scattering cross section to observed data. Oksyuk (1965) used this Fisk's potential to calculate the rotational cross sections for H_2, N_2, and O_2 in the adiabatic approximation (56). Fisk's potential, however, is of a short-range type and has neither long-range quadrupole nor polarization parts in it. For this reason, the resulting rotational cross section is somewhat smaller than what we expect from more realistic interactions.

Hara (1969a) studied recently the scattering of electrons from the space-fixed H_2 molecule, including long-range quadrupole interaction, target polarization, and electron exchange. The scattering amplitude thus obtained was then used to calculate the rotational excitation cross section in the adiabatic approximation (Hara, 1969b). His results are in good agreement with experimental results. (See Section V for comparison of theoretical results with observed data.)

B. THE DISTORTED-WAVE METHOD

The distorted-wave approximation is a perturbation method which can be obtained by assuming in the basic equations (7) that all the off-diagonal elements of the interaction are small. For details of the formulation, the readers are referred to Arthurs and Dalgarno (1960). In the zeroth approximation we have to solve a scattering problem with the central field $\langle lJ|V|lJ\rangle_\lambda$. This is the scattering of the electron in an averaged potential where the non-spherical part of the interaction has been considerably suppressed. This may not be a good starting solution, since, as we have seen already, the target molecule does not rotate appreciably during the encounter. When the non-spherical interaction is much weaker than the spherical part of interaction, the distorted-wave approximation is expected to give a cross section of a high reliability.

Takayanagi and Geltman (1965) calculated the rotational excitation cross section of H_2 and N_2 in the distorted-wave approximation. The distorted waves were calculated, for simplicity, for the spherical interaction

$$\begin{aligned} V_0(r) &= -\alpha/2r^4, &\quad r \geq r_c \\ &= -\alpha/2r_c{}^4, &\quad r < r_c. \end{aligned} \tag{61}$$

The cutoff distance r_c was chosen empirically to give the best result for the total scattering cross section. The nonspherical interaction which induces the rotational excitation was

$$\begin{aligned} &-[(Q/r^3) + (\alpha'/2r^4)]P_2(\hat{r}\cdot\hat{s}), &\quad r \geq r_c \\ &-[(Q/r_c{}^3) + (\alpha'/2r_c{}^4)]P_2(\hat{r}\cdot\hat{s}), &\quad r < r_c. \end{aligned} \tag{62}$$

A similar distorted-wave calculation, with a slightly different potential function, has been done independently by Sampson and Mjolsness (1965). Later, Geltman and Takayanagi (1966) realized that there was a short-range interaction, the nonspherical part of which was much stronger than interaction (62) in small distances. The short-range interaction is approximately given by

$$V_{SR}(\mathbf{r}) = v(r_A) + v(r_B), \tag{63}$$

where r_A and r_B are the distances from the nuclei, and $v(r)$ the electrostatic potential due to an undeformed isolated atom. They expanded this interaction (63) in terms of the Legendre polynomials $P_\nu(\hat{r}\cdot\hat{s})$ and kept the first two nonvanishing terms ($\nu = 0$ and 2). With these interactions, calculations were made for O_2 as well as H_2 and N_2. It is especially interesting to compare the result obtained for O_2 with that of N_2. At very low collision energy, corresponding to the temperature of $100°K$ or less, the excitation cross section of O_2 is nearly two orders of magnitude smaller than that of N_2. This is because the quadrupole moment of O_2 is an order of magnitude smaller than that of N_2 (see Table II). As the electron energy increases to about 0.1 eV, corresponding to the temperature of several hundred degrees, the oxygen cross section becomes considerably larger than that of N_2 (Fig. 8), which is in good agreement with an experimental finding by Mentzoni and Narasinga Rao (1965). This situation can be interpreted as follows. In these two molecules under consideration the quadrupole moment Q and the non-spherical part of the polarizability α' have different signs. Therefore, the two terms in (62) tend to cancel each other. In oxygen, however, the quantity Q is very small so that the polarization interaction soon dominates as the

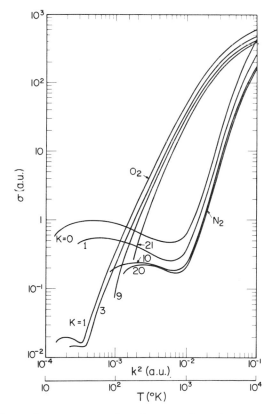

FIG. 8. The distorted-wave cross sections for several rotational transitions $K \to K+2$ in O_2 and N_2. For O_2 the rotational quantum number K differs from the total angular momentum quantum number J because of its unpaired electron spin ($S = 1$), while $K = J$ for N_2. The energy of the incident electron is scaled in terms of k^2 in a.u., so that the abscissa of the figure indicates the electron energy in the Rydberg unit (13.6 eV).

collision energy increases. There is not much cancellation in this case. Besides, the short-range interaction has the same sign as that of the polarization interaction. These are the reasons for the large increase in the cross section of O_2. For nitrogen, on the other hand, a considerable cancellation between the two terms in (62) prevents such a large increase until the collision energy is further raised.

In both N_2 and O_2, the short-range force is so strong that the distorted-wave method, in which the nonspherical interaction is assumed small, yields an unreasonably large cross section for the incident energy above 1 eV. Since the calculated cross section comes mainly from the incident p wave, we may expect from a consideration of the particle-flux conservation that the total inelastic cross section remains below $3\pi/k^2$, k being the incident electron wave number. However, the calculated value exceeds this theoretical limit in an energy range around 1 eV. Solution of a set of coupled equations, without resorting to the perturbation theory, is clearly needed.

Dalgarno and Henry (1965) made a distorted-wave calculation for hydrogen. They neglected the polarization interaction which pulls the electron into the region of stronger molecular field. Thus there is little distortion effect in their calculation and their cross section is somewhat too small in the low collision energy.

Ardill and Davison (1968) also neglected the polarization effect in their distorted-wave calculation of the rotational excitation cross section of hydrogen. However, they explicitly introduced, for the first time in this type of calculation, the effect of electron exchange and found that there is a considerable increase in the p wave partial cross section due to this effect.

C. CLOSE-COUPLING CALCULATIONS

In this approach, a finite number of sets (l, J) are retained in basic equations (7), and the resulting set of equations are solved numerically.

1. Hydrogen Molecule

Lane and Geltman (1967) applied this method to the rotational excitation of hydrogen. They expanded the short-range potential (63) in terms of $P_\nu(\hat{\mathbf{s}} \cdot \hat{\mathbf{r}})$ and retained the first three terms ($\nu = 0$, 2, and 4). The singularities at the origin of the long-range (polarization and quadrupole) interactions were removed by multiplying the cutoff factor

$$C(r) = 1 - \exp[-(r/r_c)^p], \qquad p = 6. \tag{64}$$

The parameter r_c was determined by fitting the calculated total cross section to observed data. They found values of r_c in the neighborhood of 1.8 to 2.0

a.u. most appropriate in the energy range of about 5 to 15 eV. The hydrogen molecule has only two orbital electrons and the short-range interaction is not so strong as in other molecules. The result of the close-coupling calculation thus agrees within 20% with the result of the distorted-wave calculation for the excitation $J = 0 \rightarrow 2$, provided that the same interaction potential is used. The higher order excitation processes $J = 0 \rightarrow 4$, $1 \rightarrow 5$, etc., are found to have very small probabilities.

Lane and Henry (1968) improved the calculation of Lane and Geltman by replacing the artificially cutoff polarization potential by the adiabatic polarization potential calculated by themselves. Then, in their second paper (Henry and Lane, 1969), they took account of the electron exchange. Basic equations (7) were thus replaced by a set of integrodifferential equations. By solving these equations numerically they obtained the rotational excitation cross sections of H_2, for $J = 0 \rightarrow 2$, $1 \rightarrow 3$ and the differential cross section for these excitation processes. They made calculations of the following four types: (*i*) all interactions taken into account, (*ii*) electron exchange neglected, (*iii*) target polarization neglected, (*iv*) both electron exchange and target polarization neglected. The results for (*ii*) and (*iii*) are comparable and are much larger than the result for (*iv*) (twice as large at 1 eV for the excitation $J = 0 \rightarrow 2$). The cross section (*i*) with full interactions is larger by a factor of two or three than those for (*ii*) and (*iii*). This final result (*i*) agrees well with that of Lane and Geltman (1967) with the semiempirical interactions.

2. Polar Molecules

In polar molecules, the long-range dipole interaction predominates in determining the rotational transitions, provided that the dipole moment of the molecule is not too small. The distant collision is more important than the close collision, and thus the Born approximation is expected to be fairly reliable (Section III, C, 1). To confirm this expectation, a close-coupling calculation for polar molecules was done by Itikawa and Takayanagi (1969). Two molecules, HCl and CN, were studied. One has a large and the other has a small rotational constant. The short-range interaction and the spherical part of the polarization interaction as well as the long-range dipole interaction were taken into account. The resulting cross section for the excitation $J = 0 \rightarrow 1$ agrees fairly well with the Born cross section obtained for the point–dipole interaction. Figure 9 shows the result for HCl. In this molecule, the partial contribution from the incident s and p waves is about 40% at 0.01 eV, but it decreases to 20% at 0.1 eV, 10% at 1.0 eV, and becomes only a few per cent at 10 eV. This fact reflects the situation that the transition is primarily determined by the simple dipole interaction. For CN, the relative contribution of the s and p waves is further reduced. For this molecule, the

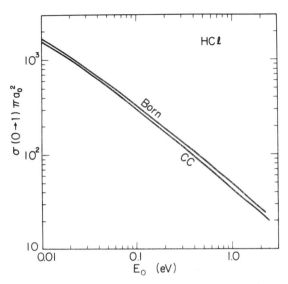

FIG. 9. Rotational cross sections $\sigma\,(0 \rightarrow 1)$ for HCl calculated using the close-coupling method (*CC*) and the Born approximation (*Born*). Incident electron energy is denoted by E_0.

dipole moment was not known at the time of calculation, so that a large value of 1 a.u. was assumed arbitrarily. Later an experimental value of 0.57 a.u. was reported (Table II). It is instructive, however, that the dipole moment as large as 1 a.u. does not invalidate the simple Born approximation for the rotational excitation. On the other hand, the elastic scattering $J = 0 \rightarrow 0$ is dominated by the s and p waves and thus is very sensitive to the choice of the short-range interaction. Itikawa (1969) then calculated the differential cross section. In the excitation $J = 0 \rightarrow 1$, the cross section is strongly peaked in the forward direction as is expected from the dominant dipole interaction. In the region of small scattering angles (0°–20°), the difference between the close coupling and the Born cross sections is very small, while at larger scattering angles ($>30°$), there is a noticeable difference, especially for the incident energy above 1 eV. This large-angle scattering is due to the close collision, where the distortion of the electron wave function is appreciable. Similar calculations for electron scattering by CO and CN have been reported recently by Allison *et al.* (1969; see also Crawford *et al.*, 1969).

It is interesting to compare the total scattering cross section with the momentum-transfer cross section [cf. Eq. (15)]:

$$\sigma_m = \sum_J \sigma_m(J_0 \rightarrow J).$$

$$\sigma_m(J_0 \rightarrow J) = 2\pi \int_0^{\pi} (1 - \cos\theta)q(J_0 \rightarrow J;\theta)\sin\theta\,d\theta. \qquad (65)$$

The small-angle scattering due to the long-range dipole interaction is suppressed by the factor $(1 - \cos \theta)$. Therefore, the momentum-transfer cross section is determined by the intermediate and short-range parts of the interaction. To study the dependence of σ_m on the dipole moment of the molecule, Itikawa (1969) adopted a simple analytic potential function

$$V = -(D/r^2)[1 - \exp\{-(r/r_c)^6\}]P_1(\hat{\mathbf{s}} \cdot \hat{\mathbf{r}}), \qquad (66)$$

with a cutoff distance $r_c = 1$ a.u. and calculated σ_m at 0.03 eV as a function of D. The rotational constant was taken to be the same as that of CN. The result is shown in Fig. 10. The pronounced peak obtained in the close-coupling calculation is very much the same as the peak obtained previously for the space-fixed finite-size dipole (Takayanagi and Itikawa, 1968), and is interpreted as a shape (potential) resonance (see Section VI). The same function (66) has been used by Itikawa and Takayanagi (1969) to study the

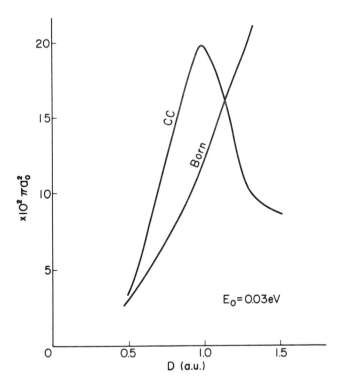

FIG. 10. Dependence of the momentum-transfer cross section σ_m on the dipole moment D. The interaction potential used for the calculation of σ_m is of the form (66) in the text. Comparison between the close-coupling calculation (*CC*) and the Born approximation (*Born*) is made for the incident electron energy 0.03 eV.

dependence of the total (elastic plus rotational transition) cross section on D. It was found that the elastic cross section alone showed a peak as a function of D, but a large contribution from the rotational transition due to the dipole interaction masked this peak and the total cross section had only a small hump as a function of D.

Crawford (1968) made a close-coupling calculation for the rotational excitation of H_2O using the interaction

$$V = f(r)P_1(\hat{s} \cdot \hat{r}),$$ (67)

$$f(r) = \begin{cases} -Dr/r_c^3, & r < r_c \\ -D/r^2, & r \geq r_c \end{cases}$$

and evaluated the momentum-transfer cross section (65). By adjusting the cutoff parameter r_c, a much better agreement with experimental data was obtained than in the Born approximation. However, the best fit is found at $r_c = \frac{1}{8}$ Å, which is too small to be interpreted as a cutoff distance. This result is probably due to the fact that he entirely neglected the short-range interaction.

V. Summary of Numerical Calculations

A. CONCLUSIONS FROM THEORETICAL CALCULATIONS

1. H_2, N_2, O_2

At very low energy (at room temperature or below), the quadrupole interaction dominates (Gerjuoy and Stein, 1955) and the Born approximation is applicable. At a little higher energy, the polarization interaction has to be considered. This has two effects: (1) the direct contribution to the rotational transition (Dalgarno and Moffett, 1963), and (2) the large distortion effect which is taken into account by the distorted-wave method (Takayanagi and Geltman, 1965; Sampson and Mjolsness, 1965). For N_2 and O_2, the distorted-wave calculations indicate the necessity of a more accurate calculation, such as a close-coupling approach. A close-coupling calculation for nitrogen has been recently reported by Burke *et al.* (1969).

In the distorted wave calculations, usually, an adjustable parameter (cutoff distance) is empirically determined. By doing so, the electron exchange effect, which has not been introduced explicitly, is effectively taken into account.

Explicit introduction of the electron exchange in the distorted-wave type calculation was started later. (Ardill and Davison, 1968). It was found that the exchange effect had a considerable effect on the excitation cross section.

Close-coupling (Henry and Lane, 1969) and adiabatic approximation

(Hara, 1969b) calculations for H_2 show that both polarization and electron exchange have to be taken into account to explain the observed data quantitatively.

For molecules other than H_2, we usually know only the molecular parameters, Q, α, α' which determine the asymptotic form of the electron–molecule interaction. Quantitative knowledge of the short-range part of the interaction is badly needed. Even for H_2, the target polarization is not completely understood yet (ambiguity in the polarization when the incident electron comes into the charge cloud of the target, nonadiabatic effects, lack of consistent formulation of scattering problem with both exchange and polarization, etc.). It is also desirable to clarify the relation between the close-coupling calculation and the adiabatic approximation.

2. Polar Molecules

As far as the rotational excitation is concerned, the Born approximation is found to be fairly accurate (Itikawa and Takayanagi, 1969; Itikawa, 1969). However, the momentum-transfer cross section for low-energy electrons shows an anomalous behavior as the dipole moment D becomes comparable with 1 a.u. (Takayanagi and Itikawa, 1968; Itikawa, 1969). Knowledge of the short-range interaction is required to predict the large-angle elastic scattering cross section accurately.

B. COMPARISON WITH EXPERIMENTAL DATA

In hydrogen only a very small number of rotational transitions are involved in the electron-swarm experiment, so that the relevant cross section can be determined fairly accurately by analyzing the experimental data. In parahydrogen, especially, the rotational cross section for $J = 0 \to 2$ can be determined uniquely since there is no other competitive inelastic processes in the energy region below the vibrational threshold. There have been two detailed investigations on this particular process: one by Engelhardt and Phelps (1963) and the other by Crompton et al. (1969). The excitation cross section obtained by the former group is somewhat larger than the result of the latter. In Fig. 11, the cross section determined by Crompton et al. is shown, together with some theoretical curves. It is seen that the close-coupling calculation by Henry and Lane (1969) agrees with the experimental data extremely well. In this calculation, both the electron exchange and the target polarization are taken into account. If either of these two effects is neglected, the resulting cross section becomes much smaller (see Section IV, C, 1). Another close-coupling calculation by Lane and Geltman (1967), based on an empirically determined potential function, is slightly below the experimental points. The

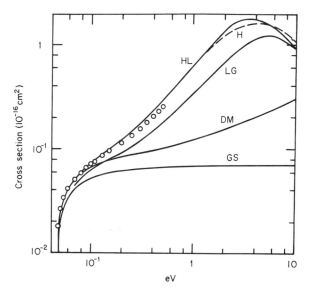

Fig. 11. The experimental and theoretical excitation cross sections $\sigma(J = 0 \to 2)$ of H_2. *HL*: Henry and Lane (1969). *H*: Hara (1969b). *LG*: Lane and Geltman (1967). *DM*: Dalgarno and Moffett (1963). *GS*: Gerjuoy and Stein (1955). Circles are experimental data obtained by Crompton *et al.* (1969).

distorted-wave method gives a cross section (not shown in the figure) not far from the close-coupling result (cf. Section IV, C, 1). Both Gerjuoy–Stein and Dalgarno–Moffett cross sections [Eqs. (32) and (38)] are satisfactory in the near-threshold region, but they fail to explain the large observed cross section at higher energies, where the distortion of the electron wave function due to the polarization interaction is important.

For the transition $J = 1 \to 3$ in H_2, Ehrhardt and Linder (1968) determined the cross section by an electron-beam experiment. Very recently, Linder (1969) presented a revised cross section. These experimental cross sections are compared with the theoretical curves in Fig. 12. The curve by Henry and Lane (1969) was obtained by a close-coupling calculation, while Hara's curve (Hara, 1969b) was obtained in the adiabatic approximation. The agreement between theoretical calculations and experimental data is fairly good, although further investigations are clearly needed to achieve a quantitative agreement. The angular distribution of inelastically scattered electrons has been calculated both by Henry and Lane and by Hara. The agreement of these calculations with experiment is again fairly satisfactory.

Comparisons of theoretical calculations with swarm experiments for other molecules are made in a review article by Phelps (1968). For nitrogen, for instance, the distorted-wave cross section obtained by Geltman and Taka-

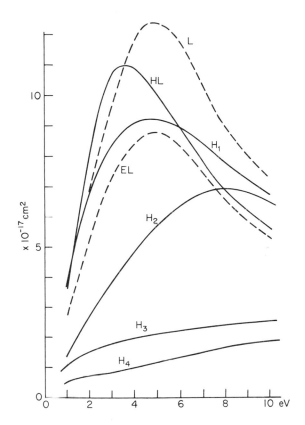

FIG. 12. The experimental and theoretical excitation cross sections $\sigma(J = 1 \rightarrow 3)$ of H_2. *HL*: Henry and Lane (1969). H_1–H_4: Hara (1969b), calculations with full interactions, without polarization, without exchange, and with static interaction only, respectively. Experimental curves are *EL* (Ehrhardt and Linder, 1968), and *L* (Linder, 1969).

yanagi (1966) is too small, as compared with experimental data, at 0.1 eV and too large at energies above 0.3 eV.

For polar molecules, Phelps (1968) has found that the Born cross section, based on the point–dipole interaction, explains the observed data fairly well.

An indirect test of theoretical calculations may be made by comparing the calculated total (elastic plus rotational) cross section with the experimental one. Itikawa and Takayanagi (1969) have calculated the elastic $(J = 0 \rightarrow 0)$ as well as rotational $(J = 0 \rightarrow 1, 2)$ cross sections of HCl in the close-coupling method. The sum of these cross sections is compared in Fig. 13 with an early experiment by Brüche (1927). The agreement between the calculated and the observed cross sections is rather good for the incident energy above about 5 eV. The peak at about 10 eV comes from the elastic cross section; it is due

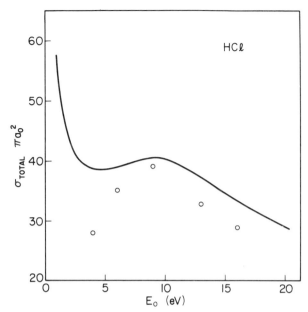

Fɪɢ. 13. Comparison of the calculated (in the close-coupling scheme) total cross section, $\sum_{J=0,1,2} \sigma(0 \to J)$, with the measured values (open circles) for electron–HCl scattering. E_0 is the incident energy in eV.

to a shape resonance. For low incident energies ($\lesssim 4$ eV) the measured cross section, when extrapolated smoothly, becomes much less than the calculated one. It is not definite, however, whether such a discrepancy really exists, because of the lack of experimental data for the incident energy less than 4 eV.

VI. Resonance Effects

As was already mentioned in Section II, B, 3, electron–molecule scattering sometimes has a resonant nature, that may affect the rotational excitation. On the other hand, the possibility of the rotational excitation itself can be the origin of a resonant scattering. That is, the rotational excitation may lead to formation of a temporary negative molecular ion when the incident energy is at an appropriate value below the rotational excitation threshold. Eventually, the electron will be reemitted, unless other processes are energetically possible.

Stabler (1963b) studied the electron capture through the rotational excitation of a positive ion of homonuclear diatomic molecule. He applied the Coulomb-modified Born approximation and assumed the point-quadrupole interaction to be responsible for the transition. However, he studied this process as the first stage of the electron–ion recombination and did not study the reemission of the electron. The electron capture by neutral polar molecules through the rotational excitation was first suggested by Turner (1966). The captured electron is in a bound state in the dipole field of the molecule. The initial state of the collision process is assumed the adiabatic wave function (53). The perturbation inducing the transition is the rotational energy operator for the molecule. He found that the capture cross section is remarkably large and suggested that this kind of process may explain the anomalously large momentum-transfer cross sections observed for some polar molecules (Hurst et al., 1963). He also calculated the rate of the reverse process (auto-detachment) and found the lifetime of the compound state to be as short as the rotational period of the molecule. Since the electron capture is soon followed by the reemission, it is more appropriate to treat these two steps as a single event rather than to study them separately. Thus Itikawa (1967) applied the second-order perturbation theory to the problem. As the pertur-bation he adopted the rotational energy operator as Turner did. It has been shown that contribution of the resonance to the momentum-transfer cross section is comparable with the nonresonant cross section (31). The width of the resonance is very broad, corresponding to the short lifetime of the inter-mediate state.

There is a lower limit to the dipole moment for which a bound state exists. The critical value has been discussed by many people (Fox and Turner, 1966; Mittleman and Myerscough, 1966; Turner and Fox, 1966; Lévy-Leblond, 1967; Fox, 1967; Brown and Roberts, 1967; Coulson and Walmsley, 1967; Crawford and Dalgarno, 1967; Crawford, 1967a; Turner et al., 1968). According to these investigations, a bound state exists only for a dipole larger than 0.639 a.u. (1.625 debye). Hence the capture mechanism cannot be applied to the polar molecules having smaller dipole moment. For example, H_2S has a moment of only 0.36 a.u., and yet Hurst et al. (1963) have observed a fairly large momentum-transfer cross section for this molecule. This example, therefore, has to be explained by another mechanism. Other mole-cules having a large observed cross section are H_2O and D_2O, where the dipole moment (0.73 a.u.) is above the critical value. However, there is another obstacle to the formation of the compound state. For a dipole whose moment is just above the critical value, a bound electron has a very small binding energy and, therefore, moves around the dipole too slowly to follow the molecular rotation. (The critical dipole was obtained for nonrotating mole-cules!) A rough estimate at $D = 0.83$ a.u. (Takayanagi and Itikawa, 1968) shows that the contribution of such a nonadiabatic effect to the bound-state energy is comparable to the binding energy itself at this value of D. Since the water molecule has a dipole moment less than this, it becomes doubtful whether the water molecule has a bound state to make the capture mechanism effective. These discussions are based on the pure dipole interaction. If other interations (of shorter ranges) are sufficiently strong to reinforce the attractive part of the dipole field, a bound state may exist even in the rotating H_2S.

The temporary capture of an electron through the rotational excitation may take place not only in the polar target but also in nonpolar molecules, provided that the electron–molecule interaction is sufficiently strong to allow for a bound state. Resonance scattering of this kind may be investigated by solving close-coupling equations with closed channels (rotational levels energetically unattainable for the given incident energy). Similar calculations have been done for atom–diatom collisions [see, e.g., Levine (1968); Levine *et al.* (1968)]. Frommhold (1968) studied experimentally the pressure-dependence of the electron-drift velocity in H_2, D_2, and N_2. By analyzing the data obtained, he suggested the temporary capture of electron through the rotational resonance process. Kouri (1968) assumed a spherical Morse-type potential between the electron and the H_2 molecule and calculated the possible energy levels of resonance arising from the process under consideration. On the other hand, the detailed close-coupling calculation of Henry and Lane (1969) for electron–hydrogen scattering showed no indication of such a resonance process. It is noted, however, that the resonance, if it exists, depends strongly on the short-range interaction, of which our knowledge is usually rather poor (especially on the electron exchange and the short-range part of the polarization force). Further studies are necessary, therefore, both theoretically and experimentally on the topic.

Another possibility of the enhancement of the cross section is the so-called potential resonance, which occurs, for instance, in scattering of a slow particle by a square-well potential with a suitable depth. This needs no temporary excitation of the target. Takayanagi and Itikawa (1968) studied the scattering of slow electrons by a two-center dipole field, namely, the field produced by a pair of point charges $+Z$, $-Z$ fixed at a distance R. The cross section obtained is averaged over the orientation of the molecule. The momentum-transfer cross section calculated for $\kappa^2 = (\frac{1}{2}kR)^2 = 0.001, 0.0001$ (k is the wave number) is shown in Fig. 14, as a function of the dipole moment $D = ZR$. The characteristic feature of the cross section is the pronounced peaks, the first of which appears in the region below 1 a.u. of the dipole moment. The position of this peak approaches slowly the critical dipole moment (mentioned above) as the incident wave number k decreases. It is interesting to see that a considerable deviation from the Born cross section starts far below the critical dipole moment. Other interactions (of shorter ranges) in real molecules may further shift the anomalous region to lower values of D. Christophorou and Christodoulides (1969) plotted the experimental momentum-transfer cross section σ_m for a large number of molecules in a σ_m-vs.-D diagram and obtained a distribution of points in favor of the

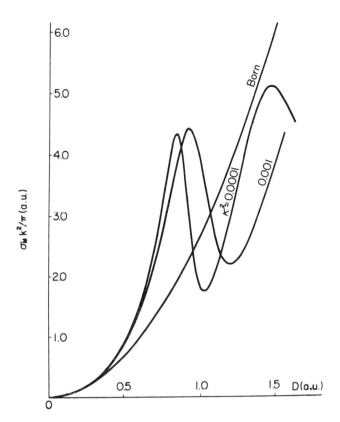

FIG. 14. Dipole-moment dependence of the momentum-transfer cross sections calculated for electron scattering by a two-center space-fixed dipole. Here, $\kappa = \frac{1}{2}kR$, k being the wave number of the incident electron and R denoting the separation of the two centers. The corresponding Born result is also shown.

anomalous cross section found theoretically. Lévy-Leblond and Provost (1967) studied the electron scattering from a space-fixed point–dipole and found a rapid increase in the momentum-transfer cross section when the dipole moment D approaches (from below) the critical value. Dalgarno et al. (1968) made a close-coupling calculation for a 0.05 eV electron scattering from a molecule with rotational constant $B = 2.4 \times 10^{-4}$ eV for three values of the dipole moment: $D = 1.600$, 1.627, and 1.650 debye. They confirmed that the momentum-transfer cross section does not show an abrupt change in the neighborhood of the critical dipole moment. All these calculations are consistent with the work by Takayanagi and Itikawa. The close-coupling calculation of Itikawa and Takayanagi (1969) also shows a similar anomaly in σ_m as a function of D (see Section IV, C, 2, especially Fig. 10). It is noted that Fig. 10 was obtained for the incident energy much above the threshold of the rotational excitation, so that the process is not the same as Turner's mechanism discussed earlier.

VII. Effects of Molecular Vibration

So far, we have assumed that the molecule is a rigid rotor. Real molecules, however, are vibrating rotors. It might be appropriate, therefore, to mention briefly possible effects of vibration on the rotational excitation. The dipole moment D, the quadrupole moment Q, the polarizabilities α, α', and other parameters characterizing the electron–molecule interaction are all dependent on the vibrational coordinates. In the rigid-rotor model, we adopt values of these parameters, say, at the mean internuclear distance. Since the electron above the thermal energy can see the instantaneous field, the proper scattering cross section may be obtained by averaging the calculated cross section over the statistical distribution of the vibrational coordinates. The amplitude of vibration, however, is not very large, especially for the zero-point vibration, so that the correction obtained in this way may not be important. But when the target molecule is in a highly excited vibrational state, the correction can be quite large.

Exceptionally important is the nonpolar molecule with infrared-active modes of vibration. Here, the mean molecular field has no long-range dipole field, while the scattered electron sees the instantaneous dipole field which may affect considerably the scattering cross section. One can study this problem in the adiabatic approximation. Function (53) has to be extended to allow for the molecular vibration, i.e.,

$$\Psi = F_s(\mathbf{r})X_{v_0}(s)Y_{J_0M_0}(\hat{\mathbf{s}}),\tag{68}$$

where X_{v_0} is the vibrational eigenfunction for the v_0th state. F_s is the solution of the electron scattering problem for a given s. For polyatomic molecules, the vector $\mathbf{s}$ is no more the internuclear distance between two nuclei, but symbolically indicates the whole set of vibrational and rotational coordinates of the molecule. Let the scattered amplitude be $f_s(\mathbf{k}_0 \to \mathbf{k})$. The amplitude for the inelastic process (simultaneous rotational and vibrational transitions) will be given by

$$f(v_0 J_0 M_0 \to vJM; \theta, \phi) = \int X_v^*(s)Y_{JM}^*(\hat{\mathbf{s}})f_s(\mathbf{k}_0 \to \mathbf{k})X_{v_0}(s)Y_{J_0M_0}(\hat{\mathbf{s}})\, d\mathbf{s}.\tag{69}$$

Abram and Herzenberg (1969) applied this formula successfully to the differential cross section of hydrogen for the simultaneous rotational and vibrational transitions. If there is an instantaneous dipole moment in the molecule, the amplitude f_s may be considerably larger than for the scattering in the averaged molecular field. However, when we put $v = v_0$ to obtain the pure rotational transition amplitude, a considerable part of the dipole scattering will disappear in the integration over s. Therefore, the effect of vibration on the pure rotational transition may not be so large as one expects at the beginning. Tice and Kivelson (1967) tried to explain a large observed cross section of CO_2 by the instantaneous dipole field. They found, however, that the calculated cross section is about 16 times too small to explain the experimental data.

Another method to study the effect of vibration on the rotational excitation is the second Born approximation. The target molecule jumps from its initial state $(v_0 J_0)$, through an intermediate state $(v'J')$, to the final state $(v_0 J)$. Again, the dipole interaction can considerably enhance the cross section, provided that both the transitions $(v_0 J_0) \rightarrow (v'J')$ and $(v'J') \rightarrow (v_0 J)$ are allowed.

VIII. Effects of Unpaired Electron Spin

Many molecules have a nonvanishing resultant spin. The total angular momentum of the molecule is given by the vector sum $\mathbf{J} = \mathbf{K} + \mathbf{S}$, where $\mathbf{K}$ is the rotational angular momentum of the molecular framework and $\mathbf{S}$ the resultant electron spin. Quantities $\mathbf{K}^2$ and $\mathbf{S}^2$ have the eigenvalues $K(K+1)$ and $S(S+1)$, respectively, and K and S are good quantum numbers. In the case of oxygen molecule in the ground $^3\Sigma$ state, these quantum numbers take the values

$$K = 1, 3, 5, \ldots; \qquad S = 1.$$

For a given value of K, the possible values of J are

$$J = K - 1, \quad K, \quad K + 1.$$

In the electron scattering, the spin vector $\mathbf{S}$ is barely changed through the electromagnetic interaction. The only possible way of changing $\mathbf{S}$ appreciably is the electron exchange. No one has ever studied the effect of the process $\Delta M_s \neq 0$ (M_s, z component of $\mathbf{S}$) on the rotational transition. If we neglect the electron exchange for the time being, we can expect that the vector $\mathbf{S}$ remains unchanged during collision while the vector $\mathbf{K}$ may undergo a transition. As a consequence, the quantum number J can be changed. However, the energy of the molecule is primarily determined by K and the dependence on J is very little, except for the lowest state of rotation ($K = 1$) for O_2. Therefore, as far as the electron energy loss in the oxygen gas is concerned, the presence of spin does not have a large influence. This has been confirmed directly, by utilizing the Born approximation, by Geltman and Takayanagi (1966). Each individual cross section σ $(K_0, J_0 \to KJ)$, however, depends considerably on J_0 and J.

Appendix

ASYMPTOTIC INTERACTION BETWEEN AN ELECTRON AND A LINEAR MOLECULE

The electrostatic interaction is calculated in terms of the charge distribution $\rho(\mathbf{r})$ inside the molecule:

$$V_{\text{static}}(\mathbf{r}) = - \int \rho(\mathbf{r}')/|\mathbf{r} - \mathbf{r}'| d\mathbf{r}'.$$

If the electron is sufficiently outside the molecule $(r > r')$, this potential may be expanded as

$$V_{\text{static}}(\mathbf{r}) = - \sum_{v=0}^{\infty} r^{-v-1} \int \rho(\mathbf{r}') r'^{v} P_v(\hat{\mathbf{r}} \cdot \hat{\mathbf{r}}') \, d\mathbf{r}'.$$

By taking the z axis along the direction of molecular axis $\hat{\mathbf{s}}$ and utilizing the formula

$$P_v(\hat{\mathbf{r}} \cdot \hat{\mathbf{r}}') = [4\pi/(2v+1)] \sum_{\lambda} Y_{v\lambda}(\hat{\mathbf{r}}) Y_{v\lambda}^{*}(\hat{\mathbf{r}}'),$$

we have

$$V_{\text{static}}(\mathbf{r}) = - \sum_{v=0}^{\infty} r^{-v-1} P_v(\hat{\mathbf{s}} \cdot \hat{\mathbf{r}}) M_v,$$

where M_v is the 2^v pole moment

$$M_v = \int \rho(\mathbf{r}') r'^{v} P_v(\hat{\mathbf{s}} \cdot \hat{\mathbf{r}}') \, d\mathbf{r}'.$$

It is noted that the higher multipole moments depend on the location of the coordinate origin. In the present article, all moments are relative to the gravity center of the molecule. At large distances, the dipole $(M_1 = D)$ and the quadrupole $(M_2 = Q)$ terms are most important:

$$V_{\text{static}}(\mathbf{r}) \xrightarrow{r \to \infty} - (D/r^2) P_1(\hat{\mathbf{s}} \cdot \hat{\mathbf{r}}) - (Q/r^3) P_2(\hat{\mathbf{s}} \cdot \hat{\mathbf{r}}) - \cdots.$$

It is important to notice that sometimes $2Q$ is called the quadrupole moment. In addition to these, there is the polarization interaction which is also important at large distances:

$$V_{\text{polariz}}(\mathbf{r}) \xrightarrow{r \to \infty} -(\alpha/2r^4) - (\alpha'/2r^4)P_2(\hat{s} \cdot \hat{r}),$$

where

$$\alpha = \tfrac{1}{3}(\alpha_{\parallel} + 2\alpha_{\perp}), \qquad \alpha' = \tfrac{2}{3}(\alpha_{\parallel} - \alpha_{\perp})$$

and $\alpha_{\parallel}$, $\alpha_{\perp}$ are the polarizabilities of the molecule in the directions parallel and perpendicular to the molecular axis. The electron-exchange effect, which cannot be written in a simple analytic form, is of a short-range nature and has no important contribution to distant collisions.

REFERENCES

Abram, R. A., and Herzenberg, A. (1969). *Chem. Phys. Letters* **3**, 187

Allison, A. C., Crawford, O. H., and Dalgarno, A. (1969). *Abstr. Papers, 6th Intern. Conf. Phys. Electron. At. Collisions*, p. 155. MIT Press, Cambridge, Massachusetts.

Altshuler, S. (1957). *Phys. Rev.* **107**, 114.

Ardill, R. W. B., and Davison, W. D. (1968). *Proc. Roy. Soc. (London)* **A304**, 465.

Arthurs, A. M., and Dalgarno, A. (1960). *Proc. Roy. Soc. (London)* **A256**, 540.

Bardsley, J. N., and Mandl, F. (1968). *Rept. Progr. Phys.* **31**, 471.

Biedenharn, L. C., Blatt, J. M., and Rose, M. E. (1952). *Rev. Mod. Phys.* **24**, 249.

Blatt, J. M., and Biedenharn, L. C. (1952). *Rev. Mod. Phys.* **24**, 258.

Boikova, R. F., and Ob'edkov, V. D. (1968). *Zh. Eksperim. i Teor. Fiz.* **54**, 1439 [*Soviet Phys. JETP* **27**, 772 (1968)].

Bridge, N. J., and Buckingham, A. D. (1966). *Proc. Roy. Soc. (London)* **A295**, 334.

Brown, W. B., and Roberts, R. E. (1967). *J. Chem. Phys.* **46**, 2006.

Brüche, E. (1927). *Ann. Physik* **82**, 25.

Burke, P. G., Faisal, F. H. M., Sinfailam, A. L., and Thompson, D. G. (1969). *Abstr. Papers, 6th Intern. Conf. Phys. Electron. At. Collisions*, p. 101. MIT Press, Cambridge, Massachusetts.

Callaway, J., and LaBahn, R. W. (1968). *Phys. Rev.* **168**, 12.

Carson, T. R. (1954). *Proc. Phys. Soc. (London)* **A67**, 909.

Chen, J. C. Y. (1966). *Phys. Rev.* **146**, 61.

Christophorou, L. G., and Christodoulides, A. A. (1969). *Proc. Phys. Soc. (London) (At. Mol. Phys.)* **2**, 71.

Condon, E. U., and Shortley, G. H. (1935). "The Theory of Atomic Spectra," Chapter 6. Cambridge Univ. Press, London and New York.

Coulson, C. A., and Walmsley, M. (1967). *Proc. Phys. Soc. (London)* **91**, 31.

Crawford, O. H. (1967a). *Proc. Phys. Soc. (London)* **91**, 279.

Crawford, O. H. (1967b). *J. Chem. Phys.* **47**, 1100.

Crawford, O. H., (1968). *Chem. Phys. Letters* **2**, 461.

Crawford, O. H., and Dalgarno, A. (1967). *Chem. Phys. Letters* **1**, 23.

Crawford, O. H., Dalgarno, A., and Hays, P. B. (1967). *Mol. Phys.* **13**, 181.

Crawford, O. H., Allison, A. C., and Dalgarno, A. (1969). *Astron. Astrophys.* **2**, 451.

Crompton, R. W., Gibson, D. K., and McIntosh, A. I. (1969). *Austral. J. Phys.* **22**, 715.

Cross, P. C., Hainer, R. M., and King, G. W. (1944). *J. Chem. Phys.* **12**, 210.

Dalgarno, A., and Henry, R. J. W. (1965). *Proc. Phys. Soc. (London)* **85**, 679.

Dalgarno, A., and Moffett, R. J. (1963). *Proc. Natl. Acad. Sci. India* **A33**, 511.

Dalgarno, A., Crawford, O. H., and Allison, A. C. (1968). *Chem. Phys. Letters* **2**, 381.
Ehrhardt, H., and Linder, F. (1968). *Phys. Rev. Letters* **21**, 419.
Engelhardt, A. G., and Phelps, A. V. (1963). *Phys. Rev.* **131**, 2115.
Fisk, J. B. (1936). *Phys. Rev.* **49**, 167.
Fox, K. (1967). *Phys. Letters* **25A**, 345.
Fox, K., and Turner, J. E. (1966). *J. Chem. Phys.* **45**, 1142.
Frommhold, L. (1968). *Phys. Rev.* **172**, 118.
Geltman, S., and Takayanagi, K. (1966). *Phys. Rev.* **143**, 25.
Gerjuoy, E., and Stein, S. (1955). *Phys. Rev.* **97**, 1671.
Hara, S. (1967). *J. Phys. Soc. Japan* **22**, 710.
Hara, S. (1969a). *J. Phys. Soc. Japan* **27**, 1009.
Hara, S. (1969b). *J. Phys. Soc. Japan* **27**, 1592.
Henry, R. J. W., and Lane, N. F. (1969). *Phys. Rev.* **183**, 221.
Herzberg, G. (1945). "Infrared and Raman Spectra." Van Nostrand, Princeton, New Jersey.
Herzberg, G. (1950). "Spectra of Diatomic Molecules." Van Nostrand, Princeton, New Jersey.
Hurst, G. S., Stockdale, J. A., and O'Kelly, L. B. (1963). *J. Chem. Phys.* **38**, 2572.
Itikawa, Y. (1967). *Phys. Letters* **24A**, 495; *Inst. Space Aeronaut. Sci. Rept.* **412**.
Itikawa, Y. (1969). *J. Phys. Soc. Japan* **27**, 444.
Itikawa, Y., and Takayanagi, K. (1969). *J. Phys. Soc. Japan* **26**, 1254.
Kleinman, C. J., Hahn, Y., and Spruch, L. (1968). *Phys. Rev.* **165**, 53.
Kołos, W., and Wolniewicz, L. (1964). *J. Chem. Phys.* **41**, 3674.
Kouri, D. J. (1968). *J. Chem. Phys.* **49**, 5205.
Lane, N. F., and Geltmen, S. (1967). *Phys. Rev.* **160**, 53.
Lane, N. F., and Henry, R. J. W. (1968). *Phys. Rev.* **173**, 183.
Levine, R. D. (1968). *J. Chem. Phys.* **49**, 51.
Levine, R. D., Johnson, B. R., Muckerman, J. T., and Bernstein, R.B. (1968). *J. Chem. Phys.* **49**, 56.
Lévy-Leblond, J.-M. (1967). *Phys Rev.* **153**, 1.
Lévy-Leblond, J.-M., and Provost, J.-P. (1967). *Phys. Letters* **26B**, 104.
Linder, F. (1969). *Abstr. Papers, 6th Intern. Conf. Phys. Electron. At. Collisions*, p. 141. MIT Press, Cambridge, Massachusetts.
Massey, H. S. W. (1931). *Proc. Cambridge Phil. Soc.* **28**, 99.
Massey, H. S. W. (1969). "Electronic and Ionic Impact Phenomena," 2nd Ed., Vol. II, Chapter 11. Oxford Univ. Press, London and New York.
Mentzoni, M. H., and Narasinga Rao, K. V. (1965). *Phys. Rev. Letters* **14**, 779.
Mittleman, M. H., and Myerscough, V. P. (1966). *Phys. Letters* **23**, 545.
Mittleman, M. H., Peacher, J. L., and Rozsnyai, B. F. (1968). *Phys. Rev.* **176**, 180.
Morse, P. M. (1953). *Phys. Rev.* **90**, 51.
Oksyuk, Yu. D. (1965). *Zh. Eksperim. i Teor. Fiz.* **49**, 1261 [*Soviet Phys. JETP* **22**, 873 (1966)].
Phelps, A. V. (1968). *Rev. Mod. Phys.* **40**, 399.
Racah, G. (1942). *Phys. Rev.* **62**, 438.
Sampson, D. H. (1965). *Phys. Rev.* **137**, A4.
Sampson, D. H., and Mjolsness, R. C. (1965). *Phys. Rev.* **140**, A1466.
Slater, J. C. (1960). "Quantum Theory of Atomic Structure," Vol. II. McGraw-Hill, New York.
Stabler, R. C. (1963a). *Phys. Rev.* **131**, 679.
Stabler, R. C. (1963b). *Phys. Rev.* **131**, 1578.

Stogryn, D. E., and Stogryn, A. P. (1966). *Mol. Phys.* **11**, 371.
Takayanagi, K. (1954). *Progr. Theoret. Phys.* (*Kyoto*) **11**, 557.
Takayanagi, K. (1966). *J. Phys. Soc. Japan* **21**, 507.
Takayanagi, K. (1967). *Progr. Theoret. Phys.* (*Kyoto*) *Suppl.* **40**, 216.
Takayanagi, K., and Geltman, S. (1965). *Phys. Rev.* **138**, A1003.
Takayanagi, K., and Itikawa, Y. (1968). *J. Phys. Soc. Japan* **24**, 160.
Takayanagi, K., and Ohno, K. (1955). *Progr. Theoret. Phys.* (*Kyoto*) **13**, 243.
Thomson, R., and Dalby, F. W. (1968). *Can. J. Phys.* **46**, 2815.
Tice, R., and Kivelson, D. (1967). *J. Chem. Phys.* **46**, 4748.
Turner, J. E. (1966). *Phys. Rev.* **141**, 21.
Turner, J. E., and Fox, K. (1966). *Phys. Letters* **23**, 547.
Turner, J. E., Anderson, V. E., and Fox, K. (1968). *Phys. Rev.* **174**, 81.
Wijnberg, L. (1966), *J. Chem. Phys.* **44**, 3864.
Wollrab, J. E. (1967). "Rotational Spectra and Molecular Structure," Chapter 2. Academic Press, New York.

NOTE ADDED IN PROOF

The following references have been published in this field of study since the manuscript was submitted.

Bottcher, C. (1969). *Chem. Phys. Letters* **4**, 320.
Chang, E. S., and Temkin, A. (1969). *Phys. Rev. Letters* **23**, 399.
Lane, N. F., and Geltman, S. (1969). *Phys. Rev.* **184**, 46.

THE DIFFUSION OF ATOMS AND MOLECULES

E. A. MASON and T. R. MARRERO[1]

Brown University, Providence, Rhode Island

[1] Present address: Department of Chemical Engineering, University of Missouri, Columbia, Missouri.

I. Introduction and Historical Sketch

Our purpose is twofold: to give an account of the physics of gaseous diffusion of atoms and molecules, and to survey the experimental methods and results obtained in the determination of gaseous diffusion coefficients. We do not consider dense gases or liquids, and in fact limit the discussion to moderate densities and lower, so that no more than binary collisions are important. Some of the free-molecule effects usually associated with lower densities are treated, since much of the current interest in gaseous diffusion is concerned with porous media, where free-molecule effects are often evident even above atmospheric pressure.

We consider only isothermal phenomena, the subject of thermal diffusion having been reviewed previously (Mason *et al.*, 1966). Although we thus ignore temperature gradients, we cannot ignore pressure gradients, which cause flow. The coupling between diffusion and flow is much more common than is generally realized, for almost all experimental arrangements involving diffusion also cause flow. This has been a great source of confusion in the past. Only in recent years has the problem been clarified, and even now some difficulties remain.

In this survey the phenomenological aspect is considered first, in Section II. Here one seeks to describe how gases mix and flow, and what the variables are that influence this behavior; ultimately such a description takes the form of some differential equations containing coefficients to be found by experiment. The emphasis is on the simplest physical explanations that are essentially correct without resort to the full mathematical machinery of the rigorous kinetic theory of gases. It is remarkable how far one can go in this way. The next two sections (III and IV) deal with the molecular aspect of diffusion, whose task is to relate the phenomenological coefficients to atomic and molecular properties. The account of the molecular theory can be fairly brief, thanks to the existence of standard references (Chapman and Cowling, 1970; Hirschfelder *et al.*, 1954; Waldmann, 1958). Some of the more recent results are summarized in Sections III and IV. Most of the interesting surprises in recent years have been on the experimental and phenomenological rather than the molecular side, which has largely developed into a tool for the study of intermolecular forces. The final two Sections (V and VI) review the determination of diffusion coefficients. These are based largely on a comprehensive survey of gaseous diffusion coefficients carried out under the National Standard Reference Data System of the U. S. National Bureau of Standards (Marrero and Mason, 1971).

Serious study of gaseous diffusion dates from the first work of Thomas Graham (1829), who studied the diffusion of various gases out of a closed vessel into the surrounding air through a small tube. No diffusion coefficients were obtained from these experiments at the time, for the mathematical statement of the law of diffusion, analogous to Fourier's law of heat conduction, was not to be given for another 26 years (Fick, 1855). It remained for Maxwell (1867) to calculate diffusion coefficients from Graham's observations, as well as from some later ones made by Graham (1863) using a closed vertical tube. These last measurements gave a diffusion coefficient for CO_2–air only a few percent different from the best modern value, and probably represent the first accurate result obtained. A few years later, Loschmidt (1870a,b) made an extensive series of measurements on ten gas pairs, using a closed vessel in the form of a long tube with a slide plate in the middle. Each half of the tube was filled with a different gas, the middle slide plate was opened for a measured time period, and the partially mixed gases then analyzed. This method, or variations of it, was the preferred experimental technique for many years, and is still frequently used. A number of different experimental techniques have been developed in the last 25 years. It is a tribute to Loschmidt's skill that most of his results have scarcely been improved upon; it is also a comment on the difficulty of measuring gaseous diffusion coefficients with high accuracy.

After his first work, Graham's interest in gaseous diffusion took a rather different turn, prompted by some observations published by Doebereiner (1823), in which it was noticed that hydrogen escaped from a jar with a slight crack in it, so that the water of the pneumatic trough rose into the jar. We discuss this work in detail here because it has been sadly misunderstood in recent times, despite its enshrinement in textbooks as "Graham's Law of Diffusion" (Mason, 1967; Mason and Kronstadt, 1967). Had it been remembered and understood, much effort and confusion in recent years might have been avoided. Graham's investigations were reported in a paper read before the Royal Society of Edinburgh in 1831, and later published in three sections (Graham, 1833). He took a calibrated glass tube with a porous plate at one end and with the other end immersed in a vessel of water (or mercury). The gas to be investigated was added to the tube by displacement of water, and its volume noted. As the gas diffused out and the air diffused in through the porous plate, the water level tended to rise or fall in the tube, depending on whether the gas was lighter or heavier than air. Since a change in the water level would have caused a pressure difference across the porous plate, Graham kept the pressure uniform by flowing water into or out of the outer vessel to keep the outer water level the same as the one inside the tube. After some time all the gas had diffused out and been replaced by the air that had diffused in. Graham noticed that the ratio of the volume of gas diffused out to the volume of air diffused in was equal to the ratio of the square root of the molecular weight of air to that of the gas. It should be noted that these experiments were *not* carried out in the free-molecule regime; the free-molecule result is called *effusion*, which was also discovered by Graham, but much later (Graham, 1846). It may also be noted that no diffusion coefficients were obtained from these experiments; they can be calculated from the rate of change of the water level (Evans *et al.*, 1969), but the mathematical analysis is more complicated than that needed for the closed-tube experiments.

The crucial point of difference between Graham's experiments and Loschmidt's is the uniform total pressure of the former. It is not so obvious that a small pressure gradient has to be present in the Loschmidt experiment in order to keep the fluxes of the two diffusing gases equal and opposite, which they clearly must be in order that the pressure does not continuously increase in one half of the tube and decrease in the other half. This pressure gradient is almost immeasurably small, except in capillary tubes, and was not detected by direct measurement until comparatively recent times (Kramers and Kistemaker, 1943). Its consquences, however, are large. Loschmidt and subsequent workers assumed that Graham's law of diffusion was applicable to experiments of the Loschmidt type; a consequence of this assumption is that the diffusion coefficients of gases 2 and 3 into a reference gas 1 should vary inversely as the square roots of the molecular weights of 2 and 3,

$$\mathscr{D}_{12}/\mathscr{D}_{13} \approx (M_3/M_2)^{1/2}, \tag{1}$$

which is actually only a crude approximation for most systems. Graham's law of diffusion thus came to be regarded as only a rough approximation (Partington, 1949), although anyone who took the trouble to examine the original data would have seen that the level of accuracy was about 1%. This particular misconception still persists (Kirk, 1967).

The importance of Graham's law of diffusion is not what it says about diffusion coefficients (it says nothing), but is its bearing on the coupling between diffusion and flow. Put another way, it concerns the boundary condition next to the wall (slip velocity) of a diffusing gas mixture, a matter of importance in any application. Having been misunderstood for over 100 years, the diffusion law was rediscovered experimentally (Hoogschagen, 1953, 1955), and has since become the subject of considerable activity under the name of "equal pressure counterdiffusion" or "isobaric diffusion" (for a partial list of references, see Mason and Kronstadt, 1967).

Turning now to the molecular theory of diffusion, we find a history of considerable difficulty compared to the theory for other transport phenomena.

A momentum-transfer theory was devised by Maxwell (1860) and independently by Stefan (1871, 1872). This leads to an expression for the binary diffusion coefficient which is independent of composition, and which in fact is the same as the first approximation later obtained in the accurate rigorous Chapman–Enskog theory. Apparently, however, Maxwell was not too happy with the momentum-transfer method, "... which led me into great confusion, especially in treating of the diffusion of gases " (Maxwell, 1873), and preferred a more rigorous method based on his equations of transfer (Maxwell, 1867). It can be shown that the momentum-transfer approach leads to the first approximation of the full Maxwell–Chapman transfer theory (see Kennard, 1938; Present, 1958). The momentum-transfer theory was neglected for many years; it was then apparently rediscovered by Frankel (1940) and by Present and de Bethune (1949; see also Present, 1958), and further elaborated by Furry (1948) and by Williams (1958). Momentum-transfer arguments can be used in a very simple way to guide the formulation of a full phenomenological description of diffusion and flow, as will be shown in Section II.

A mean-free-path theory of diffusion was devised by O. E. Meyer in 1877 (Meyer, 1899; Kennard, 1938). This predicts that the binary diffusion coefficient should depend strongly on the relative composition of the mixture, a result later found to disagree with experiment (see Kennard, 1938). Attempts to refine the theory by corrections for the "persistence of velocities" (Jeans, 1925) soon met with mathematical difficulties, one being slow convergence. It was also difficult to see how the refinements would be just sufficient to make the diffusion coefficient essentially independent of composition. Nevertheless, the first-order mean-free-path theory was attractive because of the clarity of the physical picture and the simplicity of the calculations; *ad hoc* attempts to patch up the theory by ignoring collisions between like molecules were made (Jeans, 1925). These gave the same result as the momentum-transfer theory, but were hardly justifiable. A rigorous treatment has recently been given that establishes the persistence corrections as an iterative solution of the same integral equation that appears in the Chapman–Enskog theory (Monchick and Mason, 1967).

The rigorous treatment of kinetic theory was carried out in the second decade of this century by Chapman, who took Maxwell's equations of transfer as his starting point, and by Enskog, who started with Boltzmann's equation for the distribution function. A delightful personal account of these developments has been given by Chapman (1967). Further developments involving inelastic collisions and quantum effects are discussed in Section III. This theory gives the connection between diffusion coefficients and molecular properties, particularly the intermolecular potentials, and is essentially a finished story.

The problem of coupling between flow and diffusion is in a less finished state. The general phenomenon of coupling between different modes of transport was first investigated by Maxwell (1879), who considered the stresses induced in a single gas by a temperature gradient. The extension to mixtures, and hence to diffusion, was first considered by Kramers and Kistemaker (1943), who used a simple momentum-transfer argument. Recent rigorous treatments have been mostly by Russian workers, using moment methods to solve the Boltzmann equation. A much simpler approach, yielding essentially the same conclusions, can be based on the Chapman–Enskog theory by treating the container walls or the particles of a porous medium as one component of a multicomponent mixture (Mason *et al.*, 1967a). These matters are discussed in Section IV.

II. Phenomenological Description

A. Modes of Gas Motion

In the absence of conditions causing turbulence, there are three main types of isothermal gas transport in free space or through tubes or porous media, as first clearly recognized by Graham (1833, 1846, 1849, 1863).

(1) Diffusion, in which the different species of a mixture move under the influence of composition gradients. This is a continuum phenomenon in that molecule–molecule collisions dominate over molecule–wall collisions.

(2) Transpiration, or laminar viscous flow, in which the gas acts as a continuum fluid driven by a pressure gradient. This is sometimes called convective or bulk flow. The pressure is high enough that molecule–molecule collisions dominate over molecule–wall collisions.

(3) Effusion, now more commonly called free-molecule or Knudsen flow. Here the pressure is so low that collisions between molecules can be ignored compared to molecule–wall collisions.

Corresponding to these three transport mechanisms are three transport coefficients: the continuum or normal diffusion coefficient $\mathscr{D}_{ij}$ (for a binary mixture of species i and j), the viscosity coefficient η, and the Knudsen diffusion coefficient D_{iK} (for species i). There are also three corresponding parameters characteristic of the solid medium through which the gas moves: the porosity–tortuosity ratio ε/q for continuum diffusion, the viscous flow parameter B_o, and the Knudsen flow parameter K_o. These three parameters are all simply related for a medium of simple geometry, such as a long circular capillary of diameter d:

$$\varepsilon/q = 1, \qquad B_o = d^2/32, \qquad K_o = d/4. \tag{2}$$

But for an actual porous medium the relation is complicated and generally unknown, and the three parameters are usually found from experiment rather than by calculation from an assumed geometry.

1. Continuum Diffusion

Diffusion is the most difficult of the three transport mechanisms to define satisfactorily, because it is most convenient to define a diffusion coefficient for a mixture in which the net flux is zero and in which there is also no pressure gradient (so there is no viscous flow). In this case the purely diffusive fluxes $\mathbf{J}_{1D}$ and $\mathbf{J}_{2D}$ of a binary mixture are

$$\mathbf{J}_{1D} = -\mathscr{D}_{12}\nabla n_1 \quad \text{and} \quad \mathbf{J}_{2D} = -\mathscr{D}_{21}\nabla n_2, \tag{3}$$

where $\mathbf{J}_{1D}$ and $\mathbf{J}_{2D}$ are in molecules/cm²-sec, n_1 and n_2 are in molecules/cm³, and $\mathscr{D}_{12}$ and $\mathscr{D}_{21}$ are in cm²/sec. It is easy to see that $\mathscr{D}_{21} = \mathscr{D}_{12}$, since $\mathbf{J}_{1D} + \mathbf{J}_{2D} = \mathbf{J}_D = 0$ for no net flux, and $\nabla(n_1 + n_2) = \nabla n = 0$ for no pressure gradient. Unfortunately, such a situation is very difficult to produce experimentally, since a small pressure gradient is necessary to keep $\mathbf{J}_D = 0$. This can be understood in a simple physical way as follows. Suppose we put two different gases at the same temperature and pressure into communication through a small tube (or a porous plate), as indicated in Fig. 1. The faster

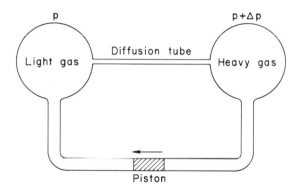

FIG. 1. Binary diffusion. If the piston is held stationary, a pressure difference develops that is just sufficient to keep the net flux zero through the diffusion tube. If the pressure is kept uniform, the piston must be moved as indicated, and there is a net flux of gas through the diffusion tube.

light molecules tend to get through the tube more quickly than the slower heavy ones, so that the pressure rises on the side of the heavy gas until the viscous flow induced back through the tube is just sufficient to make the net flux zero. The resulting steady-state pressure difference is very small if the tube has a large diameter, but can be appreciable for small capillaries or fine porous media. Thus Eqs. (3) can be used only for equal counter-current diffusion in large tubes.

To keep the pressure uniform during the diffusion, an arrangement like a side tube with an impermeable frictionless piston would be necessary, as indicated in Fig. 1. Since the piston must be moved to keep the pressure uniform, the net flux of gas is clearly not zero. This in principle is the type of experiment performed by Graham (1833), and this particular form of apparatus has been used in practice (McCarty and Mason, 1960; Mason, 1961). We can still describe this diffusion by the same diffusion coefficient $\mathscr{D}_{12}$, provided we allow for the net diffusive flux by including an extra term that apportions the net flux contribution to each species,

$$\mathbf{J}_{1D} = -\mathscr{D}_{12}\nabla n_1 + x_1\mathbf{J}_D, \tag{4a}$$

$$\mathbf{J}_{2D} = -\mathscr{D}_{21}\nabla n_2 + x_2\mathbf{J}_D, \tag{4b}$$

where $\mathbf{J}_D = \mathbf{J}_{1D} + \mathbf{J}_{2D}$, and $x_i = n_i/n$ is the mole fraction. All the fluxes are diffusive in this case because there is no pressure gradient. Another way of looking at Eqs. (4) is to say that $\mathscr{D}_{12}$ is defined in a coordinate system moving with a velocity corresponding to the average velocity associated with $\mathbf{J}_D$. It is easy to show, by addition of these two equations, that $\mathscr{D}_{21} = \mathscr{D}_{12}$ because $\nabla n = 0$.

Equations (4) thus define pure diffusive transport in the continuum region; the key point is that the pressure is uniform so that the viscous transport is zero. To find the flux when the pressure is nonuniform, we must find the viscous flux due to the pressure gradient, and find how to combine it with the diffusive flux. It turns out that the two fluxes are simply additive to a high degree of approximation (Section II, B); because of this simple additivity, diffusion at uniform pressure can be regarded as a more fundamental process than diffusion with zero net flux.

Graham (1833) discovered experimentally that the fluxes in Eqs. (4) were related,

$$(\mathbf{J}_{1D}/\mathbf{J}_{2D}) = (M_2/M_1)^{1/2}. \tag{5}$$

This looks like a free-molecule result (effusion), but is quite different. The best simple physical explanation of this diffusion law is based on a calculation of the momentum transferred to the walls by the diffusing molecules, as suggested by Hoogschagen (1953, 1955) following the ideas of Kramers and Kistemaker (1943). We briefly repeat the argument here, since the same type of argument can be used to show how continuum diffusion combines with free-molecule diffusion in the transition regime (Section II, B), and how to extend Eqs. (4) to multicomponent mixtures (Section II, C). We first define an average diffusion velocity $\bar{V}_{iD}$ for species i as

$$\bar{V}_{iD} = J_{iD}/n_i. \tag{6}$$

This is much different from the mean molecular speed $\bar{v}_i$; in a diffusing gas mixture $\bar{V}_{iD}$ is usually of the order of magnitude of 1 cm/sec, and decreases to zero as the mixture finally becomes uniform, whereas $\bar{v}_i$ is of the order of magnitude of 10^4 cm/sec at ordinary temperatures. We then argue that the net force on the walls must be zero if there is no pressure gradient in the gas, and hence that the total momentum transferred to the walls by all the molecular collisions is zero. We can calculate the average momentum transferred per unit time by the ith species as approximately equal to the average momentum per molecular impact (proportioned to $m_i \bar{V}_{iD}$, where m_i is the molecular mass) multiplied by the average number of impacts per unit time (proportional to $n_i \bar{v}_i$). The sum over all the species must be zero,

$$(m_1 \bar{V}_{1D})(n_1 \bar{v}_1) + (m_2 \bar{V}_{2D})(n_2 \bar{v}_2) = 0 \tag{7}$$

or by rearrangement,

$$J_{1D} m_1 \bar{v}_1 + J_{2D} m_2 \bar{v}_2 = 0. \tag{8}$$

Since $\bar{v}_i = (8kT/\pi m_i)^{1/2}$, Eq. (5) or Graham's law of diffusion follows immediately. The generalization to v species is obviously

$$\sum_{i=1}^{v} J_{iD} m_i \bar{v}_i = 0 \quad \text{or} \quad \sum_{i=1}^{v} J_{iD} m_i^{1/2} = 0. \tag{9}$$

This diffusion law is, from the derivation, only approximate, but its range of validity does not depend on any relation between the mean free path and the internal geometry of the experimental apparatus. The approximation arises from calculating the average of a product as the product of two averages; this happens to be quite accurate in this case. Thus Graham's diffusion law applies for any tubes or any porous media over the whole pressure range from the continuum region down to the free-molecule region.

The same line of argument, that momentum transfers are additive, is sufficient basis for discussion of combined transport and of multicomponent diffusion.

One final generalization can be made. The foregoing discussion is based on the implicit assumption that the geometry of the hole or tube through which diffusion occurs is well known, which is seldom the case for porous media. It is nevertheless found experimentally that diffusion in a porous medium can be described by transport equations of the same form as Eqs. (4), provided $\mathscr{D}_{12}$ is replaced by an effective diffusion coefficient D_{12}, whose value depends on the geometry of the medium. It is customary to take D_{12} proportional to $\mathscr{D}_{12}$,

$$D_{12} = (\varepsilon/q)\mathscr{D}_{12}, \tag{10}$$

where the constant ε/q is called the porosity–tortuosity factor. This name comes from a simple model in which a porous medium is visualized as a number of tortuous holes through a solid. The free space for diffusion is then only a fraction ε of the total apparent volume of the solid, and each hole is on the average longer by a factor q than a straight hole through the solid. Both D_{12} and $\mathscr{D}_{12}$ are inversely proportional to pressure.

2. Continuum Viscous Flow

By viscous flow we mean that portion of the gas flow in the continuum region that is caused by a pressure gradient. If there is no turbulence, the behavior of the gas is described by the coefficient of viscosity. Moreover, a mixture behaves the same as a single gas, since such bulk flow has no tendency to cause a mixture to separate into its components.

The computation of the viscous flux for gases is carried out exactly as in the case of liquids to obtain Poiseuille's law, but remembering that a gas is compressible. The basic idea of the calculation is simple: If the gas is not accelerating, the net force on any element of it must be zero, so that the viscous-drag force just balances the force due to the pressure difference across the element. The result has the plausible form

$$\mathbf{J}_{\text{visc}} = \text{Flow/Area} = -(nB_o/\eta)\nabla p, \tag{11}$$

where $\mathbf{J}_{\text{visc}}$ is the viscous flux in molecules/cm^2-sec, n is the total number density in molecules/cm^3, B_o is a constant characteristic of the hole geometry having units of cm^2, η is the viscosity in gm/cm-sec, and p is the pressure in dyne/cm^2. The compressibility arises because $n = p/kT$, making the coefficient of the gradient dependent upon position through its dependence on pressure. The boundary condition for continuum flow is that the gas velocity is zero at the side walls. Any "slip flow" at the walls is to be regarded as a free-molecule component of the flow.

For mixtures, the viscous flux of species i is proportional to its mole fraction, since the flow is nonseparative,

$$\mathbf{J}_{i\,\text{visc}} = x_i \mathbf{J}_{\text{visc}}. \tag{12}$$

3. Free-Molecule or Knudsen Diffusion

The original studies of effusion (Graham, 1846) were limited to small holes in very thin plates. This assured that the molecules made no collisions with each other during their passage through the hole, which we now know is the important criterion. When this condition is met, the flux of molecules through any kind of hole is equal to the number of molecules passing into the entrance of the hole multiplied by the probability that a molecule that enters the hole will eventually get all the way through and not bounce back out the entrance. This probability depends only on the geometry of the hole and the law of reflection for molecules hitting the inner walls of the hole. Such free-molecule flow was studied in considerable detail around 1907–1908 by Knudsen (see Knudsen, 1950), and is now often called Knudsen flow. If there is gas of molecular density n at one end of the hole and a vacuum at the other, the effusive or Knudsen flux J_K is

$$J_K = wn\bar{v}, \tag{13}$$

where w is the probability factor and $\bar{v}$ is the mean molecular speed. For an infinitesimally thin orifice $w = 1/4$. If there is a gas mixture instead of a single gas, each species acts independently of the other species, and the total flow is a sum of terms,

$$J_K = \sum_{i=1}^{v} w_i n_i \bar{v}_i. \tag{14}$$

For gas on both sides of the hole, the net flux is proportional to the difference in gas densities at the two ends; in differential form this is

$$\mathbf{J}_{iK} = -D_{iK} \nabla n_i, \tag{15}$$

for one species of a mixture, where D_{iK} is called the Knudsen diffusion coefficient. Comparing Eqs. (13) and (15), we see that D_{iK} is proportional to $\bar{v}_i$; it is therefore customary to write

$$D_{iK} = \tfrac{4}{3}K_o \bar{v}_i, \tag{16}$$

where K_o is called the Knudsen permeability constant and is characteristic of the hole geometry. It is evident that D_{iK} is independent of pressure and increases with temperatures as $T^{1/2}$.

For different gases at the same pressure on the two sides of the hole, Eqs. (15) and (16) give

$$-(\mathbf{J}_{1K}/\mathbf{J}_{2K}) = D_{1K}/D_{2K} = \bar{v}_1/\bar{v}_2 = (M_2/M_1)^{1/2}, \tag{17}$$

which is Graham's law of effusion. This has the same appearance as Graham's law of diffusion, but arises in quite a different way. The effusion law results because J_i is *directly* proportional to $\bar{v}_i$, whereas the diffusion law arises because J_i is *inversely* proportional to $m_i \bar{v}_i$, as shown by Eq. (8). The accuracies and ranges of validity of the two laws are also quite different. The effusion law is exact provided that the mean free path of the molecules is much larger than the hole diameter; it is thus valid essentially only at low pressures. The diffusion law is only approximate, according the derivation given, but holds at all pressures.

B. Combined Transport

We have already stated that diffusive flows (continuum and free-molecule) combine by additivity of momentum transfers; we must now inquire how viscous and diffusive flows combine. The two flows contribute independently to the total transport, which thus turns out to be simply a sum of diffusive and viscous fluxes with no extra terms due to coupling between the two mechanisms. For gases, this additivity can be shown to follow from rigorous kinetic theory to a high degree to approximation (Chapman and Cowling, 1941, 1970; Zhdanov et al., 1962; Mason et al., 1967a), but the conclusion is quite a general one. It depends only on the fact that the various flows are proportional to gradients (linear laws), and that quantities of different tensorial character do not couple in the linear approximation in isotropic systems. This result is sometimes known as Curie's theorem (de Groot and Mazur, 1962; Fitts, 1962; Yao, 1968).

These combining principles have simple electrical analogs: Diffusive flows combine like resistors in series (voltage drops add, as do momentum transfers), whereas diffusive and viscous flows combine like resistors in parallel (currents add, as do fluxes) (Mason and Evans, 1969).

Let us consider the diffusive flows first. We note that the momentum transferred by species i is proportional to the gradient of the partial pressure of that species ($p_i = n_i kT$), and rearrange Eqs. (4) for continuum diffusion and Eq. (15) for free-molecule diffusion to emphasize the momentum transfer rather than the flux:

$$-(1/kT)\nabla p_1 = (1/\mathscr{D}_{12})(\mathbf{J}_{1D} - x_1 \mathbf{J}_D), \tag{18}$$

and

$$-(1/kT)\nabla p_1 = (1/D_{1K})\mathbf{J}_{1K}, \tag{19}$$

with similar equations for species 2. Thus it is the ∇p_1 terms which add; moreover $\mathbf{J}_{1D}$ and $\mathbf{J}_{1K}$ are the same, just as the current through two series resistors is the same. Combining Eqs. (18) and (19) in this way, we obtain

$$-(1/kT)\nabla p_1 = (1/D_{1K})\mathbf{J}_{1D} + (1/\mathscr{D}_{12})(\mathbf{J}_{1D} - x_1 \mathbf{J}_D), \tag{20}$$

with a similar equation for species 2. This equation describes the diffusion of one component of a binary mixture (not of a multicomponent mixture) at uniform total pressure throughout the entire pressure range from the free-molecule to the continuum limits.

If a gradient of the total pressure exists, the viscous fluxes must be added to the diffusive fluxes,

$$\mathbf{J}_1 = \mathbf{J}_{1D} + \mathbf{J}_{1 \text{ visc}}, \qquad \mathbf{J}_2 = \mathbf{J}_{2D} + \mathbf{J}_{2 \text{ visc}}, \tag{21}$$

and the total flux is

$$\mathbf{J} = \mathbf{J}_1 + \mathbf{J}_2 = \mathbf{J}_D + \mathbf{J}_{\text{visc}}, \tag{22}$$

the viscous fluxes being given by Eqs. (11) and (12). The easiest way to do this is to replace each diffusive flux in Eq. (20) by

$$\mathbf{J}_{iD} = \mathbf{J}_i - x_i \mathbf{J}_{\text{visc}}, \tag{23}$$

which leads to

$$-(1/kT)\nabla p_1 = [(1/D_{1K}) + (1/\mathcal{D}_{12})]\mathbf{J}_1 - (x_1/\mathcal{D}_{12})\mathbf{J} - (x_1/D_{1K})\mathbf{J}_{\text{visc}}. \tag{24}$$

Substituting for $\mathbf{J}_{\text{visc}}$ from Eq. (11) and rearranging, we can obtain the following neat form:

$$\mathbf{J}_1 = -D_1 \nabla n_1 + x_1 \delta_1 \mathbf{J} - x_1 \gamma_1 (nB_o/\eta)\nabla p, \tag{25}$$

where

$$1/D_1 = (1/D_{1K}) + (1/\mathcal{D}_{12}),$$

$$\delta_1 = D_1/\mathcal{D}_{12} = D_{1K}/(D_{1K} + \mathcal{D}_{12}),$$

$$\gamma_1 = D_1/D_{1K} = \mathcal{D}_{12}/(D_{1K} + \mathcal{D}_{12}) = 1 - \delta_1.$$

A similar equation holds for $\mathbf{J}_2$, with corresponding definitions of D_2, δ_2, and γ_2.

The complete phenomenological description of diffusion and flow in a binary mixture over the whole pressure range is given by Eq. (25) plus the corresponding equation for the second component. The nice thing about this form of the equation is that it clearly exhibits both the free-molecule and continuum limits, as well as the behavior in the transition region, through the pressure dependences of D_1, δ_1, and γ_1. These dependences follow from the fact that D_{1K} is independent of pressure and $\mathcal{D}_{12}$ is inversely proportional to pressure. Thus at very low pressures $D_1 = D_{1K}$, $\delta_1 = 0$, and $\gamma_1 = 1$, so that Eq. (15) for free-molecule diffusion is recovered. At high pressures $D_1 = \mathcal{D}_{12}$, $\delta_1 = 1$, and $\gamma_1 = 0$ (but $n\gamma_1 = n\mathcal{D}_{12}/D_{1K} = $ constant), so that Eqs. (4) for continuum diffusion are recovered if the pressure gradient is zero. The interplay of the various terms in Eq. (25) at different pressures leads to quite interesting behavior, especially in the transition region, which has only been recently recognized and understood (Mason et al., 1967a; Gunn and King, 1969).

Although the two independent fluxes $\mathbf{J}_1$ and $\mathbf{J}_2$ are completely described by Eq. (25) and its analog for component 2, it is often convenient to take one of the independent fluxes to be the total flux $\mathbf{J}$. That is, instead of having two diffusion equations in which the pressure gradient appears in a subsidiary role, we would like to have one diffusion equation, plus one flow equation in which the pressure gradient appears in a primary role (the second diffusion equation then becomes redundant). This flow equation can be obtained by adding Eq. (24) and its analog for component 2, eliminating $\mathbf{J}_2$ by $\mathbf{J}_2 = \mathbf{J} - \mathbf{J}_1$, and substituting for $\mathbf{J}_{\text{visc}}$ from Eq. (11). The result is

$$\mathbf{J} - \beta_1 \mathbf{J}_1 = -(D_{2K}/kT)[1 + (B_o p/\eta \bar{D}_K)]\nabla p, \tag{26}$$

where

$$\beta_1 \equiv 1 - (D_{2K}/D_{1K}) = 1 - (m_1/m_2)^{1/2},$$

$$1/\bar{D}_K = (x_1/D_{1K}) + (x_2/D_{2K}).$$

Here $\beta_1 \mathbf{J}_1$ is the portion of the net flux that is driven by diffusion alone, and $(\mathbf{J} - \beta_1 \mathbf{J}_1)$ is thus the portion driven by the pressure gradient alone.

Equations (25) and (26) furnish an alternate complete description of combined diffusion and flow in a binary mixture. It is sometimes easier to integrate Eq. (26) approximately than either of the diffusion equations. It should be emphasized that the integration of any of these equations is almost impossible to effect analytically because of the pressure and composition dependence of the coefficients. Although we have previously stated that the diffusive and viscous fluxes are independent and hence are additive, this does not mean that the diffusion and viscous-flow equations can be solved independently, even though no direct coupling terms occur. For example, in a typical problem we must know the viscosity in order to solve the viscous-flow equation. But the viscosity of a gas mixture depends on the mixture composition, which must be obtained from the solution of the diffusion equation. However, the solution of the diffusion equation depends upon knowledge of the gas flow occurring, and this in turn depends upon the viscous-flow equation. Thus we come full cycle—the two equations are coupled in a very real way.

C. MULTICOMPONENT DIFFUSION

To obtain the extension to multicomponent mixtures, we first combine momentum transfers for continuum diffusion, add on the free-molecule momentum transfer, then include the viscous flux by way of Eq. (23), and finally substitute for $\mathbf{J}_{\text{visc}}$ from Eq. (11). For the first step we go back to Eq. (18) and its analog for component 2, and eliminate the total diffusive flux $\mathbf{J}_D$ between the two equations, obtaining

$$-(1/kT)\nabla p_1 = (1/\mathscr{D}_{12})(x_2 \mathbf{J}_{1D} - x_1 \mathbf{J}_{2D}) = (n_1 n_2/n\mathscr{D}_{12})(\bar{\mathbf{V}}_{1D} - \bar{\mathbf{V}}_{2D}), \quad (27)$$

where we have used the relation $\nabla p_2 = -\nabla p_1$. A similar equation exists for ∇p_2, but it is the same as Eq. (27) with reversed signs. The extension to multicomponent mixtures is now plausible: For each new species there is another momentum-transfer term on the right-hand side of the equation,

$$-(1/kT)\nabla p_1 = (n_1 n_2/n\mathscr{D}_{12})(\bar{\mathbf{V}}_{1D} - \bar{\mathbf{V}}_{2D}) + (n_1 n_3/n\mathscr{D}_{13})(\bar{\mathbf{V}}_{1D} - \bar{\mathbf{V}}_{3D}) + \cdots,$$
$$(28a)$$

$$-(1/kT)\nabla p_2 = (n_2 n_1/n\mathscr{D}_{21})(\bar{\mathbf{V}}_{2D} - \bar{\mathbf{V}}_{1D}) + (n_2 n_3/n\mathscr{D}_{23})(\bar{\mathbf{V}}_{2D} - \bar{\mathbf{V}}_{3D}) + \cdots,$$
$$(28b)$$

$$\vdots$$

etc.

For a mixture of v species, there is a total of v equations, of which only $v - 1$ are independent (any one equation is equal to the sum of the other $v - 1$ equations). This set of equations is valid for continuum diffusion at constant total pressure, and is usually called the Stefan–Maxwell equations.

Adding in the free-molecule and viscous contributions, we obtain the general result,

$$-\nabla n_1 = \frac{1}{D_{1K}}\left[\mathbf{J}_1 + x_1\left(\frac{nB_o}{\eta}\right)\nabla p\right] + \frac{1}{\mathscr{D}_{12}}(x_2\mathbf{J}_1 - x_1\mathbf{J}_2)$$

$$+ \frac{1}{\mathscr{D}_{13}}(x_3\mathbf{J}_1 - x_1\mathbf{J}_3) + \cdots, \tag{29a}$$

$$-\nabla n_2 = \frac{1}{D_{2K}}\left[\mathbf{J}_2 + x_2\left(\frac{nB_o}{\eta}\right)\nabla p\right] + \frac{1}{\mathscr{D}_{21}}(x_1\mathbf{J}_2 - x_2\mathbf{J}_1)$$

$$+ \frac{1}{\mathscr{D}_{23}}(x_3\mathbf{J}_2 - x_2\mathbf{J}_3) + \cdots, \tag{29b}$$

$$\vdots$$

etc.

For binary mixtures, these equations are equivalent to Eqs. (24) or (25).

The arguments leading to Eqs. (29) are based largely on plausibility rather than on detailed theory, and the results at this stage must be regarded as phenomenological. In particular, there is no assurance that the $\mathscr{D}_{ij}$ in Eqs. (29) are the same as the ones in the corresponding binary mixture equations. For dilute gases, in which at most only binary encounters are important, it seems reasonable that this should be so. The same form of equations might be expected to apply to dense gases and liquids, but the $\mathscr{D}_{ij}$ would then presumably depend on all the components in the mixture, not just on i and j. Put another way, Eqs. (29) as written tacitly assume that the $\mathscr{D}_{ij}$ are independent of mixture composition. Detailed theories show that this is nearly, but not quite, true (Chapman and Cowling, 1970; Hirschfelder et al., 1954; Sandler and Mason, 1968). This is an important simplification in the treatment of multicomponent gaseous diffusion.

Although it is easy enough to write down the phenomenological equations for multicomponent diffusion, it is very difficult to solve them. Multicomponent diffusion is much more complicated than binary diffusion; even ternary diffusion shows qualitatively different features than binary diffusion. The reason is that the diffusion of one pair of components can "drag along" a third component, so to speak. That is, the fluxes of the different components are coupled together, and do not depend on just their own gradients. The diffusion of a third "solvent" gas even in the absence of a composition gradient was called osmotic diffusion by Hellund (1940), who first studied it theoretically. Other related phenomena have been described by Toor (1957):

diffusion barrier, in which the flux of a component is zero even though its gradient is not zero; and *reverse diffusion*, in which a component diffuses against its gradient. All three phenomena have been observed experimentally by Duncan and Toor (1962), who showed that the results were well described by Eqs. (28).

A peculiar consequence of the coupling of different component fluxes in multicomponent diffusion is that density inversions can develop that cause the system to become gravitationally unstable. Convection can then set in, and the whole mixing process becomes extremely complex. The theoretical possibility of gravitational instability in ternary diffusion was first pointed out by Wendt (1962) in connection with liquids. Instabilities in ternary gaseous diffusion were independently and inadvertently discovered experimentally (Miller and Mason, 1966; Miller *et al.*, 1967), and were then looked for and found in liquids (Miller, 1966). Although a quantitative mathematical theory of these instabilities is difficult to construct, it is fairly easy to give a simple physical explanation of their existence. We have already pointed out in Section II, A, 1 that in binary diffusion with zero net flux there must be a lower pressure on the side of the lighter gas. If a third gas is added as an initially uniform heavy solvent, the pressure difference must still develop, and this forces the solvent towards one side, even though there is no gradient of solvent concentration. The overall density gradient can then be inverted if a sufficient quantity of heavy solvent gas is present. This is the explanation for instabilities in osmotic diffusion; similar explanations can be given for other ternary cases, such as a mixture of two gases diffusing into a third gas (Miller and Mason, 1966).

D. Special Cases

Several well-known special cases can be obtained from the foregoing general equations. We have already mentioned that Graham's law of effusion follows as the low-pressure limit of Eq. (25); here we shall briefly mention the phenomena of uniform-pressure diffusion (Graham's law of diffusion), transition diffusion, diffusion pressure effect (diffusive slip), and viscous slip.

1. Uniform-Pressure Diffusion

This is just Graham's (1833) diffusion experiment. To recover Graham's diffusion law from Eq. (25), we set $\nabla p = 0$, $\nabla n_1 = -\nabla n_2$, and combine the equation with the corresponding equation for component 2 to yield

$$(\mathbf{J}_1/D_1) + (\mathbf{J}_2/D_2) = \mathbf{J}[x_1(\delta_1/D_1) + x_2(\delta_2/D_2)]. \tag{30}$$

This reduces to

$$-(\mathbf{J}_1/\mathbf{J}_2) = D_{1K}/D_{2K} = (m_2/m_1)^{1/2}, \tag{31}$$

which also looks like the effusion law, but which has been derived without any special conditions on the pressure. It therefore applies at all pressures, and not just in the free-molecule region. As is discussed in Section II, A, 1, its validity rests on the momentum-transfer argument summarized by Eq. (7).

The diffusion law follows even more readily from Eq. (26) on setting $\mathbf{V}p = 0$. Then

$$\mathbf{J} - \beta_1 \mathbf{J}_1 = 0, \tag{32}$$

from which Eq. (31) is obtained on setting $\mathbf{J} = \mathbf{J}_1 + \mathbf{J}_2$.

It is interesting to examine how the flux varies in uniform-pressure diffusion as the total pressure is varied. In the continuum region we expect the flux to be independent of pressure, and in the free-molecule region we expect it to be directly proportional to pressure. If we set $\mathbf{V}p = 0$ in Eq. (25) and substitute for $\mathbf{J}$ from Eq. (32), we obtain

$$\mathbf{J}_1(1 - x_1\,\delta_1\beta_1) = -D_1\,\mathbf{V}n_1. \tag{33}$$

The pressure dependence is most easily seen if we make a differential approximation, $\mathbf{V}n_1 \approx n\,\Delta x_1/L$, where L is the length of the tube or porous medium through which diffusion occurs, and Δx_1 is the mole fraction difference across L. If we define flux and pressure scale factors that are constants for a given system,

$$\Phi \equiv nD_{12}/L, \tag{34}$$

$$\pi_1 \equiv pD_{12}/D_{1K}, \tag{35}$$

then Eq. (33) can be written in the form

$$J_1/\Phi = -[(p/\pi_1)\Delta x_1]/[1 + (1 - x_1\beta_1)(p/\pi_1)]. \tag{36}$$

The flux thus increases from zero linearly with pressure, and reaches a constant value at high pressure; the pressure dependence is the same as that of a Langmuir adsorption isotherm.

If the mole fraction difference Δx_1 is not small, Eq. (33) can be easily integrated for the case of one-dimensional steady-state diffusion (Evans et al., 1961a,b),

$$J_1 = (nD_{12}/\beta_1 L) \ln\{[1 - \delta_1\beta_1 x_1(L)]/[1 - \delta_1\beta_1 x_1(0)]\}, \tag{37}$$

where $x_1(0)$ and $x_1(L)$ are the mole fractions at the two ends of L. This can be written as a universal equation for J_1 in terms of the scale factors Φ and π_1,

$$J_1/\Phi = (1/\beta_1) \ln\{[1 + [1 - \beta_1 x_1(L)](p/\pi_1)]/[1 + [1 - \beta_1 x_1(0)](p/\pi_1)]\}. \tag{38}$$

At high pressures J_1 becomes constant,

$$J_1 \to (\Phi/\beta_1)\ln[[1 - \beta_1 x_1(L)]/[1 - \beta_1 x_1(0)]], \tag{39}$$

and at low pressures proportional to p,

$$J_1 \to -\Phi(p/\pi_1)[x_1(L) - x_1(0)] = -p(D_{1K}/kT)[x_1(L) - x_1(0)], \tag{40}$$

as expected. Notice that the viscosity appears nowhere in any of these equations, indicating that there is no viscous flow even though the net flux J is not zero.

A plot of Eq. (38) for He–Ar diffusion through a low-permeability porous graphite specimen is shown in Fig. 2, together with some experimental

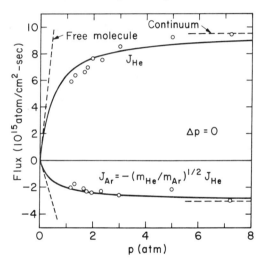

FIG. 2. Helium and argon diffusion fluxes as a function of pressure in a uniform-pressure (Graham) experiment in a low-permeability porous graphite at 25°C. The solid curves are from Eq. (38).

measurements (Evans *et al.*, 1962). The fluxes of both components were measured independently; the results conform both with Graham's law of diffusion and with the predicted pressure dependence. The only adjustable constant in the calculation is D_{12} (taken as $nD_{12} = 2.61 \times 10^{15}$ atoms/cm-sec), the value of D_{1K} having been determined ($D_{HeK} = 3.93 \times 10^{-4}$ cm²/sec) from separate permeability data (Mason *et al.*, 1967a). Notice that diffusion is still in the transition region between free-molecule and continuum behavior even at pressures of several atmospheres. This is characterized by the pressure factor π_1, equal to 0.270 atm for this system; for comparison, π_1 would be about 3×10^{-4} atm for a capillary tube of 1 mm diameter, representing a scale change of about 10^3.

2. Transition Diffusion

Here we wish to examine how the flux varies with pressure in a Loschmidt-type of experiment, where $\mathbf{J} = 0$. Again we expect the flux to be constant in the continuum region and directly proportional to pressure in the free-molecule region. Setting $\mathbf{J} = 0$ in Eq. (25) and its corresponding equation for component 2, and algebraically eliminating ∇p between the two equations, we obtain

$$\mathbf{J}_1[(1/x_1 D_1) + (1/x_2 D_2)] = -n[(1/x_1) + (1/x_2)]\nabla x_1, \tag{41}$$

where we have used the facts that $\mathbf{J}_2 = -\mathbf{J}_1$ and $\nabla x_2 = -\nabla x_1$. The pressure dependence is most easily displayed by making the differential approximation, which yields

$$J_1/\Phi = -[(p/\pi_1)\Delta x_1]/[1 - x_1\beta_1(1 - \beta_1)^{-1} + (p/\pi_1)]. \tag{42}$$

This equation shows the same general behaviour with pressure as does Eq. (36) for the uniform-pressure case.

Equation (41) is easily integrated for the one-dimensional steady-state case, and the result can be written in terms of the scale factors Φ and π_1 of Eqs. (34) and (35) as

$$\frac{J_1}{\Phi} = -\left(\frac{1 - \beta_1}{\beta_1}\right)\left(\frac{p}{\pi_1}\right) \ln \left\{\frac{\beta_1 x_1(L) + (1 - \beta_1)[1 + (p/\pi_1)]}{\beta_1 x_1(0) + (1 - \beta_1)[1 + (p/\pi_1)]}\right\}. \tag{43}$$

At high pressures J_1 becomes constant,

$$J_1 \to -\Phi[x_1(L) - x_1(0)], \tag{44}$$

and at low pressures proportional to p,

$$J_1 \to -\frac{pD_{1K} D_{2K}}{kT(D_{1K} - D_{2K})} \ln \left[\frac{\beta_1 x_1(L) + (1 - \beta_1)}{\beta_1 x_1(0) + (1 - \beta_1)}\right], \tag{45}$$

as expected.

It is curious that the viscosity appears nowhere in any of these equations, even though we expect viscous back-flow to be occurring. This is a peculiarity of the $\mathbf{J} = 0$ condition; regardless of what value the viscosity has, the pressure difference adjusts itself to keep $\mathbf{J} = 0$. Thus the viscosity affects the steady-state pressure difference, as will be seen in Section II, D, 3, but not the flux. The viscosity appears explicitly in the flux equations, however, unless either $\mathbf{J} = 0$ or $\nabla p = 0$.

A plot of Eq. (43) is shown in Fig. 3 for He–Ar diffusion through the same graphite specimen as shown in Fig. 2, together with the experimental measurements of Evans et al. (1963). No adjustable constants are involved here, the value of D_{12} having been obtained from the independent measurements at uniform pressure shown in Fig. 2.

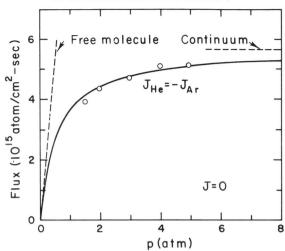

FIG. 3. Flux as a function of pressure in an equal counterdiffusion (Loschmidt) experiment for the same system shown in Fig. 2. The solid curve is from Eq. (43) and involves no adjustable constants.

3. Diffusion Pressure Effect (Diffusive Slip)

The pressure difference that results from diffusion in a closed system was first investigated by Kramers and Kistemaker (1943). They also recognized that this pressure difference was related to the fact that the mass-average velocity of the gas was not zero at the wall, or that there was a nonzero momentum flow, or slip, in the gas adjacent to the wall. The effect was later rediscovered and discussed as an analog of the Kirkendall effect in solids (McCarty and Mason, 1960; Miller and Carman, 1960; Mason, 1961), and has recently become the subject of considerable experimental activity (Waldmann and Schmitt, 1961; Wicke and Hugo, 1961; Suetin and Volobuev, 1964; Kotousov, 1965; Volobuev and Suetin, 1965, 1967; Kosov and Kurlapov, 1966).

The piston in Fig. 1 is held fixed so that $\mathbf{J} = 0$, and we ask what the pressure difference will be. To find out, we set $\mathbf{J} = 0$ and $n_1 = x_1 n$ in Eqs. (25) and (26), and eliminate $\mathbf{J}_1$ between the two equations to obtain a differential equation relating the pressure difference dp to the composition difference dx_1. This can be put in a form which explicitly exhibits the pressure dependence,

$$dp = [p(D_{2K} - D_{1K})/(A_0 + A_1 p + A_2 p^2)]\, dx_1, \qquad (46)$$

where the constants are

$$A_0 = x_1 D_{1K} + x_2 D_{2K}, \qquad (47)$$

$$A_1 = (D_{1K} D_{2K}/p D_{12}) + B_o/\eta, \qquad (48)$$

$$A_2 = [(x_1 D_{2K} + x_2 D_{1K})/p D_{12}](B_o/\eta). \qquad (49)$$

Thus the steady-state pressure difference increases from zero linearly with p at low pressures, reaches a maximum value of

$$(dp)_{max} = (D_{2K} - D_{1K})/[A_1 + 2(A_0 A_2)^{1/2}] \, dx_1, \tag{50}$$

at a pressure of

$$p_{max} = (A_0/A_2)^{1/2}, \tag{51}$$

and finally falls back to zero at high pressures as $1/p$. Equation (46) involves the viscosity, unlike Eqs. (36) and (42).

A comparison of Eq. (46) with the He–Ar measurements of Evans et al. (1963) is shown in Fig. 4, using the differential approximation of $\Delta p \approx dp$ and $\Delta x_1 \approx dx_1$ and taking mean values for x_1, η, and p. The agreement is rather good, and only the lowest pressure measurement has much experimental uncertainty. No adjustable constants are involved here, the values of the necessary parameters having been determined separately (Mason et al., 1967a).

There are, however, unexplained anomalies involving the diffusion pressure effect. For instance, Waldmann and Schmitt (1961) find a sign reversal in Δp vs. p for Ar–CO_2 diffusion.

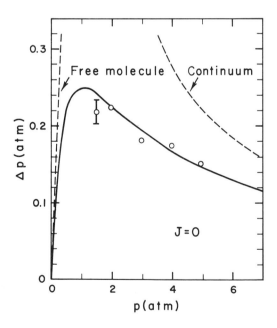

FIG. 4. The pressure difference at $\mathbf{J} = 0$ (diffusive slip) as a function of mean pressure for the same system shown in Figs. 2 and 3. The solid curve is from Eq. (46) and involves no adjustable constants.

4. Viscous Slip

This topic is not obviously related to diffusion, but can be treated as a special kind of diffusion problem through the dusty-gas model discussed in Section IV, B. In the present phenomenological context it can be considered a combination of viscous and free-molecule flow.

The slipping of a single gas over a solid surface was discovered experimentally by Kundt and Warburg (1875). It is found that a gas moves under the influence of a pressure gradient according to the equation

$$\mathbf{J} = -(K/kT)\,\nabla p, \tag{52}$$

where K is called the permeability coefficient. A plot of K against mean pressure is a straight line whose slope is inversely proportional to the gas viscosity; Kundt and Warburg found that the line did not go through the origin.

This result can be obtained from Eq. (25) by setting $x_1 = 1$ and $\mathbf{J}_1 = \mathbf{J}$, whereby Eq. (52) follows with K given as

$$K = (pB_o/\eta) + D_K. \tag{53}$$

This is the basis for the determination of gas viscosity by capillary flow, or the determination of the parameters B_o and K_o for a porous medium by permeability measurements with a known gas.

An interesting anomaly occurs in K at very low pressures. There D_K (or K_o) is slightly pressure-dependent and increases slightly as p approaches zero. Thus the intercept obtained by extrapolation from higher pressures is less than the true zero-pressure limit of $K_o = d/4$, and in fact the curve may have a small minimum at very low pressures (see Knudsen, 1950). The extrapolated intercept is lower than the true intercept by a factor of approximately $3\pi/16$ (Maxwell, 1879; see Present, 1958).

III. Molecular Theory of Continuum Diffusion

A. Elementary Theories

These theories have already been discussed to some extent in Section I, and we add only a few remarks here. The momentum-transfer theory is quite good; the main objection seems to be the lack of a convincing connection between the transfer of momentum and the transfer of molecules. Such a connection only seems to emerge from the elaborate Maxwell–Chapman theory, where it becomes quite clear (see p. 191 of Kennard, 1938). A lesser objection is that the theory gives no account of viscosity or thermal conductivity, which gives it an *ad hoc* air.

Mean-free-path theory gives a reasonable account of viscosity and thermal conductivity (the numerical coefficients are poor, but the other parts are not bad), but fails badly in its predictions of diffusion phenomena. It gives an erroneous account of the composition dependence of the diffusion coefficient, fails to predict even the existence of thermal diffusion, and has been severely criticized on that account (Chapman, 1928; Furry, 1948).

The mean-free-path theory for many years coexisted with the rigorous Chapman–Enskog theory, yet seemingly inconsistent with it, even though there was an intuitive feeling that a connection had to exist between the two theories. The connection was found by Monchick (1962), who solved the integral equation for the second Chapman–Enskog approximation to the velocity distribution function by a Liouville–Neumann iteration scheme. He showed that the Nth iteration corresponded to a free-path or free-flight theory in which the trajectory of a particle had been followed through $N + 1$ collisions. The extension to mixtures (Monchick and Mason, 1967) showed that the higher iterations did indeed lead to a correct prediction of the composition dependence of $\mathscr{D}_{12}$ and the existence of thermal diffusion, although the rate of convergence was painfully slow.

Thus the connections among all the theories are now known. The momentum-transfer theory appears as the first approximation in the Chapman solution of Maxwell's equation of transfer, and the mean-free-path theory appears as the first iteration in a solution of the linearized Boltzmann equation. Maxwell's equations of transfer can be obtained from the Boltzmann equation by forming moments, thereby completing the interconnection.

All the theories, even in their simplest forms, rigorously and correctly predict that $\mathscr{D}_{12}$ is inversely proportional to density or pressure. This is a consequence of the binary collision assumption: The flux of a species is proportional to the number density of the species, but inversely proportional to the number density of the molecules with which the diffusing species collides. The two effects exactly cancel, and the total flux of diffusing species is therefore independent of total pressure. In a similar way, the fluxes of momentum and energy carried by the molecules are also pressure-independent, so that the coefficients of viscosity and thermal conductivity of a gas are independent of gas pressure as long as the binary collision mechanism dominates. The reason the diffusion coefficient is inversely proportional to pressure rather than independent of it is that, for essentially historical reasons, the flux is taken proportional to the gradient of the species number density, $\mathbf{J}_1 = -\mathscr{D}_{12}\,\nabla n_1$. Since ∇n_1 obviously is proportional to pressure, $\mathscr{D}_{12}$ must be inversely proportional if $\mathbf{J}_1$ is to be independent. Had the flux been taken proportional to the gradient of mole fraction ∇x_1, as is perhaps more sensible, then the coefficient of proportionality would have been pressure-independent, as are the coefficients of viscosity and thermal conductivity.

B. Chapman–Enskog Theory

All transport phenomena arise by deviations, however slight, from the Maxwell equilibrium velocity distribution function. The basic problem of rigorous kinetic theory is to solve the Boltzmann equation for the nonequilibrium distribution function. The fluxes are then simple integrals over the distribution function; the diffusive flux, for example, is the integral of molecular velocity over the distribution function. Comparison of the flux expressions with the phenomenological equations given in Section II identifies the diffusion coefficients and gives explicit expressions for them in terms of molecular collisions.

It should be pointed out that the form of the phenomenological transport equations does *not* follow automatically from a mathematical solution of the Boltzmann equation. Such a situation would be theoretically very satisfying, but the complex mathematical structure of the equation has so far prevented any such derivation. Instead, one takes for granted that the phenomenological equations are correct, and seeks "normal" solutions of the Boltzmann equation that lead to these equations, ignoring the many other possible types of solutions. The solutions are, in other words, forced into the desired form.

An outline of the Chapman–Enskog procedure is as follows. First, the distribution function is written as the equilibrium distribution plus a perturbation term. The perturbation term is assumed small, whereby the nonlinear Boltzmann equation can be transformed to a linearized equation for the perturbation function. To solve this equation, it is *assumed* that the perturbation is proportional to the various gradients of composition, pressure, etc., the coefficients of proportionality being unknown complicated functions of molecular velocities. This assumption is clearly sufficient to generate the phenomenological equations: on forming fluxes by integration over the distribution function, we automatically find them to be directly proportional to the gradients, and the transport coefficients to be integrals over the unknown functions which were the coefficients of proportionality between the distribution-function perturbation and the gradients.

The next step is to find these unknown functions, each one of which satisfies a linear integrodifferential equation obtained from the linearized Boltzman equation. If we tried to solve these equations by an iterative procedure, we would obtain a free-path or free-flight type of theory (Monchick, 1962; Monchick and Mason, 1967); the convergence is unfortunately quite poor. The Chapman–Enskog method proceeds by expansion of the functions as power series in the components of the molecular velocities. For the expansions it is convenient, but not necessary, to use orthogonal functions to simplify the subsequent calculations. The orthogonal functions usually used are Sonine or Laguerre polynomials. When these expansions are substituted back into the integral expressions for the transport coefficients, it turns out (because of the orthogonality) that each transport coefficient is exactly equal to just one (or two) of the expansion coefficients.

A mathematical summary of the foregoing procedure applied to binary diffusion, ignoring all external forces and gradients other than of composition, is as follows. The distribution function for species i is written as

$$f_i = f_i^{(0)}(1 + \Phi_i), \tag{54}$$

where $f_i^{(0)}$ is the Maxwell distribution (normalized to unity), and Φ_i is the perturbation. The diffusive flux is

$$\mathbf{J}_{iD} = n_i \overline{\mathbf{V}}_i = n_i \int \mathbf{V}_i f_i \, d\mathbf{V}_i = n_i \int \mathbf{V}_i \Phi_i f_i^{(0)} \, d\mathbf{V}_i, \tag{55}$$

where $\mathbf{V}_i$ is the molecular velocity in the coordinate system for which $\mathbf{J}_D = 0$. It is *assumed* that Φ_i has the form

$$\Phi_i = n \sum_{\substack{j=1 \\ j \neq i}}^{v} C_i^{j} \, \mathbf{V}_i \cdot \nabla x_j, \tag{56}$$

where C_i^{j} is an unknown (scalar) function of molecular velocities. For a binary system there is only one term in the sum, and $\nabla x_j = -\nabla x_i$. Substituting Eq. (56) into Eq. (55) and comparing the resulting expression with the phenomenological definition of the diffusion coefficients given in Eq. (3), we find

$$\mathscr{D}_{ij} = \tfrac{1}{3} n_i \int C_i^{j} V_i^2 f_i^{(0)} \, d\mathbf{V}_i. \tag{57}$$

The functions C_i^{j} are then expanded in terms of Sonine polynomials,

$$C_i^{j} - C_i^{k} = \sum_{m=0}^{M} c_{im}^{jk} S_{3/2}^{(m)}(x), \tag{58}$$

where the c_{im}^{jk} are expansion coefficients, and the argument of the Sonine polynomials is $x = m_i V_i^2 / 2kT$. This is the general expansion for the multicomponent case; only one term occurs on the left for the binary case. It is assumed that the expansion becomes exact as M approaches infinity. Because of the orthogonality properties of the $S_{3/2}^{(m)}$, Eq. (57) becomes

$$\mathscr{D}_{ij} = (n_i kT/m_i) c_{i0}^{ji}. \tag{59}$$

The problem now is to find the expansion coefficient c_{i0}^{ji}. To do this, the expansion of Eq. (58) is substituted back into the linear integrodifferential equations for the C_i^{j}, which are then solved by a moment method. The result is an infinite set of algebraic equations for all the c_{im}^{ji} as unknowns, and the coefficients of these unknowns are complicated multiple integrals over molecular velocities. These integrals result from the moment formation; most of the integrations can be carried out explicitly, but not all, until the intermolecular potential is specified.

The diffusion coefficient is thus equal to a single unknown in an infinite set of algebraic equations, which cannot be solved exactly except in very special cases. An approximation procedure is used in which the set is systematically truncated in some plausible way; the simplest truncation gives the first approximation to $\mathscr{D}_{ij}$, the next step gives the second approximation, and so on. Two truncation schemes are commonly used, one due to Chapman and Cowling, and the other to Kihara (see Mason, 1957a,b, for an exposition of these schemes). In the first approximation $\mathscr{D}_{ij}$ is independent of composition; the second and higher approximations introduce composition dependence. The convergence is very rapid.

The Chapman–Enskog method depends on the following conditions:

(i) *Binary collisions*. This is inherent in the Boltzman equation itself.

(ii) *Small perturbation*. Departures from the equilibrium distribution are small, so that the fluxes are linear in the gradients.

(iii) *Classical mechanics*. Historically, classical mechanics was necessarily used by Boltzmann, Chapman, and Enskog. The generalization to quantum mechanics is easy, however, and is discussed in Section III, D.

(iv) *Elastic collisions*. The extension to inelastic collisions is discussed in Section III, E.

The first Chapman–Enskog approximation for the binary diffusion coefficient is

$$[\mathscr{D}_{12}]_1 = \tfrac{3}{16}(2\pi kT/\mu_{12})^{1/2}(1/n\overline{\Omega}_{12}^{(1,1)}), \tag{60}$$

where k is Boltzmann's constant, T is the absolute temperature, and $\mu_{12} = m_1 m_2/(m_1 + m_2)$ is the reduced mass of a pair of molecules. The diffusion collision integral $\overline{\Omega}_{12}^{(1,1)}$ has units of area and is dependent on the temperature and the intermolecular potential,

$$\overline{\Omega}^{(1,1)}(T) = \tfrac{1}{2}(kT)^{-3}\int_0^\infty e^{-E/kT}E^2 S^{(1)}(E)\, dE, \tag{61}$$

where E is the initial relative kinetic energy of a pair of colliding molecules, and $S^{(1)}(E)$ is the diffusion (momentum-transfer) cross section,

$$S^{(1)}(E) = 2\pi \int_0^\infty (1 - \cos \chi)b\, db, \tag{62}$$

where χ is the deflection angle for a collision with impact parameter b.

The expression for $[\mathscr{D}_{12}]_1$ in practical units is

$$[\mathscr{D}_{12}]_1 = 0.008258[(M_1 + M_2)/2M_1M_2]^{1/2}(T^{3/2}/p\overline{\Omega}_{12}^{(1,1)}), \tag{63}$$

where M_1 and M_2 are the molecular weights in grams per mole, T is in degrees Kelvin, p is in atmospheres, and $\overline{\Omega}_{12}^{(1,1)}$ is in square Ångstroms.

The second approximation for $\mathscr{D}_{12}$ can be written as

$$[\mathscr{D}_{12}]_2 = [\mathscr{D}_{12}]_1/(1 - \Delta_{12}) \approx [\mathscr{D}_{12}]_1(1 + \Delta_{12}), \tag{64}$$

where

$$\Delta_{12} = \tfrac{1}{10}(6C_{12}^* - 5)^2$$
$$\times \;[(x_1{}^2 P_1 + x_2{}^2 P_2 + x_1 x_2 P_{12})/(x_1{}^2 Q_1 + x_2{}^2 Q_2 + x_1 x_2 Q_{12})]. \tag{65}$$

The P's and Q's are complicated algebraic expressions containing various types of collision integrals, and C_{12}^* is a dimensionless ratio of collision integrals. The correction term Δ_{12} is temperature dependent and contains the small composition dependence of $\mathscr{D}_{12}$; it is discussed in more detail in Section III, I.

The third approximation for $\mathscr{D}_{12}$ is considerably more complicated. Because of the rapid convergence, the third approximation is seldom needed for atoms and molecules, but the necessary integrals have been evaluated (Mason, 1957a).

Other cross sections and collision integrals occur in the higher approximations and in the expressions for other transport coefficients. The generalized definition is

$$\bar{\Omega}^{(l,\,s)}(T) = [(s + 1)!(kT)^{s+2}]^{-1} \int_0^\infty e^{-E/kT} E^{s+1} S^{(l)}(E)\, dE, \tag{66}$$

$$S^{(l)}(E) - 2\pi \left[1 - \frac{1 + (-1)^l}{2(1 + l)} \right]^{-1} \int_0^\infty (1 - \cos^l \chi) b\, db. \tag{67}$$

Given the intermolecular potential, collision integrals can be calculated, but tedious numerical integration is required for realistic forms of the potential.

It is convenient in calculations to use dimensionless collision integrals, defined as

$$\Omega^{(l,\,s)*} \equiv \bar{\Omega}^{(l,\,s)}/\pi\sigma^2, \tag{68}$$

where σ is an arbitrary molecular size or range-of-potential parameter. If σ is chosen in a reasonable way, the numerical values of the reduced collision integrals are usually about unity, the deviations from unity reflecting the "softness" of the potential in comparison to an ideal rigid sphere of diameter σ. The following ratios of collision integrals occur repeatedly in higher approximations:

$$A^* \equiv \Omega^{(2,\,2)*}/\Omega^{(1,\,1)*}, \tag{69}$$

$$B^* \equiv [5\Omega^{(1,\,2)*} - 4\Omega^{(1,\,3)*}]/\Omega^{(1,\,1)*}, \tag{70}$$

$$C^* \equiv \Omega^{(1,\,2)*}/\Omega^{(1,\,1)*}, \tag{71}$$

$$E^* \equiv \Omega^{(2,\,3)*}/\Omega^{(2,\,2)*}. \tag{72}$$

C. Convergence of Approximations

How close $[\mathscr{D}_{12}]_1$ is to $\lim_{M \to \infty}[\mathscr{D}_{12}]_M$ depends on composition, molecular masses, and the intermolecular potential. The accuracy may be assessed by numerical comparison of $[\mathscr{D}_{12}]_1$, $[\mathscr{D}_{12}]_2$, $[\mathscr{D}_{12}]_3$, etc., for a series of special cases (Mason, 1957b). For the case of nearly equal molecular masses, $[\mathscr{D}_{12}]_1$ is probably accurate within 2% regardless of the composition or the inter-molecular potential. If the molecular masses are very unequal and the heavy component is in trace concentration, then $[\mathscr{D}_{12}]_1$ is accurate to within 1%. If the light component is the trace, then $[\mathscr{D}_{12}]_1$ may be inaccurate. The worst case known is a mixture of rigid spheres, for which $[\mathscr{D}_{12}]_1$ is low by about 13% and $[\mathscr{D}_{12}]_2$ low by about 5%. In practical cases it is probably safe to regard $[\mathscr{D}_{12}]_1$ as accurate within about 5% for all gas pairs, and $[\mathscr{D}_{12}]_2$ as accurate within 2%; most cases are usually even better.

D. Quantum Effects

The only modification necessary for quantum effects is the replacement of the integration over impact parameters by one over the differential cross section. The general transport cross sections must thus be written as

$$S^{(l)}(E) = \{1 - [1 + (-1)^l/2(1 + l)]^{-1}\} \int_0^{2\pi} d\phi \int_0^{\pi}$$
$$\times (1 - \cos^l \chi)I(\chi, \phi, E)\sin \chi \, d\chi, \qquad (73)$$

where $I(\chi, \phi, E)$ is the differential cross section for scattering through polar angle χ and azimuth angle ϕ.

Quantum effects become significant when the de Broglie wavelength, $\lambda = h/\mu v$, approaches the size parameter σ. Thus λ/σ is a measure of quantum effects, and gases behave classically for $\lambda/\sigma \ll 1$. In kinetic theory it is common practice to use the de Boer parameter Λ^*,

$$\Lambda^* \equiv h/[\sigma(2\mu\varepsilon)^{1/2}], \qquad (74)$$

which is simply λ/σ for a colliding pair of molecules with reduced mass μ and kinetic energy equal to the depth ε of the intermolecular potential well. The larger the value of Λ^*, the more important are the quantum effects at a given value of the reduced temperature $T^* \equiv kT/\varepsilon$. This is illustrated in Table I, based on calculations for the Lennard-Jones (12-6) potential (Munn et al., 1965a). Typical values of the de Boer parameter are as follows: 0.35 for Ne–Ar, 1.3 for He–Ne, 1.5 for H_2–D_2, and 2.9 for ^{3}He–^{4}He. Quantum deviations can be quite large for light gases at low temperatures.

TABLE I

QUANTUM EFFECTS ON DIFFUSION COEFFICIENTS IN TERMS OF THE
DE BOER PARAMETER Λ^* AND THE REDUCED TEMPERATURE $T^* = kT/\varepsilon$ [a]

| T^* \ Λ^* | $[\mathscr{D}_{12}]_1$(quantal)/$[\mathscr{D}_{12}]_1$(classical) | | | | |
	0.5	1.0	1.5	2.0	3.0
1.0	1.009	1.032	1.105	1.224	1.444
1.5	1.008	1.031	1.080	1.150	1.269
2.0	1.006	1.025	1.060	1.105	1.182
3.0	1.004	1.016	1.035	1.060	1.101
5.0	1.001	1.007	1.015	1.028	1.047
10.0	1.000	1.003	1.004	1.009	1.016

[a] Calculated for a Lennard–Jones (12–6) potential by Munn *et al.* (1965a).

Quantum deviations for diffusion are further illustrated in Figs. 5 and 6, where the diffusion coefficients for H_2–D_2 and ^{3}He–^{4}He are shown as a function of temperature and compared with theoretical calculations based on the Lennard–Jones (12–6) potential (Monchick *et al.*, 1965; Diller and Mason, 1966). There are no adjustable constants involved, the potential parameters having been determined from second virial coefficient data. It can be seen

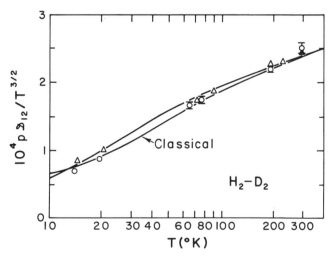

FIG. 5. Quantum effects in H_2–D_2 diffusion, from Diller and Mason (1966). Here $\Lambda^* = 1.5$ and $\varepsilon/k = 36°$K.

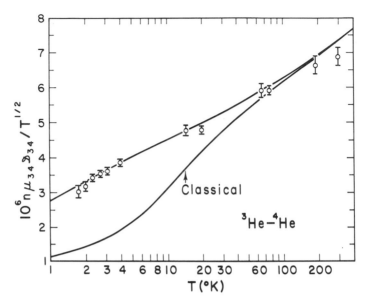

FIG. 6. Quantum effects in ^{3}He–^{4}He diffusion, from Monchick *et al.* (1965). Here $\Lambda^* = 2.9$ and $\varepsilon/k = 10°$K.

that the quantum effects are only of the order of the (rather large) experimental uncertainty for H_2–D_2, but are much larger for ^{3}He–^{4}He.

It is often desired to adjust measurements of $\mathscr{D}_{12}$ for one set of isotopes to another molecular weight basis. The mass dependence of $\mathscr{D}_{12}$ may involve three factors. First, the main dependence is the inverse proportionality to the square root of the reduced mass of the gas pair, as shown in Eqs. (60) and (63). A second mass dependence occurs in the correction term Δ_{12}, but this is almost always negligible. The third dependence is in the diffusion collision integral, which depends on mass through the de Broglie wavelength or de Boer parameter. This mass dependence is a quantum effect, and depends on the temperature and the form of the intermolecular potential.

E. INELASTIC COLLISIONS

To extend the foregoing results to polyatomic and polar molecules, it is necessary to take inelastic collisions into account. This requires a reformulation of the Boltzmann equation. The usual method involves a semiclassical point of view due to Wang Chang and Uhlenbeck and to de Boer (Wang Chang *et al.*, 1964), in which the translational molecular motion is treated classically and the internal molecular motions are treated quantummechanically. Each internal molecular state is considered a separate chemical species, and inelastic collisions considered as chemical reactions. There is thus a separate Boltzmann equation for each internal state, just as for multicomponent mixtures. This set can be solved by the Chapman–Enskog method outlined in Section III, B, with the following modifications. The expansion for the perturbation Φ_i given in Eq. (56) must include terms involving the angular velocities of the molecules as well as their translational velocities (Kagan and Afanas'ev, 1962; Waldmann, 1963). These extra " spin polarization" terms appear to have only a slight effect on the numerical value of $\mathscr{D}_{12}$, as judged by calculations for the model of loaded spheres (Sandler and Dahler, 1967; Sandler and Mason, 1967). The expansion for the functions $C_i{}^j$ given in Eq. (58) must also include terms involving the internal energy as well as the translational energy, even if spin polarization is ignored, and Eq. (59) for $\mathscr{D}_{ij}$ then contains two expansion coefficients instead of one.

A classical analog of the Wang Chang–Uhlenbeck–de Boer theory has been given by Taxman (1958) and, in a more general form, by She and Sather (1967).

The foregoing treatments were applied first to a single gas, as might be expected, and so could not include diffusion. The extension to mixtures has been made by Monchick et al. (1963, 1966, 1968) and by Alievskiĭ and Zhdanov (1969). The external appearance of the expression for $[\mathscr{D}_{ij}]_1$, as given by Eq. (60) or (63), remains the same but the explicit expression for the collision integrals is much more complicated. The general expressions for the diffusion and viscosity collision integrals (without spin polarization) are found to be

$$\overline{\Omega}_{qq'}^{(l', s)}(T) = 2[(s + 1)! Z_q Z_{q'}]^{-1} \sum_{ijkl} \exp(-\varepsilon_{qi} - \varepsilon_{q'j})$$

$$\times \int_0^\infty \gamma^{2s+3} \exp(-\gamma^2) S_{ij}^{(l')kl}(E)\, d\gamma, \tag{75}$$

where

$$\gamma^2 S_{ij}^{(1)kl}(E) = \int_0^{2\pi} d\phi \int_0^\pi (\gamma^2 - \gamma\gamma' \cos \chi) I_{ij}^{kl}(\chi, \phi, E) \sin \chi \, d\chi, \tag{76}$$

$$\gamma^4 S_{ij}^{(2)kl} = \tfrac{3}{2} \int_0^{2\pi} d\phi \int_0^\pi [\gamma^2(\gamma^2 - \gamma'^2 \cos^2 \chi) - \tfrac{1}{6}(\gamma^2 - \gamma'^2)^2] I_{ij}^{kl}(\chi, \phi, E) \sin \chi \, d\chi, \tag{77}$$

and

$$\gamma^2 - \gamma'^2 = \varepsilon_{qk} + \varepsilon_{q'l} - (\varepsilon_{qi} + \varepsilon_{q'j}), \tag{78}$$

$$\gamma^2 = E/kT, \qquad \gamma'^2 = E'/kT, \tag{79}$$

in which E' is the relative kinetic energy after collision and the species are denoted by q and q'. The ε's are the energies of the internal quantum states of the species, divided by kT. The normalization factors Z_q and $Z_{q'}$ are the internal partition functions,

$$Z_q = \sum_i \exp(-\varepsilon_{qi}), \qquad Z_{q'} = \sum_j \exp(-\varepsilon_{q'j}). \qquad (80)$$

The differential scattering cross section $I_{ij}^{kl}(\chi, \phi, E)$ describes collisions between two molecules initially in internal states i and j which emerge from the collision in final states k and l at the angle χ, ϕ.

No evaluations of Eqs. (76) and (77) have yet been made for realistic models, but the following argument suggests that inelastic effects on diffusion coefficients are small. Inelastic collisions enter $\bar{\Omega}^{(1,1)}$ only through the term $\gamma\gamma' \cos \chi$; to a first approximation $\gamma' \approx \gamma$ and the inelastic collisions have no effect. For a second approximation γ' can be written as γ plus some terms in $\Delta\varepsilon = \gamma^2 - \gamma'^2$; the inelastic correction terms are then of the form $\gamma(\Delta\varepsilon) \cos \chi$. For isotropic scattering such terms vanish on integration, and even for non-isotropic scattering the inelastic contribution is probably small unless there is some special correlation between χ and $\Delta\varepsilon$. Similar arguments hold for $\bar{\Omega}^{(2,2)}$. At any rate it is an empirical fact that the diffusion coefficients and viscosities of polyatomic gases can be well correlated by potential models that ignore inelastic collisions.

Perhaps the most important conclusion of the formal kinetic-theory results for inelastic collisions concerns the expression for the mixture viscosity, from which $\mathscr{D}_{12}$ can be determined (Weissman and Mason, 1962; Weissman, 1964). The algebraic expressions appear mathematically the same whether the molecular collisions are elastic or inelastic, and the only real effect comes from the value that is used for the collision integral ratio A^*. A first-order expansion for A^* indicates only a small correction for inelastic collisions, but good approximations are not yet available.

F. Combined Transport

The phenomenological independence of the diffusive and viscous fluxes has already been mentioned in Section II, B. In the rigorous kinetic theory this independence follows from treating the perturbation distribution function Φ_i as a first-order correction. When the perturbation is followed to the second order, however, the flux equations become much more complicated and are no longer independent (Chapman and Cowling, 1941, 1970). Three major changes occur. The first is that the resultant macroscopic flux equations now contain second derivatives and products of first derivatives of the macroscopic variables, and not just first derivatives as in the linear approximation. Second, terms corresponding to the viscous transfer of momentum appear in the diffusion equation, and vice versa. The third change is that additional boundary conditions are needed since more derivatives appear in the equations.

Fortunately for everyone's sanity, almost all of the new terms can be shown to be unnecessary, either because they are small compared to the first-order terms in all cases of experimental interest, or because retaining them involves inconsistencies in the use of asymptotic series. Only one of the new terms must be kept, a term proportional to the gradient of the rate-of-distortion tensor. However, this second-order term can be manipulated such that the form of the diffusion equation remains the same but the coefficient of pressure diffusion is modified.

The modified pressure diffusion coefficient was explicitly evaluated for a special case by Chapman and Cowling (1941, 1970), for the general case by Zhdanov *et al.* (1962) for monatomic mixtures, and by Alievskiĭ and Zhdanov (1969) for polyatomic mixtures. No practical cases are presently known where this modification is significant, however. It might be important in sound propagation through a gas mixture in a narrow tube, where pressure diffusion is a major contributor to the absorption and dispersion of the sound wave.

G. Multicomponent Diffusion

Solution of the linearized Boltzmann equation, in the absence of pressure and temperature gradients, for a multicomponent mixture of v species yields the following set of $v - 1$ independent diffusion equations, (Zhdanov *et al.*, 1962; Monchick *et al.*, 1966; Alievskiĭ and Zhdanov, 1968):

$$-\nabla x_i = \sum_{j=1}^{v} \{x_i x_j (1 - \Delta_{ij})/[\mathscr{D}_{ij}]_1\}[(\mathbf{J}_{iD}/n_i) - (\mathbf{J}_{jD}/n_j)]. \tag{81}$$

This result is comparable to the phenomenological Eqs. (28), and gives just the continuum diffusion portion of the total transport. The $[\mathscr{D}_{ij}]_1$ are exactly the binary coefficients given by Eq. (60), and the Δ_{ij} are the corrections obtained in the second approximation. However, the Δ_{ij} do not depend only on the species i and j, but on all the species in the mixture in a complicated manner. Since Δ_{ij} is fairly small for most binary mixtures, it is reasonable to ignore the behavior of the multicomponent Δ_{ij} in Eq. (81), and to replace the group $[\mathscr{D}_{ij}]_1/(1 - \Delta_{ij})$ by the binary $\mathscr{D}_{ij}$.

The only explicit calculations of the multicomponent Δ_{ij} have been for the special case of the diffusion of a trace species through a stagnant binary mixture (as in Blanc's law for ion mobilities). Numerical calculations showed that the multicomponent effects are usually small, of the order of experimental errors, but that systems showing large effects are possible (Sandler and Mason, 1968).

Equations (81) are sometimes inverted to give one equation for each flux in terms of all the gradients (Hirschfelder *et al.*, 1954; Waldmann, 1958); the coefficients of the gradients are then multicomponent diffusion coefficients, which are complicated functions of composition and the binary diffusion coefficients. Because of the auxiliary condition

$$\sum_{j=1}^{v} \nabla x_j = 0, \tag{82}$$

the multicomponent diffusion coefficients are not uniquely defined. Hirschfelder *et al.* (1954) resolved the lack of uniqueness by setting the diagonal coefficients equal to zero; this leads to an unsymmetric matrix of coefficients, for which it is not obvious that the Onsager reciprocal relations are true (although they turn out to be). Curtiss (1968) has shown how to resolve the lack of uniqueness in such a way as to make the coefficients symmetric, and thus clearly consistent with the reciprocal relations. His results appear to be equivalent to those obtained earlier by Waldmann (1958).

H. Density Dependence

All the molecular theory discussed thus far gives $\mathscr{D}_{12}$ as inversely proportional to density, because of the binary collision assumption. In recent years there has been a great deal of work on the extension from dilute to moderately dense gases. A comprehensive list of references is given in the review article of Ernst *et al.* (1969); another recent review has been given by Cohen (1969). The results have been surprising, since divergences have appeared. No review of diffusion should fail to mention this work. Moreover, the results have implications even for the low-density transition region between free-molecule and continuum behavior, according to the dusty-gas model discussed in Section IV, B.

By analogy with the virial expansion for the equation of state of a gas, it was expected that the density dependence of the transport coefficients could be expressed as a power series in the number density,

$$n\mathscr{D}_{12} = n\mathscr{D}_{12}^0/(1 + \alpha_1 n + \alpha_2 n^2 + \cdots), \tag{83}$$

where $n\mathscr{D}_{12}^0$ gives the binary collision limit, and $\alpha_1, \alpha_2, \ldots$ are "transport virial coefficients." The coefficients α_i turned out to be all divergent above a certain term. In three dimensions the first divergent coefficient is α_2, and in two dimensions it is α_1. The divergences have been shown to be logarithmic and to be caused by long-range correlation resulting from recollisions between pairs of molecules. The first divergent term should thus be of the form

$$\alpha_i n^i \to \alpha_i' n^i (1 + \alpha_i'' \ln n). \tag{84}$$

Simple mean-free-path arguments suggest that a double series expansion in terms of n and ln n is needed (Cohen, 1969).

At present it is not known whether the logarithmic terms are of sufficient numerical importance to be needed in fitting experimental data (Hanley et al., 1969), or even whether there are still further contributions with another density dependence.

I. Composition Dependence

The small composition dependence of $\mathscr{D}_{12}$ resides in the correction term Δ_{12}, given by Eq. (65) as

$$\Delta_{12} = \tfrac{1}{10}(6C_{12}^* - 5)^2\left(\frac{x_1{}^2 P_1 + x_2{}^2 P_2 + x_1 x_2 P_{12}}{x_1{}^2 Q_1 + x_2{}^2 Q_2 + x_1 x_2 Q_{12}}\right), \tag{65}$$

where

$$P_1 = \frac{2M_1{}^2}{M_2(M_1 + M_2)}\left(\frac{2M_2}{M_1 + M_2}\right)^{1/2}\left[\frac{\overline{\Omega}_{11}^{(2,\,2)}}{\overline{\Omega}_{12}^{(1,\,1)}}\right], \tag{85}$$

$$P_{12} = 15\left(\frac{M_1 - M_2}{M_1 + M_2}\right)^2 + \frac{8M_1 M_2 A_{12}^*}{(M_1 + M_2)^2}, \tag{86}$$

$$Q_1 = \frac{2}{M_2(M_1 + M_2)}\left(\frac{2M_2}{M_1 + M_2}\right)^{1/2}\left[\frac{\overline{\Omega}_{11}^{(2,\,2)}}{\overline{\Omega}_{12}^{(1,\,1)}}\right]$$
$$\times [(\tfrac{5}{2} - \tfrac{6}{5}B_{12}^*)M_1{}^2 + 3M_2{}^2 + \tfrac{8}{5}M_1 M_2 A_{12}^*], \tag{87}$$

$$Q_{12} = 15\left(\frac{M_1 - M_2}{M_1 + M_2}\right)^2(\tfrac{5}{2} - \tfrac{6}{5}B_{12}^*) + \frac{4M_1 M_2 A_{12}^*}{(M_1 + M_2)^2}(11 - \tfrac{12}{5}B_{12}^*)$$
$$+ \frac{8(M_1 + M_2)}{5(M_1 M_2)^{1/2}}\left[\frac{\overline{\Omega}_{11}^{(2,\,2)}}{\overline{\Omega}_{12}^{(1,\,1)}}\right]\left[\frac{\overline{\Omega}_{22}^{(2,\,2)}}{\overline{\Omega}_{12}^{(1,\,1)}}\right]. \tag{88}$$

The expressions for P_2 and Q_2 are obtained from those for P_1 and Q_1 by the interchange of subscripts, which refer to molecular interactions. That is, the subscript "11" denotes interactions between two species 1 molecules, and so on. These are the Chapman and Cowling expressions; the Kihara expressions can be obtained by setting $B_{12}^* = 5/4$. Since $\overline{\Omega}_{12}^{(1,1)}$ is related to $\mathscr{D}_{12}$, and $\overline{\Omega}_{11}^{(2,2)}$ is related to the viscosity of pure species 1, it is clear that the P's and Q's can largely be expressed in terms of experimental quantities.

The foregoing expressions are for elastic collisions only. Formal expressions for the case of inelastic collisions are available (Zhdanov et al., 1962; Monchick et al., 1966), but no use has yet been made of them. They are complicated to use, and Δ_{12} is only a small correction, usually less than 0.05.

Even the elastic collision expressions for Δ_{12} are rather tedious to use for so small a correction, and attempts have been made to obtain simplified but adequate approximations (Wilke and Lee, 1955; Amdur and Schatzki, 1958; Mason et al., 1964). Perhaps the best of these semiempirical results relates Δ_{12} to values of the thermal diffusion factor α_T, which are often known experimentally,

$$\Delta_{12} \approx \tfrac{1}{5}\zeta x_1(\alpha_T)^2[x_1(Q_1/S_1) - x_2(Q_2/S_2)], \tag{89}$$

in which ζ is a numerical constant whose value probably lies between 1 and 2, and

$$S_1 = (M_1/M_2)[2M_2/(M_1 + M_2)]^{1/2}[\overline{\Omega}_{11}^{(2,2)}/\overline{\Omega}_{12}^{(1,1)}]$$
$$- [4M_1M_2 A_{12}^*/(M_1 + M_2)^2] - [15M_2(M_2 - M_1)/2(M_1 + M_2)^2], \tag{90}$$

the expression for S_2 being obtained by interchange of subscripts. Here it is important to note that species 1 is the heavier one. The advantage of Eq. (89) is that the sensitive portions of the expression for Δ_{12} are taken from experiment, and the portions that must be calculated theoretically, at least in part (the Q's and S's), are comparatively insensitive to intermolecular forces and temperature (Mason et al., 1964).

The composition dependence of Δ_{12} can be quite accurately approximated by (Mason et al., 1964).

$$\Delta_{12} \approx (\text{constant})[x_1/(1 + cx_1)], \tag{91}$$

where x_1 is the mole fraction of the heavier species. The constants in this equation of course depend on temperature, molecular masses, and intermolecular forces. Marrero and Mason (1971) have found that the constants can be predicted with acceptable accuracy by the following formulas:

$$\Delta_{12} \approx \zeta(6C_{12}^* - 5)^2[ax_1/(1 + cx_1)], \tag{92}$$

where ζ is a numerical constant between 1 and 2 as in Eq. (89), C_{12}^* is the collision integral ratio of Eq. (71) and must be found from some model of the intermolecular potential (e.g., the 12–6 or exp–6 models), and

$$a = [0.17/(1 + 1.8m)^2][\overline{\Omega}_{12}^{(1,1)}/\overline{\Omega}_{22}^{(2,2)}], \tag{93}$$

$$c + 1 = 10a(1 + 1.8m + 3m^2), \tag{94}$$

$$m = M_2/M_1. \tag{95}$$

The collision integrals in a may be obtained either from experiment or by calculation from a potential model. The quantities a and c vary only weakly with temperature, and can usually be taken as constant.

To give some idea of the usefulness of Eqs. (92) through (95), Table II

TABLE II

COMPOSITION DEPENDENCE AT 295°K OF $\mathscr{D}_{12}$
ACCORDING TO EXPERIMENT[a] AND TO
EQS. (92)–(95)[b]

System	$\dfrac{\mathscr{D}_{12}(x_1 = 1/2)}{\mathscr{D}_{12}(x_1 = 0)}$	ζ
He–Ne	1.030	1.64
He–Ar	1.039	1.67
He–Kr	1.044	1.65
He–Xe	1.049	1.78
Ne–Ar	1.009	1.2
Ne–Kr	1.013	1.01
Ne–Xe	1.020	1.25
Ar–Kr	1.003	1.4
Ar–Xe	1.003	1.8
Kr–Xe	1.001	1.8

[a] van Heijningen et al. (1968).
[b] The value of ζ should lie between 1 and 2 if the equations are satisfactory. A Lennard–Jones (12–6) potential was used in the calculations.

gives a comparison with the accurate experimental results of van Heijningen et al. (1968) at 295°K. The composition dependence is adequately given by Eq. (92); the comparison of the constants a and c is made by showing the value of ζ needed to reproduce the experimental value of the ratio $D_{12}(x_1 = \frac{1}{2})/\mathscr{D}_{12}(x_1 = 0)$. It can be seen that ζ does lie between 1 and 2, and that the magnitude of the composition variation of $\mathscr{D}_{12}$ is small.

J. TEMPERATURE DEPENDENCE

Almost the entire temperature dependence of $\mathscr{D}_{12}$ is given by $[\mathscr{D}_{12}]_1$, or at constant pressure by the factor $T^{3/2}/\overline{\Omega}^{(1,1)}$. The higher approximations have only a slight effect, and in the following discussion we neglect the temperature dependence of the small correction term Δ_{12}. The temperature dependence of $\overline{\Omega}^{(1,1)}$ can be calculated from the intermolecular potential. Calculations for various plausible potential models have shown that the derivative $d \ln \overline{\Omega}^{(1,1)}/d \ln T$ usually lies between 0 and $-1/2$, so that the derivative $(\partial \ln \mathscr{D}_{12}/\partial \ln T)_p$ should lie between 3/2 and 2. This is found to be the case experimentally.

The general characteristics of $(\partial \ln \mathscr{D}_{12}/\partial \ln T)_p$ are as follows. At very low temperatures the dominant molecular interaction is the long-range r^{-6}

London dispersion energy, which causes $\overline{\Omega}^{(1,1)}$ to vary as $T^{-1/3}$. At very high temperatures the dominant interaction is the (roughly) exponential short-range repulsion, which causes $\overline{\Omega}^{(1,1)}$ to have a weaker temperature dependence than at low temperatures. Thus $(\partial \ln \mathscr{D}_{12}/\partial \ln T)_p$ is equal to 11/6 at very low temperatures, and equal to a smaller value (~ 1.7) at high temperatures, the high-temperature value being slightly dependent on temperature. At intermediate temperatures, where both the attractive and repulsive interactions are significant, the transition in $(\partial \ln \overline{\Omega}_{12}/\partial \ln T)_p$ is not monotonic but exhibits a maximum.

These general features are shown in Fig. 7, obtained from experimental data and from potential model calculations. The derivative $(\partial \ln \mathscr{D}_{12}/\partial \ln T)_p$

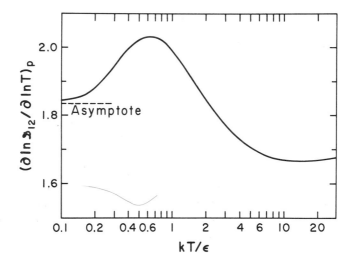

FIG. 7. Qualitative temperature dependence of diffusion coefficients at constant pressure.

is always appreciably greater than the constant value of 3/2 shown by rigid spheres.

More quantitative information about the temperature dependence of $\mathscr{D}_{12}$ requires additional details about intermolecular potentials. It is well known that molecules attract each other at large separation distances, and repel each other at small separations. In principle, quantum theory provides a method for calculating the interaction between a pair of molecules. The long-range dispersion energy can in fact be calculated fairly accurately. However, when molecules are close enough for their electron clouds to overlap, the resulting repulsion energy is too complicated to be calculated in any simple way. The various interactions and their effect on $\mathscr{D}_{12}$ are considered below.

1. Short-Range Interactions

Quantum-mechanical calculations indicate that the repulsive potential can be approximated by an exponential function, and over a more limited range by an inverse power. Both give completely erroneous accounts of the potential at large separations, where attractive forces dominate. The simple algebraic form of a single-term exponential or inverse power potential permits the collision integrals to be calculated numerically.

The exponential potential can be written as

$$\varphi(r) = \varphi_o e^{-r/\rho}, \tag{96}$$

where φ_o and ρ are constants. Monchick (1959) evaluated the $\overline{\Omega}^{(l,s)}$ over a wide temperature range for this potential, and found the temperature dependence to be approximately

$$\overline{\Omega}^{(1,\,1)}(T) \propto \rho^2 [\ln(\varphi_o/kT)]^2. \tag{97}$$

The inverse-power potential can be written as

$$\varphi(r) = K/r^n, \tag{98}$$

where K and n are constants. For this potential it can be shown that the temperature dependence is exactly (Hirschfelder et al., 1954)

$$\overline{\Omega}^{(1,\,1)} \propto (nK/kT)^{2/n}. \tag{99}$$

2. Long-Range Interactions

These interactions behave asymptotically as (neglecting retardation effects)

$$\varphi(r) = -C/r^6, \tag{100}$$

where C is the London constant. The collision integral has the form,

$$\overline{\Omega}^{(1,\,1)} \propto (C/T)^{1/3}, \tag{101}$$

according to classical mechanics. Thus as $T \to 0$, $\mathscr{D}_{12} \propto T^{11/6}$ classically, but quantum corrections become important at sufficiently low temperatures. A general expression for the quantum-mechanical $\overline{\Omega}^{(1,1)}$ at very low temperatures is not presently available.

3. Intermediate-Range Interactions

At intermediate separations the potential is not dominated by either attractive or repulsive forces, and has a "well" whose detailed shape is not precisely known. Descriptive approximations of this region are often given by semi-empirical expressions like the Lennard–Jones (n–6) or the exp–6 models. Such potential models interpolate between functions derived for solely attractive or repulsive interactions, and cannot be expected to be very accurate in general. The temperature dependence of the collision integral is complicated for such potentials, and no precise expression can be given corresponding to intermediate temperatures (from about 200° to 1000°K for most gas pairs). However, Sutherland (1894) showed how to make a correction in simple mathematical form for weak attractive forces. He considered the model of rigid spheres with weak attractive forces, and showed that

$$\Omega^{(1,\,1)*} = 1 + (S/T), \tag{102}$$

where S is a positive constant. The temperature dependence of $\mathscr{D}_{12}$ is then

$$\mathscr{D}_{12} \propto T^{3/2}/[1 + (S/T)], \tag{103}$$

which is remarkably good for fitting experimental results over moderate temperature ranges. This form has also been shown to give an accurate representation (within 0.2 % for $1.4 < kT/\varepsilon < 3.5$) of the numerically calculated collision integrals for the Lennard–Jones (12–6) potential (Kim and Ross, 1967). Another form of comparable quality was later suggested by Reinganum (1900),

$$\Omega^{(1,\,1)*} = e^{S/T}, \tag{104}$$

$$\mathscr{D}_{12} \propto T^{3/2}e^{-S/T}, \tag{105}$$

which reduces to the Sutherland form for small S/T.

4. Empirical Representation

Marrero and Mason (1971) have found that the temperature dependence of $\mathscr{D}_{12}$ at constant pressure and composition can be fitted within experimental uncertainty over a very wide temperature range by means of an empirical composite of the foregoing mathematical forms,

$$\ln(p\mathscr{D}_{12}) = \ln A + s \ln T - \ln[\ln(\varphi_o/kT)]^2 - (S/T) - (S'/T^2), \tag{106}$$

where A, s, φ_o, S, and S' are constants. The value of φ_o is taken from independent molecular-beam experiments; its precise value is not critical. The parameter $s \gtrsim 3/2$ is determined from the diffusion data and can compensate for errors in φ_o. The parameters A, S, and S' are also adjustable; S and S' allow for the attractive forces, but S' is needed only when accurate data are available, and can often be taken as zero.

An advantage of Eq. (106) is that it is linear in all the adjustable parameters, so least-squaring is easy. A disadvantage is that the Sutherland–Reinganum terms for the effect of the long-range attractive interactions do not permit the equation to be used at very low temperatures, where the London dispersion energy dominates. In terms of Fig. 7 for the temperature dependence of $\mathscr{D}_{12}$, Eq. (106) is usable only for $kT/\varepsilon > 1$.

K. Calculations from Intermolecular Potentials

One frequently needs a reliable value of $\mathscr{D}_{12}$ for a system where no experimental data exist. The value of $\mathscr{D}_{12}$ can be calculated if the intermolecular potential can be estimated. Such estimates can come from quantum-mechanical calculations, spectroscopic data, virial coefficients, viscosities, molecular scattering experiments, and so on (Mason and Monchick, 1967). Even after the potential is known, however, fairly laborious numerical calculations are needed to obtain $\mathscr{D}_{12}$. In this section we indicate how such calculations can be made, and give a brief summary of available numerical tabulations.

1. Computational Procedures

If a classical–mechanical calculation will suffice, the first step is to determine the deflection angle χ as a function of energy and impact parameter by evaluation of the integral,

$$\chi(b, E) = \pi - 2b \int_{r_c}^{\infty} \{1 - (b/r)^2 - [\varphi(r)/E]\}^{-1/2}(dr/r^2), \qquad (107)$$

where r_c is the distance of closest approach in a collision, given by

$$1 - (b/r_c)^2 - [\varphi(r_c)/E] = 0. \qquad (108)$$

This usually requires numerical integration, unless $\varphi(r)$ has an especially simple form. The values of χ are then integrated (numerically) over all impact parameters to give the cross sections $S^{(l)}(E)$, as indicated by Eq. (67), and the cross sections are integrated over all energies according to Eq. (66) to give the collision integrals. Efficient computer programs for effecting these integrations have been described by O'Hara and Smith (1970).

If a classical calculation is not adequate (see Section III, D), then the differential cross section must be obtained by a scattering phase-shift analysis,

$$I(\chi) = |f(\chi)|^2 \qquad (109)$$

$$f(\chi) = (2i\kappa)^{-1} \sum_{l=0}^{\infty} (2l + 1)[\exp(2i\,\delta_l) - 1]P_l(\cos \chi), \qquad (110)$$

where δ_l is the phase shift, l is the angular momentum quantum number, $\kappa = 2\pi/\lambda$ is the wave number of relative motion, and $P_l(\cos \chi)$ is a Legendre polynomial. The δ_l must be found by (numerical) integration of the radial wave equation. When Eqs. (109) and (110) are substituted back into Eq. (73) for the cross sections, and the integrations over χ are carried out, the following expressions in terms of the phase shifts are obtained (Hirschfelder *et al.*, 1954):

$$S^{(1)}(E) = (4\pi/\kappa^2) \sum_{l=0}^{\infty} (l+1) \sin^2 (\delta_{l+1} - \delta_l), \tag{111}$$

$$S^{(2)}(E) = (4\pi/\kappa^2) \sum_{l} \{[(l+1)(l+2)]/(2l+3)\} \sin^2 (\delta_{l+2} - \delta_l). \tag{112}$$

The summations are over all integral values of l from 0 to ∞ for distinguishable particles, but only over the even or odd integral values for indistinguishable particles (in which case the summation is multiplied by a normalization factor of 2). For observable processes $S^{(1)}$ must always refer to distinguishable particles, but $S^{(2)}$ can refer to either. The cross sections are then integrated over all energies just as in the classical calculation.

Efficient procedures for the calculation of the phase shifts have been described by Munn *et al.* (1964). Care must be used in using semiclassical approximations in transport calculations, because these usually turn out to be exactly the same as the classical results.

2. Available Tabulations

Most of the available tabulations of collision integrals are for the classical limit only, but this suffices for most gas pairs. We have already mentioned the exponential potential, Eq. (96), and the inverse-power potential, Eq. (98). The collision integrals for both repulsive and attractive inverse-power potentials have been summarized by Kihara *et al.* (1960) and by Higgins and Smith (1968). Those for the repulsive exponential potential have been given by Monchick (1959) and by Higgins and Smith (1968), and those for the attractive exponential by Munn *et al.* (1965b).

A screened Coulomb potential is often used as a model for the interaction of neutral atoms at high energies (or of ions in a plasma),

$$\varphi(r) = \pm [\varphi_o e^{-r/\rho}/(r/\rho)], \tag{113}$$

where φ_o and ρ are constants. Collision integrals have been tabulated by Smith *et al.* (1965b) and by Mason *et al.* (1967b).

If the potential has both attractive and repulsive components so that it shows a potential well, it is often represented as a sum of two (or more) exponential or inverse-power terms. The simplest such combination gives the bireciprocal potential, often called the Lennard–Jones (n–m) potential, which is conveniently written in terms of the depth of the potential well (ε) and the separation distance for which the potential is zero (σ) or is at a minimum (r_m):

$$\varphi(r) = [n\varepsilon/(n - m)](n/m)^{m/(n-m)}[(\sigma/r)^n - (\sigma/r)^m], \tag{114}$$

$$\varphi(r) = [n\varepsilon/(n - m)][(m/n)(r_m/r)^n - (r_m/r)^m]. \tag{115}$$

Quantum-mechanical theory requires that $m = 6$ for the long-range attraction between neutral molecules and $m = 4$ for the long-range attraction between an ion and a neutral molecule (Hirschfelder et al., 1954). The value of n is not known from theory; empirical values ranging from 8 upwards have been used. Collision integrals for the (8–4) potential have been calculated by Hassé and Cook (1929, 1931), and for the (12–4) potential by Mason and Schamp (1958). The (12–6) potential has been quite popular and has been treated a number of times; the early results have been summarized by Hirschfelder et al. (1954), but more accurate calculations are now available. According to Smith and Munn (1964) the most accurate results are those of Monchick and Mason (1961) up to about $kT/\varepsilon = 20$, but at higher temperatures the results of Itean et al. (1961) are more accurate. The quantum-mechanical collision integrals for the (12–6) potential have been given by Imam-Rahajoe et al. (1965) and by Munn et al. (1965a). The (9–6) potential has been treated by Smith et al. (1965a) and by Shih and Ibele (1968), and the (16–6) and (18–6) potentials by Dymond et al. (1966). A very extensive and accurate set of collision integrals for (n–6) potentials has been given by Klein and Smith (1968) for $n = 9$, 12, 15, 18, 21, 24, 30, 40, 50, and 75. Schramm (1968) has given similar tabulations for $n = 18$, 24, and 48; and Lin and Hsu (1969) for $n = 8$, 10, 12, 14, 16, 18, and 20.

A model of rigid spheres with inverse-power attractions is sometimes called the Sutherland potential; it appears in the limit as an (∞–m) potential. The (∞–4) potential has been treated by Hassé (1926) and by Hassé and Cook (1927); the (∞–6) potential by Kotani (1942), whose results are also tabulated by Hirschfelder et al. (1954).

A (28–7) potential was proposed by Hamann and Lambert (1954) as an approximation to the average interaction between quasispherical molecules such as CF_4, SF_6, and $C(CH_3)_4$. Collision integrals have been calculated by McCoubrey and Singh (1959) and by Smith et al. (1965a). Another model which attempts to mimic polyatomic molecules with a pseudocentral potential is the rigid core model of Kihara (1953, 1963), in which the energy of interaction depends only on the shortest distance between the surfaces of the two cores. Collision integrals for a (12–6) potential outside a rigid-sphere core have been given by Barker et al. (1964). The same physical idea that underlies the Kihara core model is the basis for the spherical-shell potential (De Rocco and Hoover, 1962), in which each surface element on one spherical shell interacts by a (12–6) potential with each surface element on another spherical shell. Collision integrals for this potential have been tabulated by De Rocco *et al.* (1968) and by Schramm (1969).

The exponential–six (exp–6) potential is similar to the (n–6) potential, but uses an exponential instead of an inverse-power repulsion. This has some basis in quantum theory. The (exp–6) potential is

$$\varphi(r) = [\alpha\varepsilon/(\alpha - 6)]\{(6/\alpha) \exp[\alpha(1 - r/r_m)] - (r_m/r)^6\}, \qquad (116)$$

where ε and r_m have the same meaning as for the (n–6) potential, and the dimensionless third parameter α is analogous to n. Collision integrals for the (exp–6) potential have been calculated by Mason (1954), and some values extrapolated to higher α have been given by Mason and Rice (1954) and by Umanskii and Bogdanova (1968).

Another useful potential is obtained by taking two exponential terms, one repulsive and one attractive. This was first proposed by Morse (1929) to account for the vibrational spectra of diatomic molecules, so that it is the potential between two atoms,

$$\varphi(r) = \varepsilon\{\exp[-2(c/\sigma)(r - r_m)] - 2 \exp[-(c/\sigma)(r - r_m)]\}, \qquad (117)$$

where ε, σ, and r_m have the usual significance. The dimensionless parameter c is related to σ and r_m by

$$c = \sigma \ln 2/(r_m - \sigma). \qquad (118)$$

Tables of collision integrals for the Morse potential have been given by Smith and Munn (1964) and by Samoïlov and Tsitelauri (1964).

There are sometimes theoretical reasons for using three-term potentials. Mason and Schamp (1958) have given the collision integrals for a (12–6–4) potential,

$$\varphi(r) = \tfrac{1}{2}\varepsilon[(1 + \gamma)(r_m/r)^{12} - 4\gamma(r_m/r)^6 - 3(1 - \gamma)(r_m/r)^4], \qquad (119)$$

where γ measures the relative strength of the r^{-6} energy term. This model, as well as the (8–4) and (∞–4) models, were originally suggested to represent the interaction of an ion and a neutral molecule, but the mathematical form may be useful for representing potentials between neutral atoms and molecules as well.

The interaction energy of two dipoles varies as r^{-3}, and of two quadrupoles as r^{-5}. Terms representing such interactions have been added to (12–6) potentials, giving

$$\varphi(r) = 4\varepsilon_o[(\sigma_o/r)^{12} - (\sigma_o/r)^6 \pm \delta(\sigma_o/r)^3], \qquad (120)$$

$$\varphi(r) = 4\varepsilon_o[(\sigma_o/r)^{12} - (\sigma_o/r)^6 \pm q(\sigma_o/r)^5]. \qquad (121)$$

The parameters ε_o and σ_o have their usual interpretation in the limit as the dimensionless third parameters δ and q go to zero. These parameters should really be orientation-dependent, but it is extremely difficult to calculate collision integrals for a true orientation-dependent potential. Such potentials have been treated on the assumption that they are effectively central, the (12–6–3) potential by Monchick and Mason (1961) and the (12–6–5) potential by Smith et al. (1967).

3. Combination Rules

It frequently happens that no direct information on the intermolecular potential for a particular gas pair of interest is available, but that the potentials for the separate pure species are known. Various semiempirical combination rules are available for predicting the parameters for a 1–2 interaction from those for the 1–1 and 2 2 interactions. While far from perfect, such rules seem to work well enough to allow the prediction of diffusion coefficients to a level of accuracy of perhaps 10%. (This is only a rough rule of thumb, and the accuracy may be considerably better or worse, depending on circumstances.)

The most commonly used potential is probably the (12–6), and the most commonly used combination rules for this model are

$$\sigma_{12} = \tfrac{1}{2}(\sigma_{11} + \sigma_{22}), \tag{122}$$

$$\varepsilon_{12} = (\varepsilon_{11}\varepsilon_{22})^{1/2}. \tag{123}$$

The justification for the σ rule is the analogy with rigid spheres; the ε rule is based on a similar geometric-mean rule for the London dispersion coefficient, which has some theoretical justification. These two rules are often used as reasonable first approximations for any potential having an attractive well; for convenience the σ rule may be replaced or supplemented by an analogous rule for r_m,

$$(r_m)_{12} = \tfrac{1}{2}[(r_m)_{11} + (r_m)_{22}]. \tag{124}$$

Supplementary rules are needed for the three-parameter potentials. The simplest choices for the (n–6) and Morse potentials would be Eqs. (122) and (123) plus

$$n_{12} = \tfrac{1}{2}(n_{11} + n_{22}), \tag{125}$$

or

$$c_{12} = \tfrac{1}{2}(c_{11} + c_{22}). \tag{126}$$

For the (exp–6) potential the simplest choice would be Eqs. (123) and (124) plus (Mason, 1955)

$$\alpha_{12} = \tfrac{1}{2}(\alpha_{11} + \alpha_{22}). \tag{127}$$

Combination rules for the parameters γ, δ, and q of the (12–6–4), (12–6–3), and (12–6–5) potentials are seldom needed, because they depend on values of the polarizabilities, dipole moments, and quadrupole moments, respectively, which are often known independently. If this is not the case, geometric combination rules are probably satisfactory.

The rules are even simpler for single-term attractive and repulsive potentials. Theory indicates a geometric-mean rule for the London dispersion coefficient,

$$C_{12} = (C_{11}C_{22})^{1/2}. \tag{128}$$

This has been tested by Barker (1963) and found to be quite accurate, and good refinements are even available (Wilson, 1965, 1968; Crowell, 1968).

Theory likewise suggests, but more weakly, a geometric-mean combination for the repulsive potential (Mason and Monchick, 1967). For the exponential repulsive potential of Eq. (96) this procedure yields

$$(\varphi_o)_{12} = [(\varphi_o)_{11}(\varphi_o)_{22}]^{1/2}, \tag{129}$$

$$\rho_{12}^{-1} = \tfrac{1}{2}(\rho_{11}^{-1} + \rho_{22}^{-1}). \tag{130}$$

For the inverse-power repulsive potential of Eq. (98) this yields Eq. (125) for the parameter n, and

$$K_{12} = (K_{11}K_{22})^{1/2} \tag{131}$$

for the parameter K. These rules have been directly tested by means of high-velocity molecular beam scattering, and seem to work well (Amdur *et al.*, 1954; Amdur and Mason, 1956; Kamnev and Leonas, 1966).

More elaborate combination rules than the simple rules of Eqs. (122) through (127) have been devised, usually based on application of Eqs. (128) through (131) to individual terms in a potential model, or on some other theoretical consideration. But the simple rules work well enough for most purposes, and the reader interested in refinements must refer to the literature (Hirschfelder *et al.*, 1954; Mason, 1955; Srivastava, 1958; Mason and Monchick, 1962, 1967; Saran, 1963; Saxena and Gambhir, 1963; Konowalow, 1969; Hiza and Duncan, 1969).

IV. Diffusion in Rarefied Gases

The phenomenological description of diffusion in rarefied gases has been given in Section II. Here we review the molecular theory of such diffusion, supplementing the molecular theory of continuum diffusion discussed in Section III.

A. SLIP CALCULATIONS

At pressures somewhat less than those for which a gas behaves entirely as a continuum, the finite mean free path manifests itself as corrections to the boundary conditions. These corrections take the forms of various apparent discontinuities in macroscopic variables at the boundaries, and are known by such names as viscous slip, thermal creep, diffusive slip, and temperature jump (see, for example, Kennard, 1938).

The historical background is briefly as follows. Maxwell (1879) gave an extensive analysis of stresses arising in a moving, heated single gas consisting of structureless molecules, and discussed the phenomena of viscous slip, temperature jump, and thermal creep. Except for simplifications of the reasoning (Kennard, 1938), no substantial changes have been made in this approach. Much later, Kramers and Kistemaker (1943) gave an elementary discussion of the analogous phenomenon of diffusive slip, and demonstrated the effect experimentally. A more elaborate theoretical treatment of diffusive slip, along the lines of the simplified Maxwell approach, was later given by Kucherov (1957). A new approach was indicated by Grad (1949), who developed a moment solution of the Boltzmann equation, usually applied in the form of a thirteen-moment approximation. Kucherov and Rikenglaz (1959) attempted to treat all four phenomena by the thirteen-moment method; they considered only binary mixtures and did not evaluate all of the terms in the expression for the slip velocity. The latter defect was later remedied by Zhdanov (1967), who obtained a complete expression for the slip velocity of a binary mixture of monatomic gases due to gradients of velocity, temperature, composition, and pressure.

The first attempt to treat a polyatomic gas was made by Zhdanov (1968), who developed a seventeen-moment approximation for handling the internal degrees of freedom, and obtained the slip equations for a single polyatomic gas. This was generalized to polyatomic gas mixtures by Annis and Mason (1970), who used the method of the dusty-gas model, discussed in Section IV, B.

There is also a large literature on the related phenomena associated with the motion of aerosol particles in nonuniform gases, and on treatments of slip and creep that are based on mathematical models of the Boltzmann equation rather than on the Boltzmann equation itself, but we do not attempt to review these here.

Here we discuss only the isothermal phenomena of viscous slip and diffusive slip, referring the reader to the original papers for discussions of thermal creep and temperature jump. For a binary mixture the mass-average velocity is defined as

$$\mathbf{V}_o \equiv (n_1 m_1 \overline{\mathbf{V}}_1 + n_2 m_2 \overline{\mathbf{V}}_2)/(n_1 m_1 + n_2 m_2), \tag{132}$$

where $\overline{\mathbf{V}}_1$ and $\overline{\mathbf{V}}_2$ are the average molecular *velocities* (not speeds) in the laboratory-fixed coordinate system. Then there is a slip velocity at a wall parallel to the z direction given by (Zhdanov, 1967),

$$\begin{aligned}
V_{oz}|_{y=0} = {} & [1/(x_1 m_1^{1/2} + x_2 m_2^{1/2})]\{(\eta/p)(\pi k T/2)^{1/2}(\partial V_{oz}/\partial y) \\
& + \{[\mathscr{D}_{12}]_2/(x_1 m_1 + x_2 m_2)\}\{(m_2 m_1^{1/2} - m_1 m_2^{1/2}) + \tfrac{1}{5}(x_1 m_1^{3/2} \\
& + x_2 m_2^{3/2})[\alpha_T]_1 - (m_1 + m_2)(m_1^{1/2} - m_2^{1/2})[\Delta_{12}/(6C_{12}^* - 5)]\} \\
& \times [(\partial x_1/\partial z) + x_1 x_2 \alpha_p(\partial \ln p/\partial z)]\},
\end{aligned} \tag{133}$$

where the y-axis is perpendicular to the wall. The first term on the right is the viscous slip; it has exactly the same form as originally found by Maxwell for a single gas, except that η now refers to the viscosity of the mixture. (All molecules are assumed to be reflected from the wall in a diffuse scattering pattern.) The second term is the diffusive slip, and involves the Chapman–Enskog second approximation to the diffusion coefficient $[\mathscr{D}_{12}]_2$, the Chapman–Enskog first approximation to the thermal diffusion factor $[\alpha_T]_1$, the correction term for the diffusion coefficient Δ_{12}, and the pressure diffusion factor α_p. The term involving α_p is of a higher order in p^{-1} than the other terms, and for consistency should not appear in a first-order slip correction. Zhdanov, in fact, neglects this term in an application of Eq. (133) to the isothermal flow of a gaseous mixture in a long capillary.

The diffusive slip term in Eq. (133) consists of three parts. The first part is the largest, and gives rise to Graham's law of diffusion; it is the same as the expression obtained earlier by Kramers and Kistemaker. The second and third parts, involving $[\alpha_T]_1$ and Δ_{12}, are in the nature of correction terms; they can give rise to deviations from Graham's diffusion law.

The extension to polyatomic gases adds nothing significant to Eq. (133); the interesting polyatomic effects occur in thermal creep, which we are here ignoring. The dusty-gas approach used by Annis and Mason does, however, suggest that the $[\alpha_T]_1$ part of the diffusive slip term originates in the gas–dust diffusion corrections, Δ_{ij}. It also shows that the extension to multicomponent mixtures would be very laborious for diffusive slip, but straightforward for viscous slip,

$$V_{oz}|_{y=0} = \frac{(\eta/p)(\pi k T/2)^{1/2}}{x_1 m_1^{1/2} + x_2 m_2^{1/2} + x_3 m_3^{1/2} + \cdots} \frac{\partial V_{oz}}{\partial y}, \tag{134}$$

where η is the viscosity of the multicomponent mixture. (The multicomponent generalization of thermal creep is also straightforward.)

B. Dusty-Gas Model

A clever model for gas flow and diffusion in porous media was first suggested by Maxwell (1860), in which the porous medium is visualized as a random collection of large particles fixed in space ("dust"). This was apparently forgotten for many years and then re-invented by Deriagin (also spelled Derjaguin) and Bakanov (1957a,b), who proposed treating the particles of the porous medium as giant molecules by means of Chapman–Enskog theory. They calculated flow of a single gas near the free-molecule region, and obtained an explicit expression for the Knudsen flow parameter K_o mentioned in Section II. This approach leads to complex calculations and is valid only for dilute dusty gases, that is, very porous solids. Most of the complication comes from treating the details of the gas–dust collisions. The model was shortly thereafter independently re-invented yet once more by Evans *et al.* (1961b), who avoided the complicated part of the problem, namely the explicit calculation of the parameters K_o, B_o, and ε/q, and used the model only to determine the flux equations. By formal variation of the mole fraction of the "dust" particles, the whole pressure range from the free-molecule to the continuum region could be covered. In this form, the model has been applied to a variety of diffusion and flow problems (Mason *et al.*, 1967a, and papers referred to therein).

The results obtained by the dusty-gas model are essentially the same as the phenomenological results already discussed on the basis of momentum transfer in Section II, but have a better theoretical pedigree and supply more detail. A number of previously unsuspected relations have come to light in this way; however, most of these involve nonisothermal phenomena, which we are ignoring in this review. This is the closest approach at present to a rigorous kinetic theory in the transition region, but there are uncomfortable loose ends around. For instance, the work on the kinetic theory of dense gases (Section III, H) suggests that the dusty-gas model should also show logarithmic density terms in the transition region (Weijland and van Leeuwen, 1968). All that can be said at present is that the limited experimental data do not seem to indicate that logarithmic terms are appreciable, even though they should theoretically exist.

Other loose ends are occasional experimental indications of deviations from the $m^{-1/2}$ dependence of Graham's diffusion law, and here and there an apparently spectacular failure like the diffusion pressure effect in Ar–CO_2, which starts out normally at very low pressures but reverses sign in the transition region and ends up backwards in the continuum region (Waldmann and Schmitt, 1961). This is suspected to be an effect that appears in the dusty-gas theory through the Δ_{ij} correction terms for the diffusion coefficients (Zhdanov, 1968; Annis and Mason, 1970).

V. Determination of Diffusion Coefficients

The purpose of this section is to describe the principal experimental methods for the determination of gaseous diffusion coefficients, and to indicate some of their limitations and uncertainties. Emphasis is given to the methods that have produced the most reliable results. It is doubtful whether any diffusion coefficient has been determined with an accuracy better than 1%, although reproducibility in a given apparatus may be better. In the vast majority of work, an accuracy level of about 2% is considered commendable, and even this is not easy to achieve. This level of reliability implies that accurate diffusion measurements are relatively difficult, even with the best of modern instrumentation.

A few limited surveys of experimental methods have appeared previously (Jost, 1952; Present, 1958; Waldmann, 1958; Westenberg, 1966). A more recent comprehensive survey under the U. S. National Standard Reference Data System has been prepared (Marrero and Mason, 1971), upon which this section and the next are largely based.

A. Outline of Methods

Figure 8 presents the main methods that have been used to yield reliable gaseous diffusion coefficients, classified according to the overall apparatus

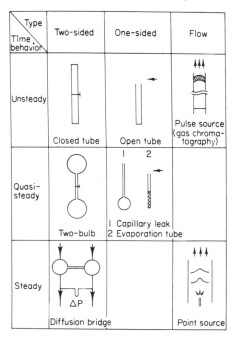

Fig. 8. Major experimental methods for diffusion coefficients.

geometry and the time behavior of the diffusion process. The methods indicated under the first two columns are without carrier gas flow in the part of the apparatus in which diffusion occurs. In the methods under the third column diffusion occurs within a flowing gas stream.

Table III lists the experimental methods as an outline of the descriptions that follow. The major methods are those that have been most frequently

TABLE III

CLASSIFICATION OF EXPERIMENTAL METHODS

Name	Primary Investigator(s)	Reliability
Major		
Closed tube	Loschmidt (1870a, b)	Good
Evaporation tube	Stefan (1873)	Poor
Two-bulb apparatus	Ney and Armistead (1947)	Good
Point source	Walker and Westenberg (1958a)	Average
Gas chromatography	Giddings and Seager (1960)	Average
Minor		
Open tube	von Obermayer (1882); Waitz (1882)	Average
Back diffusion	Harteck and Schmidt (1933)	Average
Capillary leak	Klibanova *et al.* (1942)	Poor
Unsteady evaporation	Arnold (1944)	Fair
Diffusion bridge	Bendt (1958)	Average
Dissociated gases	Wise (1959); Krongelb and Strandberg (1959)	Poor
Miscellaneous		
Droplet evaporation	Langmuir (1918); Katan (1969)	?
Dufour effect	Waldmann (1944)	?
Thermal separation rate	Nettley (1954)	?
Kirkendall effect	McCarty and Mason (1960)	?
Sound absorption	Holmes and Tempest (1960)	?
Cataphoresis	Hogervorst and Freudenthal (1967)	?
Resonance methods	See text	?

employed; they are also often, but not always, the methods that have in practice yielded the most reliable results. Each of the minor methods has been used only a few times, but might well be developed into a major method with further work. A number of miscellaneous methods, worth noting for their general applicability or their experimental ingenuity, are also listed; this list does not pretend to be complete, since it seems both futile and boring to enumerate all the miscellaneous methods by which a diffusion coefficient has been determined at some time or other. Not listed in Table III are the indirect methods, in which a diffusion coefficient is extracted from some other measurement with the aid of kinetic theory. These methods are discussed at the end of this section.

Historically, the classic techniques are the closed-tube and the evaporation-tube methods. With a few minor exceptions, all diffusion coefficients measured prior to World War II were obtained by one of these two methods, and covered only a very limited temperature range. During and after World War II there was a great revival of interest in diffusion, inspired in part by interest in isotope separation and by problems involving high temperatures in flames. The ready availability of isotopic tracers was also a great stimulus, and made so-called self-diffusion coefficients experimentally accessible. It was also gradually recognized that diffusion coefficients could be a good source of information on forces between unlike molecules, provided they could be measured over a substantial temperature range. Important new work was done with two-bulb, point-source, gas-chromatographic, and diffusion-bridge apparatus, as well as with dissociated gases. Many miscellaneous methods, often quite ingenious, have been tried but not used extensively. The existence of an accurate kinetic-molecular theory has also made possible several indirect methods, in which diffusion coefficients are deduced from measurements of other properties that at first sight seem to have no particular connection with diffusion.

B. CLOSED TUBE

This method is usually associated with the name of Loschmidt (1870a,b), who is surely entitled to major credit for its development and exploitation. However, as was remarked in Section I, the invention of the method should really be credited to Graham (1863). Loschmidt covered a maximum temperature range of about $-20°$ to $20°$C. During the 1880's von Obermayer (1887; and earlier papers referred to therein) studied many systems up to a maximum temperature of about $60°$C. However, many of his results are, for some unknown reason, systematically low by about 5%. In the early 1900's a series of investigations was carried out at the University of Halle to test the composition dependence of $\mathscr{D}_{12}$, a crucial theoretical point; this work is summarized by Lonius (1909). Boardman and Wild (1937) and Coward and Georgeson (1937) devised a rotating-plate version of the apparatus, which has been much copied. Since the war this type of apparatus has been adapted for continuous composition analysis by means of radioactive tracers (Amdur and Beatty, 1965; and earlier papers by Amdur and coworkers referred to therein) and index of refraction (Boyd *et al.*, 1951). Ivakin *et al.* (1968; and earlier papers) have coupled this technique with a Schlieren analysis that gives a continuous reading of the composition *gradient*. The lowest temperature attained with a closed-tube apparatus is $-78°$C, and the highest is $200°$C.

Reproducibility is often better than 1%, but absolute accuracies are no better than 2%; a major effort involving careful variation of many experimental parameters would probably be necessary to achieve much improvement.

C. EVAPORATION TUBE

The rate of evaporation of a liquid placed in the bottom of a long tube is controlled by the diffusion of the vapor through the surrounding gas, and can be used to measure the vapor–gas diffusion coefficient (Stefan, 1873). The method has been very widely used, but is restricted to volatile liquids and limited temperature ranges. Good modern accounts of the method have been given by Schwertz and Brow (1951) and by Lee and Wilke (1954), among others. The accuracy achieved has been generally disappointing.

D. TWO-BULB APPARATUS

A two-bulb configuration was first used by Graham (1829), and its mathematical analysis was given by Maxwell (1860). It was re-invented by Ney and Armistead (1947), who measured the self-diffusion coefficient of UF_6. It is easily adapted to cover large temperature ranges (Winn, 1950), and with moderate care gives a level of accuracy about the same as the closed-tube apparatus. It has been widely used in recent years; two recent papers, from which earlier work can be traced, are by Malinauskas and Silverman (1969) and by Srivastava and Saran (1966). Especially noteworthy work with this type of apparatus has been carried out by van Heijningen *et al.* (1966, 1968). By paying meticulous attention to details, they have apparently obtained the best absolute accuracy to date (1%), over a maximum temperature range from 65° to 400°K.

E. POINT SOURCE

This flow method was devised by Walker and Westenberg (1958a) to measure diffusion coefficients at high temperatures. One component is introduced through a fine hypodermic tube (the point source) into a slowly flowing stream of the other component, and spreads by diffusion as it moves downstream (see Fig. 8). The diffusion coefficient can be determined by sampling the gas with a fine probe and measuring its composition at different downstream positions. Temperatures up to 1200°K were obtained without great difficulty, and diffusion coefficients were determined for a large number of important gas pairs (Walker and Westenberg, 1958b, 1959, 1960; Westenberg and Frazier, 1962). Temperatures up to 1800°K have been obtained by generating the flowing stream from the burned gas of a flat flame, stabilized above a porous plate (Pakurar and Ferron, 1966; and earlier papers). The reproducibility of the point-source method is 1–2%, and absolute accuracy probably better than 5%.

F. GAS CHROMATOGRAPHY

This is another flow method, in which a pulse of one component is injected into a stream of the other component flowing through a long tube (i.e., an unpacked gas chromatograph column). The diffusion coefficient can be determined by measuring the amount of spreading of the gas pulse as it emerges from the end of the column (see Fig. 8). The theory having been available for several years, four different groups independently published experimental investigations almost simultaneously: Giddings and Seager (1960), Boheman and Purnell (1961), Bournia *et al.* (1961), and Fejes and Czárán (1961). Of these groups, only Giddings has continued to publish diffusion measurements (Giddings and Mallik, 1967; and previous papers). The method is readily adapted to routine operation (Giddings and Mallik, 1967), and to fairly wide temperature ranges (Wasik and McCulloh, 1969). Reproducibility and accuracy are comparable to that of the point-source technique.

G. MINOR METHODS

Here we briefly discuss six minor methods, in the chronological order of their development. Some are probably worth further exploitation.

1. Open Tube

If the top of a cylindrical container of gas is opened, the gas will diffuse into the surrounding atmosphere; analysis of the composition of the remaining gas after a known time permits the diffusion coefficient to be determined. The theory of the method was given by Stefan (1871), but the first measurements were made independently by von Obermayer (1882) and by Waitz (1882). The method has recently been revived and improved by Frost (1967).

2. Back Diffusion

The interdiffusion of ortho- and parahydrogen was measured down to $20°K$ by an ingenious steady-state flow method in which one component diffuses upstream against the second flowing component (Harteck and Schmidt, 1933). The composition at one or more upstream points can be used to determine the diffusion coefficient. A good description is given by

Jost (1952). It is surprising that this method has not been widely adopted for diffusion measurements at extreme temperatures, either low or high.

3. Capillary Leak

The configuration for the capillary-leak apparatus (see Fig. 8) is simply an adaptation of the method first used by Graham (1829), mentioned in Section I. Having no moving parts, it is suitable for measurements at extreme temperatures; the first diffusion coefficients at high temperatures were in fact determined in this way (Klibanova et al., 1942). Further high-temperature work has been carried out by Kosov and coworkers, who have unfortunately published their results in obscure places (Kosov, 1957; Vyshenskaya and Kosov, 1961; Kosov and Karpushin, 1966). The method has been used at low temperatures by De Paz et al. (1967), who let the gas diffusing out of the capillary enter directly into a mass spectrometer for analysis. Both the reproducibility and the accuracy at high temperatures have been poor, but the method should be capable of much improvement.

4. Unsteady Evaporation

An alternate evaporation-tube method (Section V, C) based on the unsteady-state vaporization of a liquid into a gas was developed by Arnold (1944). It has been used largely by chemical engineers; full accounts are given by Fairbanks and Wilke (1950) and by Nelson (1956). Diffusion coefficients determined by this method have been more precise and were obtained in less time than those determined by the steady-state vaporization method. A somewhat similar technique was used earlier by Mackenzie and Melville (1932, 1933) with bromine vapor.

5. Diffusion Bridge

This is a steady-state flow method, but is really Graham's classic work all over again, because it is usually operated at uniform total pressure. As shown in Fig. 8, two gas streams flow across opposite ends of a capillary tube or opposite faces of a porous septum, and the emerging streams are analyzed. The flow rates can be controlled by valves in the lines, and adjusted to produce any desired pressure difference across the capillary or septum. This technique was first used by Wicke and Kallenbach (1941), and was later used in the rediscovery of Graham's diffusion law (Hoogschagen, 1953, 1955). It has been widely used to study gaseous diffusion in porous media; almost everyone who uses it sooner or later rediscovers, but does not recognize, Graham's law of diffusion (Mason and Kronstadt, 1967). It has been used only once with a capillary, to obtain absolute values of diffusion coefficients down to very low temperatures (Bendt, 1958). The method is readily adaptable to operation over wide temperature ranges, since it is a steady-state method without moving mechanical parts.

6. Dissociated Gases

Direct measurements of the diffusion of highly reactive species, such as free radicals and valence-unsaturated atoms, are very difficult but are needed for basic understanding of many phenomena in chemical reactions and at high temperatures. Wise (1959) and Krongelb and Strandberg (1959) were the first to make such measurements, and a variety of direct techniques have now been used to measure the diffusion of H, N, and O atoms in different gases. Recent papers, containing references to earlier work, are by Young (1961), Walker (1961), Yolles and Wise (1968), and Khouw *et al.* (1969). As might be expected, scatter and consistency are not too good (10% or more in many cases). The best results, in fact, have been obtained by indirect methods such as mixture viscosities (Browning and Fox, 1964) and combined molecular-beam scattering and semiempirical quantal calculations (Yun *et al.*, 1962).

H. MISCELLANEOUS METHODS

Here are recorded a few of the more noteworthy miscellaneous methods that have on occasion been used to measure a diffusion coefficient.

1. Droplet Evaporation

Observations of the rate of evaporation of a small sphere of volatile material can be used to determine the vapor–gas diffusion coefficient (Langmuir, 1918). The applicable theory is similar to that for evaporation in a tube. A recent paper containing references to earlier work is by Bradley and Waghorn (1951).

A recent droplet-evaporation method proposed by Katan (1969) is a clever combination of Graham's uniform-pressure experiment and Stefan's evaporation technique. A bead of volatile liquid is placed in a long glass tube with a porous membrane at one end, and the bead is driven down the tube by the accumulation of vapor between the bead and the membrane. The vapor and outside gas interdiffuse through the membrane, as in Graham's experiment, and the motion of the bead can be used to determine the diffusion coefficient.

2. Dufour Effect

When two gases interdiffuse, a small transient temperature gradient is set up; this is called the Dufour effect or the diffusion thermoeffect. The asymptotic time decay of the temperature gradient can be used to determine the diffusion coefficient (Waldmann, 1944). The method has not been extensively used (Waldmann, 1947, and previous papers; Mason *et al.* 1967c).

3. Rate of Thermal Separation

The inverse of the Dufour effect is thermal diffusion, in which an imposed temperature gradient causes a composition difference to occur. The speed with which an initially uniform gas mixture separates under an imposed temperature gradient can also be used to determine the diffusion coefficient. The first quantitative experiments of this type were performed by Nettley (1954). The method has been reviewed in connection with thermal diffusion (Mason *et al.*, 1966).

4. Kirkendall Effect

In solids, the net drift of inert markers placed near a diffusion interface is called the Kirkendall effect. A similar effect exists in gases (Miller and Carman, 1960; McCarty and Mason, 1960; Mason, 1961), and the speed of the marker motion can be used to determine the diffusion coefficient. Although it was not realized at the time, the Kirkendall effect in gases is analogous to the rise or fall of the water level in Graham's diffusion experiments, which can also be used to determine diffusion coefficients (Evans *et al.*, 1969).

As elaborated in Section II, the diffusion pressure effect is closely related to uniform-pressure diffusion. It could therefore also be used to determine diffusion coefficients, but this has not yet been done.

5. Sound Absorption

The passage of a sound wave through a gas mixture produces a local partial separation of the components, caused mostly by pressure diffusion. The remixing by diffusion is out of phase with the sound wave, and thus furnishes a mechanism for absorption of sound energy. In other words, the absorption of an ultrasonic wave in a gas mixture is stronger than in either pure component, and the excess absorption depends on the diffusion coefficient, which could in principle thus be determined (Holmes and Tempest, 1960). This method has been suggested for use at very high temperatures, where most other methods fail (Carnevale *et al.*, 1967).

6. Cataphoresis

A dc discharge in a gas mixture causes a partial separation of components. The phenomenon, which also occurs in solutions, is called cataphoresis. The separation disappears by diffusion after the discharge is stopped, and the diffusion coefficient can be calculated from the rate of disappearance of the separation. Hogervorst and Freudenthal (1967) have followed this disappearance mass spectrometrically in Ne–Ar mixtures from 300° to 650°K, and obtained results in good agreement with results obtained by other methods.

7. Resonance Methods

The principle of all resonance methods is to "tag" some of the molecules in a gas or gas mixture, and then follow their dispersion due to diffusion. The tags used have been such things as the orientation of a nuclear spin (Lipsicas, 1962; Luszczynski et al., 1962), the population of magnetic sublevels in the ground state (Franzen, 1959), or a metastable excited electronic state (McCoubrey and Matland, 1956). The literature on such methods is somewhat scattered, but most of it can be traced through the foregoing references.

I. INDIRECT METHODS

1. Mixture Viscosity and Thermal Conductivity

Kinetic theory shows that the viscosity of a gas mixture depends on the composition, the viscosities and molecular weights of the pure components, and two mixture quantities. One of these is $\mathcal{D}_{12}$, and the other is the collision-integral ratio A^*_{12}, which depends only weakly on temperature and intermolecular forces. For simple gases it is fairly easy to get an independent theoretical estimate of A^*_{12}, and then $\mathcal{D}_{12}$ can be calculated by simple algebra if the viscosities are known (Weissman and Mason, 1962). Errors due to neglect of theoretical higher correction terms are quite small (Storvick and Mason, 1966), and the method gives very satisfactory values of $\mathcal{D}_{12}$ if the viscosities are measured with high accuracy (Kestin and Yata, 1968; and previous papers). As mentioned in Section III, E, the procedure is valid in principle even for polyatomic gases, although a reliable estimate of A^*_{12} is more difficult to obtain (Weissman, 1964).

A similar procedure should also work for thermal conductivity, but is less satisfactory because of the lower accuracy of the starting data (Weissman, 1965). Moreover, it is limited to monatomic gases, for the internal degrees of freedom and inelastic collisions affect the thermal conductivity much more than they affect the viscosity.

2. Thermal Diffusion

Kinetic theory also reveals several nonobvious relations between $\mathscr{D}_{12}$ and the thermal diffusion factor, which describes how a gas mixture separates under the influence of a temperature gradient. Measurements of the rather strong composition dependence of the thermal diffusion factor can be used to determine absolute values of $\mathscr{D}_{12}$ (Mason and Smith, 1966). The accuracy is only fair, however, because of the uncertainties in the measurements. A better method relates the temperature dependence of $\mathscr{D}_{12}$ to that of the thermal diffusion factor. A single isothermal measurement of $\mathscr{D}_{12}$ can thus be combined with thermal diffusion measurements to produce diffusion coefficients over a wide temperature range (Annis *et al.*, 1968; Humphreys and Mason, 1970). The final accuracy is good, because rather large uncertainties in the measurements appear only as much smaller uncertainties in the calculated diffusion coefficients.

3. Intermolecular Forces

A lengthy discussion was given in Section III, K on the calculation of $\mathscr{D}_{12}$ from intermolecular forces. At ordinary temperatures the best direct sources of information on forces between unlike molecules involve $\mathscr{D}_{12}$, and the method is useless, except for making estimates on gas pairs for which no direct measurements happen to be available. However, the short-range forces can be measured by the scattering of fast molecular beams (a comprehensive review has been given by Amdur and Jordan, 1966). This furnishes the basis for calculation of diffusion coefficients at very high temperatures (Amdur and Mason, 1958), where direct experiments are still nonexistent. Agreement between calculation and direct experiment is excellent in the small temperature range where the two overlap (Walker and Westenberg, 1959). Moreover, the long-range dispersion forces are accurately known for many systems through a combination of quantum-mechanical theory and experimental data, mostly optical and electrical (see Dalgarno, 1967, for a review). These would permit the calculation of diffusion coefficients at very low temperatures, where measurements are scarce.

VI. Experimental Results

In this section we attempt to summarize the most reliable experimental results on binary gaseous diffusion coefficients. Systems for which only limited or uncertain data exist are not included in this summary. A reader who wishes to track down experimental measurements for a system not listed in this section, or who wishes to consult the original experimental papers on a system that is listed, can locate all the pertinent references in the review of Marrero and Mason (1971), on which this section is based.

A. UNCERTAINTY LIMITS

The assignment of limits of uncertainty to experimental data always involves a large measure of subjective judgment. One does his best with such things as reproducibility and internal consistency, external consistency for different types of measuring apparatus, different workers in different laboratories, and so on; the final decisions are nevertheless based heavily on the judgments of the evaluators. We have tried to be conservative, in order that there shall be a high probability that the "true" value of a diffusion coefficient lies within the uncertainty range assigned. We have also tried to be fair and not arbitrarily downgrade good measurements, but it is quite possible that a particular $\mathscr{D}_{12}$ may be more accurate than is implied by our quoted uncertainty limits.

The reader interested in the judgments made on particular systems must consult the detailed review of Marrero and Mason (1971).

The gas pairs included can be grouped into three main categories, as shown in Fig. 9. That is, a system in Group I has an uncertainty of only $\pm 1\%$ in $\mathscr{D}_{12}$ at 300°K; the uncertainty rises to $\pm 5\%$ at 1000° K, and to $\pm 10\%$ at 10,000° K, where the results are based on molecular beam experiments. The

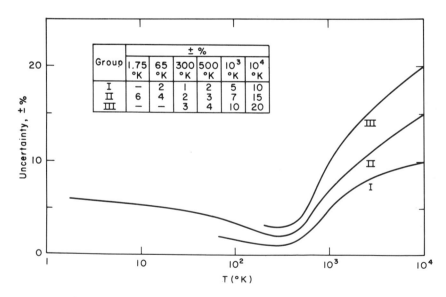

FIG. 9. Estimated uncertainty limits of diffusion coefficients as a function of temperature.

temperature dependences of the uncertainty limits are also shown in Fig. 9. Table IV lists the systems assigned to each group; borderline systems are assigned to the higher group, but are noted by a question mark. The listing is given in terms of one common member in a series of gas pairs; this leads to some duplication, but is helpful for quick reference. Also included is a group of miscellaneous systems that do not qualify for inclusion in the summary on the basis of reliability, but have been included because of their possible special interest; they include systems with dissociated gases, water vapor, or carbon dioxide. Table V lists the estimated uncertainty limits for systems of this miscellaneous group.

The sources of reliable data for the gas pairs of Groups I–III are roughly as follows. For all gas pairs the most accurate results are at approximately 300° K, because of the existence of a large number of independent measurements by the most reliable experimental methods. Usually both closed-tube and two-bulb measurements are available over the range of approximately 200°

TABLE IV

GROUPING OF GAS PAIRS ACCORDING TO UNCERTAINTY LIMITS
OF DIFFUSION COEFFICIENTS

Group I	$He–(Ne, Ar, Kr, Xe)$
	$Ne–(He, Ar, Kr, Xe)$
	$Ar–(He, Ne, Kr, Xe)$
	$Kr–(He, Ne, Ar, Xe?)$
	$H_2–N_2?$
Group II	$^3He–^4He$
	$He–(H_2, N_2, CO, O_2?, air, CO_2)$
	$H_2–(He, Ne?, Ar, Kr?, D_2, CO, air, CO_2)$
	$N_2–(Ar?, CO, CO_2)$
Group III	$Ar–(CH_4, CO, O_2, air, CO_2, SF_6)$
	$H_2–(Xe, CH_4, O_2, SF_6)$
	$CH_4–(He, Ar, H_2, N_2, O_2, air, SF_6)$
	$N_2–(Ne, Kr, Xe, CH_4, O_2, SF_6)$
	$CO–(Ar, Kr, O_2, air, CO_2, SF_6)$
	$O_2–(Ar, H_2, CH_4, N_2, CO, CO_2, SF_6)$
	$CO_2–(Ar, CO, O_2, air, N_2O, SF_6)$
	$SF_6–(He, Ar, H_2, CH_4, N_2, CO, O_2, air, CO_2)$
Misc.	$H–(He, Ar, H_2)$
	$N–N_2$
	$O–(He, Ar, N_2, O_2)$
	$H_2O–(N_2, O_2, air, CO_2)$
	$CO_2–(Ne, H_2O, C_3H_8).$

TABLE V

UNCERTAINTY LIMITS FOR GAS PAIRS
OF THE MISCELLANEOUS GROUP

System	$T(^\circ K)$	Uncertainty ($\pm \%$)
H–H$_2$	~300	5
	>1000	30
N–N$_2$, O–N$_2$, O–O$_2$	~300	10
	>1000	25
H–He, H–Ar, ⎫	~300	15
O–He, O–Ar ⎬	>1000	30
H$_2$O–N$_2$	282–373	4
H$_2$O–O$_2$	282–1070	7
H$_2$O–air	282–1070	5–10
H$_2$O–CO$_2$	296–1640	10–7
CO$_2$–Ne	195–625	3–5
CO$_2$–C$_3$H$_8$	298–550	3–5

to 500° K, with a few two-bulb measurements available at lower temperatures. Temperature limits have been extended in both directions by use of data on mixture viscosities or, in a few instances, thermal diffusion factors. For several gas pairs data are available to about 1000° K from the point-source technique. From 1000° to 10,000° K the results are derived largely from molecular beam measurements.

B. CORRELATION OF TEMPERATURE DEPENDENCE

The diffusion coefficients for all seventy-four of the gas pairs can be correlated by an empirical equation of the form discussed in Section III, J, 4, which was

$$\ln(p\mathscr{D}_{12}) = \ln A + s \ln T - \ln[\ln(\varphi_o/kT)]^2 - (S/T) - (S'/T^2). \quad (135)$$

In many cases, however, the data are not sufficiently precise to require the use of the double logarithm term, and an adequate representation is given by

$$\ln(p\mathscr{D}_{12}) = \ln A + s \ln T - (S/T). \quad (136)$$

The parameters for systems described by Eq. (135) are given in Table VI, and for those described by Eq. (136) in Table VII. It should be remarked that these parameters were obtained not by a mindless least-squares fitting of all available data, but are based on a critical evaluation and selection of what were thought to be the most reliable results. For details the original compilation must be consulted (Marrero and Mason, 1971). All results in Tables VI

TABLE VI

CORRELATION PARAMETERS OF EQ. (135) FOR DIFFUSION COEFFICIENTS

System	$10^3 A$ atm-cm^2 / sec($^\circ$K)^s	s	$10^{-8}\,\varphi_o/k$ ($^\circ$K)	S ($^\circ$K)	S' ($^\circ$K)^2	T ($^\circ$K)	Group
^{3}He–^{4}He	32.4	1.501	0.0448	−0.9630	1.894	1.74–10^4	II
He–Ne	25.41	1.509	0.212	1.87	—	65–10^4	I
He–Ar	15.21	1.552	0.410	1.71	—	77–10^4	I
He–Kr	10.61	1.609	1.42	−32.65	2036.	77–10^4	I
He–Xe	7.981	1.644	4.02	−68.87	5416.	169–10^4	I
He–H$_2$	27.0	1.510	0.0534	—	—	90–10^4	II
He–N$_2$	15.8	1.524	0.265	—	—	77–10^4	II
He–CO	15.8	1.524	0.265	—	—	77–10^4	II
Ne–Ar	8.779	1.546	1.94	1.82	1170.	90–10^4	I
Ne–Kr	8.520	1.555	6.73	20.4	—	112–10^4	I
Ne–Xe	6.747	1.584	19.0	10.1	—	169–10^4	I
Ar–Kr	5.346	1.556	13.0	47.3	—	169–10^4	I
Ar–Xe	5.000	1.563	36.8	59.9	—	169–10^4	I
Ar–H$_2$	23.5	1.519	0.488	39.8	—	242–10^4	II
Kr–Xe	2.933	1.608	128	52.7	—	169–10^4	I
Kr–H$_2$	18.2	1.564	1.69	26.4	—	77–10^4	II
H$_2$–D$_2$	24.7	1.500	0.0636	6.072	38.10	14–10^4	II
H$_2$–N$_2$	15.39	1.548	0.316	−2.80	1067.	65–10^4	I
H$_2$–CO	15.39	1.548	0.316	−2.80	1067.	65–10^4	II
N$_2$–CO	4.40	1.576	1.57	−36.2	3825.	78–10^4	II

and VII have been adjusted to refer to equimolar mixtures, except for mixtures involving air, which refer to trace diffusion through a large excess of air. It is not strictly correct to treat air as a single species, but is a very accurate approximation. Most of the diffusion coefficients involving air in Table VII were generated from the corresponding diffusion coefficients for N$_2$ and O$_2$ according to Blanc's law (Section III, G); even when direct measurements were available, they were always cross checked with Blanc's law.

The order of listing in Table VI and VII is as follows: (i) mixtures of noble gases with noble gases, arranged according to the atomic weight of the lighter component; (ii) mixtures of noble gases with other gases arranged according to the atomic weight of the noble gas component; (iii) other mixtures, arranged according to the molecular weight of the lighter component; (iv) dissociated gases. Except for ^{3}He–^{4}He and H$_2$–D$_2$, isotopic mixtures are not included, since the self-diffusion coefficient is merely proportional to the viscosity.

The results given in Tables VI and VII cannot be extrapolated to low temperatures, for the form of Eqs. (135) and (136) is unsuitable when the

TABLE VII

CORRELATION PARAMETERS OF EQ. (136) FOR DIFFUSION COEFFICIENTS

System	$10^5 A$ atm-cm^2 / sec($°$K)^s	s	S ($°$K)	T ($°$K)	Group
He–CH$_4$	3.13	1.750	—	298–10^4	III
He–O$_2$	4.37	1.710	—	244–10^4	II
He–air	3.78	1.729	—	244–10^4	II
He–CO$_2$	3.31	1.720	—	200–530	II
He–SF$_6$	3.87	1.627	—	290 10^4	III
Ne–H$_2$	5.95	1.731	—	90–10^4	II
Ne–N$_2$	1.59	1.743	—	293–10^4	III
Ne–CO$_2$	1.07	1.776	—	195–625	misc.
Ar–CH$_4$	0.784	1.785	—	307–10^4	III
Ar–N$_2$	0.904	1.752	—	244–10^4	II
Ar–CO	0.904	1.752	—	244–10^4	III
Ar–O$_2$	0.977	1.736	—	243–10^4	III
Ar–air	0.917	1.749	—	244–10^4	III
Ar–CO$_2$	1.74	1.646	89.1	276–1800	III
Ar–SF$_6$	1.48	1.596	145.4	328–10^4	III
Kr–N$_2$	0.653	1.766	—	248–10^4	III
Kr–CO	0.653	1.766	—	248–10^4	III
Xe–H$_2$	3.68	1.712	16.9	242–10^4	III
Xe–N$_2$	0.470	1.789	—	242–10^4	III
H$_2$–CH$_4$	3.13	1.765	—	293–10^4	III
H$_2$–O$_2$	4.17	1.732	—	252–10^4	III
H$_2$–air	3.64	1.750	—	252–10^4	II
H$_2$–CO$_2$	3.14	1.750	11.7	200–550	II
H$_2$–SF$_6$	7.82	1.570	102.3	298–10^4	III
CH$_4$–N$_2$	1.00	1.750	—	298–10^4	III
CH$_4$–O$_2$	1.68	1.695	44.2	294–10^4	III
CH$_4$–air	1.03	1.747	—	298–10^4	III
CH$_4$–SF$_6$	1.10	1.657	69.2	298–10^4	III
N$_2$–O$_2$	1.13	1.724	—	285–10^4	III
N$_2$–H$_2$O	0.187	2.072	—	282–373	misc.
N$_2$–CO$_2$	3.15	1.570	113.6	288–1800	II
N$_2$–SF$_6$	1.66	1.590	119.4	328–10^4	III
CO–O$_2$	1.13	1.724	—	285–10^4	III
CO–air	1.12	1.730	—	285–10^4	III
CO–CO$_2$	0.577	1.803	—	282–473	III
CO–SF$_6$	1.76	1.584	139.4	297–10^4	III
O$_2$–H$_2$O	0.189	2.072	—	282–450	misc.
	2.78	1.632	—	450–1070	misc.
O$_2$–CO$_2$	1.56	1.661	61.3	287–1083	III
O$_2$–SF$_6$	2.65	1.522	129.0	297–10^4	III
air–H$_2$O	0.187	2.072	—	282–550	misc.
	2.75	1.632	—	450–1070	misc.
air–CO$_2$	2.70	1.590	102.1	280–1800	III
air–SF$_6$	1.83	1.576	121.1	328–10^4	III
H$_2$O–CO$_2$	9.24	1.500	307.9	296–1640	misc.
CO$_2$–N$_2$O	0.281	1.866	—	195–550	III
CO$_2$–C$_3$H$_8$	0.177	1.896	—	298–550	misc.
CO$_2$–SF$_6$	0.140	1.886	—	328–472	III
H–He	14.2	1.732	—	275–10^4	misc.
H–Ar	1.45	1.597	—	275–10^4	misc.
H–H$_2$	11.3	1.728	—	190–10^4	misc.
N–N$_2$	1.32	1.774	—	280–10^4	misc.
O–He	4.68	1.749	—	280–10^4	misc.
O–Ar	0.751	1.841	—	280–10^4	misc.
O–N$_2$	1.32	1.774	—	280–10^4	misc.
O–O$_2$	1.32	1.774	—	280–10^4	misc.

long-range London dispersion energy dominates the interaction (Section III, J, 4). In such a case the diffusion coefficient has the asymptotic form (classically),

$$p\mathscr{D}_{12} = AT^{11/6}, \tag{137}$$

as discussed in Section III, J, 2. The constant A is easily calculated if the London constant C is known. Values are given in Table VIII for a number of the gas pairs of Tables VI and VII for which C is known (Dalgarno, 1967). The uncertainty in A caused by uncertainty in C should not exceed 3% and is usually much smaller, with the possible exception of mixtures involving Xe.

TABLE VIII

CLASSICAL LOW-TEMPERATURE ASYMPTOTIC DIFFUSION
COEFFICIENTS, EQ. (137)

System	$\dfrac{10^6 A}{\text{sec}(°\text{K})^{11/6}}$ atm-cm^2	$\dfrac{c^a}{e^2 a_o^5}$	Λ^{*b}	ε/k^b (°K)
He–Ne	31.2	3.0	1.32	23.7
He–Ar	20.3	9.6	0.86	40.2
He–Kr	17.9	13	0.80	39.0
He–Xe	15.6	19	0.68	46.5
He–CH$_4$	19.0	14	0.89	37
He–N$_2$	20.4	10	0.96	31
Ne–Ar	8.26	20	0.35	61.7
Ne–Kr	6.79	27	0.28	69.8
Ne–Xe	5.84	38	0.26	69.1
Ne–H$_2$	30.1	8.2	1.35	34
Ne–N$_2$	8.69	21	0.37	57
Ar–Kr	3.51	91	0.14	145
Ar–Xe	2.93	130	0.11	178
Ar–H$_2$	19.5	28	0.87	64
Ar–CH$_4$	5.27	98	0.22	130
Ar–N$_2$	4.93	69	0.21	107
Kr–Xe	2.00	190	0.08	197
Kr–H$_2$	17.1	40	0.75	80
Kr–N$_2$	3.91	96	0.16	132
Xe–H$_2$	15.1	58	0.67	87
Xe–N$_2$	3.29	140	0.14	145
H$_2$–CH$_4$	17.5	43	0.82	68
H$_2$–N$_2$	19.3	30	0.87	62.9
CH$_4$–N$_2$	5.54	100	0.23	120

[a] Dalgarno (1967).

[b] Based on the 12–6 potential; parameters for noble gas pairs and for H$_2$–N$_2$ from van Heijningen et al. (1966, 1968), and for other gas pairs from Hirschfelder et al. (1954).

The range of validity of the results in Table VIII is difficult to estimate accurately, but an approximate range may be found as follows. The high-temperature limit is set by the fact that the dispersion energy ceases to dominate; from Fig. 7 we would estimate that Table VIII can be used only for $kT/\varepsilon < 0.2$. The low-temperature limit is set by quantum effects. As discussed in Section III, D and illustrated in Table I and Figs. 5 and 6, quantum effects are quite large at $T^* < 0.2$ for systems with large values of Λ^*. Thus there is no hope of applying Eq. (137) to systems such as ^3He–^4He, He–H_2, and H_2–D_2. Numerical calculations for a 12–6 potential (Munn *et al.*, 1965a) in fact suggest that Eq. (137) is of only qualitative value for $\Lambda^* > 1$ and $T^* < 0.2$, and even for $\Lambda^* = 1$ is useful only to a 10% level of accuracy down to $T^* = 0.1$. However, for $\Lambda^* = 0.5$, Eq. (137) is probably accurate within about 3% down to $T^* = 0.02$, and of course is even better for $\Lambda^* < 0.5$. Thus the quantum deviations from Eq. (137) depend very strongly on Λ^* and T^* in a complex fashion. No simple rule for the low-temperature limit of usefulness of Eq. (137) seems possible; in cases of doubt it is probably best to consult the available numerical calculations for the 12–6 potential.

C. CORRELATION OF COMPOSITION DEPENDENCE

The weak composition dependence of $\mathscr{D}_{12}$ is readily correlated by the empirical equation discussed in Section III, I, which was

$$\Delta_{12} \approx \zeta(6C_{12}^* - 5)^2[ax_1/(1 + cx_1)], \tag{138}$$

where x_1 is the mole fraction of the heavier species.

Values of the parameters of Eq. (138) are given in Table IX, using the same order of listing as used for Tables VI and VII. A number of systems in which D_2 replaces H_2 are also included. There are only fourteen systems (not counting D_2 systems) for which enough accurate experimental data exist to justify assigning the parameter ζ a value other than 1.0. These are the 10 noble gas pairs (van Heijningen *et al.*, 1968), H_2–N_2 (van Heijningen *et al.*, 1966) and He–N_2, Ar–H_2, and H_2–CO_2 (Mason *et al.*, 1964). The value of C_{12}^* in Eq. (138) is to be calculated for the 12–6 potential with the values of ε_{12}/k as given in Table IX; the results are not too sensitive to the precise value of ε_{12}/k used. Several systems are omitted from Table IX because the molecular weights of the gases are so close that the composition dependence is negligible: He–D_2, Ar–CO_2, N_2–CO, N_2–O_2, CO–O_2, CO–air, CO_2–N_2O, and CO_2–C_3H_8.

Table IX is convenient for making rapid estimates of the composition dependence of $\mathscr{D}_{12}$, reliable to within the uncertainties of experimental measurement.

TABLE IX

CORRELATION PARAMETERS FOR THE COMPOSITION DEPENDENCE OF $\mathscr{D}_{12}$
ACCORDING TO EQ. (138)

System	ζ	ε/k [a] (°K)	a	c	System	ζ	ε/k [a] (°K)	a	c
^{3}He–^{4}He	1.0	10.2	0.031	0.26	Xe–H$_2$	1.0	87	0.25	1.53
He–Ne	1.64	23.7	0.098	0.45	Xe–D$_2$	1.0	87	0.23	1.43
He–Ar	1.67	40.2	0.18	1.17	Xe–N$_2$	1.0	145	0.10	0.56
He–Kr	1.65	39.0	0.23	1.56	H$_2$–D$_2$	1.0	33	0.042	0.12
He–Xe	1.78	46.5	0.29	2.08	H$_2$–CH$_4$	1.0	68	0.15	0.94
He–H$_2$	1.0	18.4	0.033	−0.11	H$_2$–N$_2$	1.00	62.9	0.17	0.89
He–CH$_4$	1.0	37	0.14	0.25	H$_2$–CO	1.0	61	0.16	0.88
He–N$_2$	1.80	31	0.17	1.22	H$_2$–O$_2$	1.0	61	0.16	0.81
He–CO	1.0	34	0.16	1.19	H$_2$–air	1.0	57	0.16	0.87
He–O$_2$	1.0	34	0.17	1.11	H$_2$–CO$_2$	1.84	80	0.21	1.33
He–air	1.0	31	0.17	1.19	H$_2$–SF$_6$	1.0	93	0.33	2.33
He–CO$_2$	1.0	44	0.23	1.74	D$_2$–CH$_4$	1.0	68	0.11	0.81
He–SF$_6$	1.0	51	0.39	3.09	D$_2$–N$_2$	1.00	62.9	0.13	0.76
Ne–Ar	1.2	61.7	0.059	0.57	D$_2$–CO	1.0	61	0.13	0.74
Ne–Kr	1.01	69.8	0.12	0.87	D$_2$–O$_2$	1.0	61	0.13	0.66
Ne–Xe	1.25	69.1	0.17	1.31	D$_2$–air	1.0	57	0.13	0.74
Ne–H$_2$	1.0	34	0.10	0.26	D$_2$–CO$_2$	1.84	80	0.18	1.20
Ne–D$_2$	1.0	34	0.078	0.16	D$_2$–SF$_6$	1.0	93	0.31	2.26
Ne–N$_2$	1.0	57	0.043	0.65	CH$_4$–N$_2$	1.0	120	0.035	0.05
Ne–CO$_2$	1.0	82	0.081	0.98	CH$_4$–O$_2$	1.0	124	0.038	0.00
Ar–Kr	1.4	145	0.051	0.30	CH$_4$–air	1.0	120	0.035	0.05
Ar–Xe	1.8	178	0.086	0.57	CH$_4$–SF$_6$	1.0	188	0.12	0.50
Ar–H$_2$	1.73	64	0.17	0.85	N$_2$–H$_2$O	1.0	266	0.020	−0.32
Ar–D$_2$	1.73	64	0.14	0.74	N$_2$–CO$_2$	1.0	132	0.041	0.38
Ar–CH$_4$	1.0	130	0.046	0.02	N$_2$–SF$_6$	1.0	154	0.14	1.04
Ar–N$_2$	1.0	107	0.029	0.10	CO–CO$_2$	1.0	145	0.041	0.38
Ar–CO	1.0	117	0.029	0.10	CO–SF$_6$	1.0	169	0.14	1.06
Ar–O$_2$	1.0	118	0.026	0.15	O$_2$–H$_2$O	1.0	296	0.033	−0.03
Ar–air	1.0	109	0.029	0.11	O$_2$–CO$_2$	1.0	147	0.037	0.44
Ar–SF$_6$	1.0	179	0.12	1.07	O$_2$–SF$_6$	1.0	171	0.14	1.14
Kr–Xe	1.8	197	0.039	0.33	air–H$_2$O	1.0	274	0.020	−0.34
Kr–H$_2$	1.0	80	0.21	1.14	air–CO$_2$	1.0	136	0.040	0.39
Kr–D$_2$	1.0	80	0.19	1.07	air–SF$_6$	1.0	159	0.14	1.06
Kr–N$_2$	1.0	132	0.066	0.28	H$_2$O–CO$_2$	1.0	384	0.060	0.34
Kr–CO	1.0	145	0.066	0.28	CO$_2$–SF$_6$	1.0	222	0.088	0.60

[a] Based on the 12–6 potential; parameters for noble-gas pairs and for H$_2$–N$_2$ from van Heijningen *et al.* (1966, 1968), and for others from Hirschfelder *et al.* (1954).

D. EXAMPLES

To give some impression of the experimental situation, we show a collection of deviation plots in Figs. 10–16 for He–Ar and H_2–N_2. These are two of the

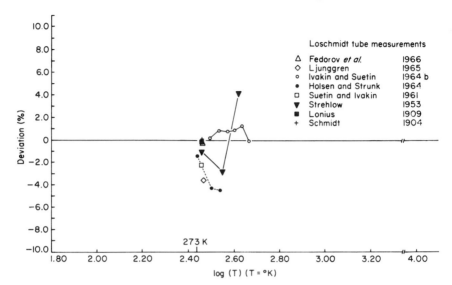

FIG. 10. Deviations of measured diffusion coefficients from the correlation function, Eq. (135). Loschmidt-tube measurements on He–Ar.

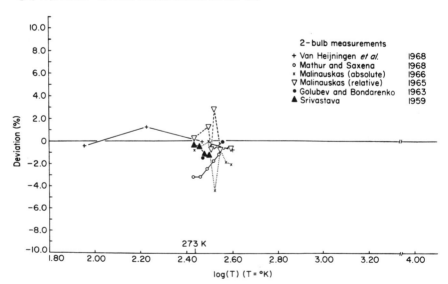

FIG. 11. Same as Fig. 10. Two-Bulb measurements on He–Ar.

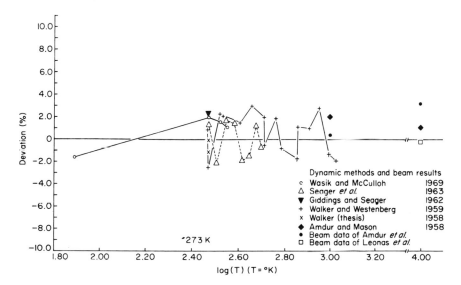

FIG. 12. Same as Fig. 10. Flow methods and beam-scattering results (Amdur, Leonas) on He–Ar. Note scale break between 10^3 and 10^4 °K.

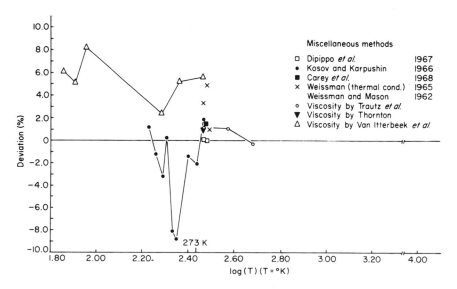

FIG. 13. Same as Fig. 10. Miscellaneous measurements on He–Ar.

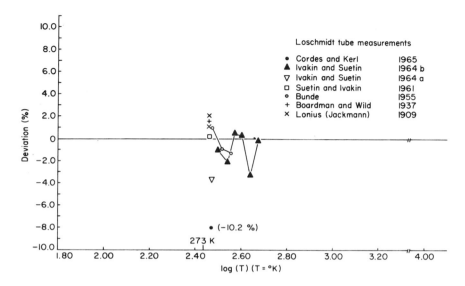

FIG. 14. Same as Fig. 10. Loschmidt-tube measurements on H_2–N_2.

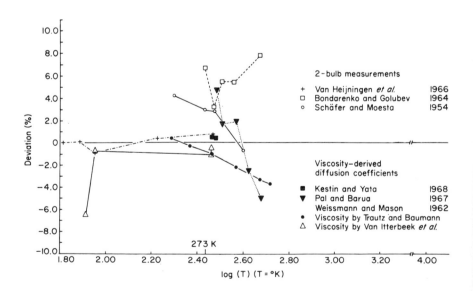

FIG. 15. Same as Fig. 10. Results for H_2–N_2 by the two-bulb method and from mixture viscosities.

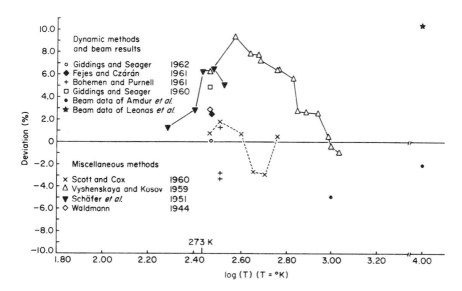

FIG. 16. Same as Fig. 10. Results for H_2–N_2 by flow methods, beam scattering (Amdur, Leonas), and miscellaneous methods. Note scale break between 10^3 and 10^4 °K.

most extensively measured gas pairs, one a noble gas pair and the other a diatomic gas pair. The temperature range extends from 65° to 10,000° K, and the plots show deviations from the correlation function of Eq. (135). The highest temperature points are from molecular-beam scattering measurements; the He–Ar potentials were based on direct measurements (Amdur et al., 1954; Kamnev and Leonas, 1965), but the H_2–N_2 potentials were calculated by geometric combination rules from He–H_2, He–N_2, and He–He data (Amdur and Smith, 1968; Amdur et al., 1957; Amdur and Harkness, 1954), and from H_2–H_2 and N_2–N_2 data (Belyaev and Leonas, 1967a,b).

Several general features are apparent in Figs. 10–16. First, the overall consistency of the data is rather good, although some reported results show considerable scatter. Second, the Loschmidt-tube and two-bulb measurements are more consistent than others, and show no evidence of any systematic disagreement. Third, careful appraisal of the experimental data is necessary to obtain the most reliable estimate of $\mathscr{D}_{12}$. A random selection of a value of $\mathscr{D}_{12}$ from the literature could easily yield a result having an uncertainty of 5%, although the original paper would probably claim much less. Finally, it would appear that heroic efforts are needed to achieve an uncertainty level of 1% or less.

ACKNOWLEDGMENT

This work was supported in part by the U.S. Army Research Office, Durham, under Grant DA–ARO–D–31–124–G925.

REFERENCES

Alievskiĭ, M. Ya., and Zhdanov, V. M. (1969). *Soviet Phys.—JETP (English Transl.)* **28**, 116 [*Zh. Eksperim. Teor. Fiz.* **55**, 221 (1968)].
Amdur, I., and Beatty, J. W., Jr. (1965). *J. Chem. Phys.* **42**, 3361.
Amdur, I., and Harkenss, A. L. (1954). *J. Chem. Phys.* **22**, 664.
Amdur, I., and Jordan, J. E. (1966). *Advan. Chem. Phys.* **10**, 29.
Amdur, I., and Mason, E. A. (1956). *J. Chem. Phys.* **25**, 632.
Amdur, I., and Mason, E. A. (1958). *Phys. Fluids* **1**, 370.
Amdur, I., and Schatzki, T. F. (1958). *J. Chem. Phys.* **29**, 1425.
Amdur, I., and Smith, A. L. (1968). *J. Chem. Phys.* **48**, 565.
Amdur, I., Mason, E. A., and Harkness, A. L. (1954). *J. Chem. Phys.* **22**, 1071.
Amdur, I., Mason, E. A., and Jordan, J. E. (1957). *J. Chem. Phys.* **27**, 527.
Annis, B. K., and Mason, E. A. (1970). *Phys. Fluids* **13**, to be published.
Annis, B. K., Humphreys, A. E., and Mason, E. A. (1968). *Phys. Fluids* **11**, 2122.
Arnold, J. H. (1944). *Trans. Am. Inst. Chem. Engrs.* **40**, 361.
Barker, J. A. (1963). *Ann. Rev. Phys. Chem.* **14**, 229.
Barker, J. A., Fock, W., and Smith, F. (1964). *Phys. Fluids* **7**, 897.
Belyaev, Yu. N., and Leonas, V. B. (1967a). *Soviet Phys. "Doklady" (English Transl.)* **11**, 866 [*Dokl. Akad. Nauk SSSR* **170**, 1039 (1966)].
Belyaev, Yu. N., and Leonas, V. B. (1967b). *Soviet Phys. "Doklady" (English Transl.)* **12**, 233 [*Dokl. Akad. Nauk. SSSR* **173**, 306 (1967)].
Bendt, P. J. (1958). *Phys. Rev.* **110**, 85.
Boardman, L. E., and Wild, N. E. (1937). *Proc. Roy. Soc.* **A162**, 511.
Bohemen, J., and Purnell, J. H. (1961). *J. Chem. Soc.* p. 360.
Bondarenko, A. G., and Golubev, I. F. (1964). *Gaz. Prom.* **9**, 50.
Bournia, A., Coull, J., and Houghton, G. (1961). *Proc. Roy. Soc.* **A261**, 227.
Boyd, C. A., Stein, N., Steingrimsson, V., and Rumpel, W. F. (1951). *J. Chem. Phys.* **19**, 548.
Bradley, R. S., and Waghorn, G. C. S. (1951). *Proc. Roy. Soc.* **A206**, 65.
Browning, R., and Fox, J. W. (1964). *Proc. Roy. Soc.* **A278**, 274.
Bunde, R. E. (1955). *Univ. Wisconsin, Naval Res. Lab. Rept.* **CM-850**.
Carey, C., Carnevale, E. H., and Uva, S. (1968). Private communication.
Carnevale, E. H., Lynnworth, L. C., and Larson, G. S. (1967). *J. Chem. Phys.* **46**, 3040.
Chapman, S. (1928). *Phil. Mag.* **5**, 630.
Chapman, S. (1967). *In* "Lectures in Theoretical Physics, Vol. 9C: Kinetic Theory" (W. E. Britten, ed.), pp. 1–13. Gordon and Breach, New York.
Chapman, S., and Cowling, T. G. (1941). *Proc. Roy. Soc.* **A179**, 159.
Chapman, S., and Cowling, T. G. (1970). "The Mathematical Theory of Non-Uniform Gases," 3rd ed. Cambridge Univ. Press, London and New York.
Cohen, E. G. D. (1969). *In* "Transport Phenomena in Fluids" (H. J. M. Hanley, ed.), pp. 157–207. Dekker, New York.

Cordes, H., and Kerl, K. (1965). *Z. Physik. Chem. (Frankfurt)* **45**, 369.

Coward, H. F., and Georgeson, E. H. M. (1937). *J. Chem. Soc.* p. 1085.

Crowell, A. D. (1968). *J. Chem. Phys.* **49**, 3324.

Curtiss, C. F. (1968). *J. Chem. Phys.* **49**, 2917.

Dalgarno, A. (1967). *Advan. Chem. Phys.* **12**, 143.

de Groot, S. R., and Mazur, P. (1962). "Non-Equilibrium Thermodynamics," pp. 5, 57–64. North-Holland Publ., Amsterdam.

De Paz, M., Turi, B., and Klein, M. L. (1967). *Physica*, **36**, 127.

Deriagin, B. V., and Bakanov, S. P. (1957a). *Soviet Phys. "Doklady" (English Transl.)* **2**, 326 [*Dokl. Akad. Nauk. SSSR* **45**, 267 (1957)].

Deriagin, B. V., and Bakanov, S. P. (1957b). *Soviet Phys.—Tech. Phys. (English Transl.)* **2**, 1904 [*Zh. Tekh. Fiz.* **27**, 2056 (1957)].

De Rocco, A. G., and Hoover, W. G. (1962). *J. Chem. Phys.* **36**, 916.

De Rocco, A. G., Storvick, T. S., and Spurling, T. H. (1968). *J. Chem. Phys.* **48**, 997.

Diller, D. E., and Mason, E. A. (1966). *J. Chem. Phys.* **44**, 2604.

Di Pippo, R., Kestin, J., and Oguchi, K. (1967). *J. Chem. Phys.* **46**, 4986.

Doebereiner, J. W. (1823). *Ann. Chim. Phys.* **24**, 332.

Duncan, J. B., and Toor, H. L. (1962). *Am. Inst. Chem. Engrs. J.* **8**, 38.

Dymond, J. H., Rigby, M., and Smith, E. B. (1966). *Phys. Fluids* **9**, 1222.

Ernst, M. H., Haines, L. K., and Dorfman, J. R. (1969). *Rev. Mod. Phys.* **41**, 296.

Evans, R. B., III, Truitt, J., and Watson, G. M. (1961a). *J. Chem. Eng. Data* **6**, 522.

Evans, R. B., III, Watson, G. M., and Mason, E. A. (1961b). *J. Chem. Phys.* **35**, 2076.

Evans, R. B., III, Watson, G. M., and Truitt, J. (1962). *J. Appl. Phys.* **33**, 2682.

Evans, R. B., III, Watson, G. M., and Truitt, J. (1963). *J. Appl. Phys.* **34**, 2020.

Evans, R. B., III, Love, L. D., and Mason E. A. (1969). *J. Chem. Educ.* **46**, 423.

Fairbanks, D. F., and Wilke, C. R. (1950). *Ind. Eng. Chem.* **42**, 471.

Fedorov, E. B., Ivakin, B. A., and Suetin, P. E. (1966). *Soviet Phys.—Tech. Phys. (English Transl.)* **11**, 424 [*Zh. Tekh. Fiz.* **36**, 569 (1966)].

Fejes, P., and Czárán, L. (1961). *Acta Chim. Hung.* **29**, 171.

Fick, A. (1855). *Ann. Physik* **94**, 59; *Phil. Mag.* **10**, 30.

Fitts, D. D. (1962). "Nonequilibrium Thermodynamics," pp. 35–36. McGraw-Hill, New York.

Frankel, S. P. (1940). *Phys. Rev.* **57**, 661.

Franzen, W. (1959). *Phys. Rev.* **115**, 850.

Frost, A. C. (1967). Ph.D. Thesis, Columbia Univ., New York.

Furry, W. H. (1948). *Am. J. Phys.* **16**, 63.

Giddings, J. C., and Mallik, K. L. (1967). *Ind. Eng. Chem.* **59** (4), 18.

Giddings, J. C., and Seager, S. L. (1960). *J. Chem. Phys.* **33**, 1579.

Giddings, J. C., and Seager, S. L. (1962). *Ind. Eng. Chem., Fundamentals* **1**, 277.

Golubev, I. F., and Bondarenko, A. G. (1963). *Gaz. Prom.* **8**, 46.

Grad, H. (1949). *Commun. Pure Appl. Math.* **2**, 331.

Graham, T. (1829). *Quart. J. Sci.* **2**, 74. Reprinted in "Chemical and Physical Researches," pp. 28–35. Edinburgh Univ. Press, Edinburgh (1876).

Graham, T. (1833). *Phil. Mag.* **2**, 175, 269, 351. Reprinted in "Chemical and Physical Researches," pp. 44–70. Edinburgh Univ. Press, Edinburgh (1876).

Graham T. (1846). *Phil Trans. Roy. Soc.* **136**, 573. Reprinted in "Chemical and Physical Researches," pp. 88–161. Edinburgh Univ. Press, Edinburgh (1876).

Graham, T. (1849). *Phil. Trans. Roy. Soc.* **139**, 349. Reprinted in "Chemical and Physical Researches," pp. 162–210. Edinburgh Univ. Press, Edinburgh (1876).

Graham T. (1863). *Phil. Trans. Roy. Soc.* **153**, 385. Reprinted in "Chemical and Physical Researches," pp. 210–234. Edinburgh Univ. Press, Edinburgh (1876).

Gunn, R. D., and King, C. J. (1969). *Am. Inst. Chem. Engrs. J.* **15**, 507.

Hamann, S. D., and Lambert, J. A. (1954). *Australian J. Chem.* **7**, 1.

Hanley, H. J. M., McCarty, R. D., and Sengers, J. V. (1969). *J. Chem. Phys.* **50**, 857.

Harteck, P., and Schmidt, H. W. (1933). *Z. Phys. Chem.* **B21**, 447.

Hassé, H. R. (1926). *Phil. Mag.* **1**, 139.

Hassé, H. R., and Cook, W. R. (1927). *Phil. Mag.* **3**, 977.

Hassé, H. R., and Cook, W. R. (1929). *Proc. Roy. Soc.* **A125**, 196.

Hassé, H. R., and Cook, W. R. (1931). *Phil. Mag.* **12**, 554.

Hellund, E. J. (1940). *Phys. Rev.* **57**, 737.

Higgins, L. D., and Smith, F. J. (1968). *Mol. Phys.* **14**, 399.

Hirschfelder, J. O., Curtiss, C. F., and Bird, R. B. (1954). "Molecular Theory of Gases and Liquids." Wiley, New York.

Hiza, M. J., and Duncan, A. G. (1969). *Phys. Fluids* **12**, 1531.

Hogervorst, W., and Freudenthal, J. (1967). *Physica* **37**, 97.

Holmes, R., and Tempest, W. (1960). *Proc. Phys. Soc. (London)* **75**, 898.

Holsen, J. N., and Strunk, M. R. (1964). *Ind. Eng. Chem., Fundamentals* **3**, 143.

Hoogschagen, J. (1953). *J. Chem. Phys.* **21**, 2096.

Hoogschagen, J. (1955). *Ind. Eng. Chem.* **47**, 906.

Humphreys, A. E., and Mason, E. A. (1970). *Phys. Fluids* **13**, 65.

Imam-Rahajoe, S., Curtiss, C. F., and Bernstein, R, B. (1965). *J. Chem. Phys.* **42**, 530.

Itean, E. C., Glueck, A. R., and Svehla, R. A. (1961). *NASA Tech. Note.* **TN D-481**.

Ivakin, B. A., and Suetin, P. E. (1964a). *Soviet Phys.—Tech. Phys. (English Transl.)* **8**, 748 [*Zh. Tekh. Fiz.* **33**, 1007 (1963)].

Ivakin, B. A., and Suetin, P. E. (1964b). *Soviet Phys.—Tech. Phys. (English Transl.)* **9**, 866 [*Zh. Tekh. Fiz.* **34**, 1115 (1964)].

Ivakin, B. A., Suetin, P. E., and Plesovskikh, V. P. (1968). *Soviet Phys.—Tech. Phys. (English Transl.)* **12**, 1403 [*Zh. Tekh. Fiz.* **37**, 1913 (1967)].

Jeans, J. H. (1925). "The Dynamical Theory of Gases," 4th ed. Cambridge Univ. Press, London and New York. Reprinted by Dover, New York (1954).

Jost, W. (1952). "Diffusion in Solids, Liquids, and Gases." Academic Press, New York.

Kagan, Yu., and Afanas'ev, A. M. (1962). *Soviet Phys. JETP (English Transl.)* **14**, 1096 [*Zh. Eksperim. Teor. Fiz.* **41**, 1536 (1961)].

Kamnev, A. B., and Leonas, V. B. (1965). *Soviet Phys. "Doklady" (English Transl.)* **10**, 529. [*Dokl. Akad. Nauk SSSR* **162**, 798 (1965)].

Kamnev, A. B., and Leonas, V. B. (1966). *Soviet Phys. "Doklady" (English Transl.)* **10**, 1202 [*Dokl. Akad. Nauk. SSSR* **165**, 1273 (1965)].

Katan, T. (1969). *J. Chem. Phys.* **50**, 233.

Kennard, E. H. (1938). "Kinetic Theory of Gases." McGraw-Hill, New York.

Kestin, J., and Yata, J. (1968). *J. Chem. Phys.* **49**, 4780.

Khouw, B., Morgan, J. E., and Schiff, H. I. (1969). *J. Chem. Phys.* **50**, 66.

Kihara, T. (1953). *Rev. Mod. Phys.* **25**, 831.

Kihara, T. (1963). *Advan. Chem. Phys.* **5**, 147.

Kihara, T., Taylor, M. H., and Hirschfelder, J. O. (1960). *Phys. Fluids* **3**, 715.

Kim, S. K., and Ross, J. (1967). *J. Chem. Phys.* **46**, 818.

Kirk, A. D. (1967). *J. Chem. Educ.* **44**, 745.

Klein, M., and Smith, F. J. (1968). *J. Res. Natl. Bur. Std.* **A72**, 359.

Klibanova, Ts. M. Pomerantsev, V. V., and Frank-Kamenetskiĭ, D. A. (1942). *Zh. Tekh. Fiz.* **12**, 14.

Knudsen, M. (1950). "The Kinetic Theory of Gases," 3rd ed. Wiley, New York.

Konowalow, D. D. (1969). *J. Chem. Phys.* **50**, 12.

Kosov, N. D. (1957). *Issled. Fiz. Osnov Rabochego Protsessa Topok i Pechei (Alma-Ata: Akad. Nauk Kazakh. SSR) Sb.* pp. 285–290.

Kosov, N. D., and Karpushin, A. G. (1966). *Nekotorye Vopr. Obshchie i Prikladnoi Fiz.*, *Tr. Gorodskoi Konf.*, *Alma-Ata* pp. 94–96.

Kosov, N. D., and Kurlapov, L. I. (1966). *Soviet Phys.—Tech. Phys.* (*English Transl.*) **10**, 1623 [*Zh. Tekh. Fiz.* **35**, 2120 (1965)].

Kotani, M. (1942). *Proc. Phys.-Math. Soc. Japan* **24**, 76.

Kotousov, L. S. (1965). *Soviet Phys.—Tech. Phys.* (*English Transl.*) **9**, 1679 [*Zh. Tekh. Fiz.* **34**, 2178 (1964)].

Kramers, H. A., and Kistemaker, J. (1943). *Physica* **10**, 699.

Krongelb, S., and Strandberg, M. W. P. (1959). *J. Chem. Phys.* **31**, 1196.

Kucherov, R. Ya. (1957). *Soviet Phys.—Tech. Phys.* (*English Transl.*) **2**, 2001 [*Zh. Tekh. Fiz.* **27**, 2158 (1957)].

Kucherov, R. Ya., and Rikenglaz, L. É. (1959). *Soviet Phys. JETP* (*English Transl.*) **9**, 1253 [*Zh. Eksperimen. Teor. Fiz.* **36**, 1758 (1959)].

Kundt, A., and Warburg, E. (1875). *Ann. Phys.* **155**, 337, 525; *Phil. Mag.* **50**, 53.

Langmuir, I. (1918). *Phys. Rev.* **12**, 368.

Lee, C. Y., and Wilke, C. R. (1954). *Ind. Eng. Chem.* **46**, 2381.

Lin, S. T., and Hsu, H. W. (1969). *Am. Inst. Chem. Engrs. J.* **14**, 328.

Lipsicas, M. (1962). *J. Chem. Phys.* **36**, 1235.

Ljunggren, S. (1965). *Arkiv. Kemi* **24**, 1.

Lonius, A. (1909). *Ann. Phys.* **29**, 664.

Loschmidt, J. (1870a). *Sitzber. Akad. Wiss. Wien* **61**, 367.

Loschmidt, J. (1870b). *Sitzber. Akad. Wiss. Wien* **62**, 468.

Luszczynski, K., Norberg, R. E., and Opfer, J. E. (1962). *Phys. Rev.* **128**, 186.

McCarty, K. P., and Mason, E. A. (1960). *Phys. Fluids* **3**, 908.

McCoubrey, A. O., and Matland, C. G. (1956). *Phys. Rev.* **101**, 603.

McCoubrey, J. C., and Singh, N. M. (1959). *Trans. Faraday Soc.* **55**, 1826.

Mackenzie, J. E., and Melville, H. W. (1932). *Proc. Roy. Soc. Edinburgh* **52**, 337.

Mackenzie, J. E., and Melville, H. W. (1933). *Proc. Roy. Soc. Edinburgh* **53**, 255.

Malinauskas, A. P. (1965). *J. Chem. Phys.* **42**, 156.

Malinauskas, A. P. (1966). *J. Chem. Phys.* **45**, 4704.

Malinauskas, A. P., and Silverman, M. D. (1969). *J. Chem. Phys.* **50**, 3263.

Marrero, T. R., and Mason, E. A. (1971). "Gaseous Diffusion Coefficients" *NSRDS-NBS*. To be published.

Mason, E. A. (1954). *J. Chem. Phys.* **22**, 169.

Mason, E. A. (1955). *J. Chem. Phys.* **23**, 49.

Mason, E. A. (1957a). *J. Chem. Phys.* **27**, 75.

Mason, E. A. (1957b). *J. Chem. Phys.* **27**, 782.

Mason, E. A. (1961). *Phys. Fluids* **4**, 1504.

Mason, E. A. (1967). *Am. J. Phys.* **35**, 434.

Mason, E. A., and Evans, R. B., III (1969). *J. Chem. Educ.* **46**, 358.

Mason, E. A., and Kronstadt, B. (1967). *J. Chem. Educ.* **44**, 740.

Mason, E. A., and Monchick, L. (1962). *J. Chem. Phys.* **36**, 2746.

Mason, E. A., and Monchick, L. (1967). *Advan. Chem. Phys.* **12**, 329.

Mason, E. A., and Rice, W. E. (1954). *J. Chem. Phys.* **22**, 843.

Mason, E. A., and Schamp, H. W., Jr. (1958). *Ann. Phys.* (*N.Y.*) **4**, 233.

Mason, E. A., and Smith, F. J. (1966). *J. Chem. Phys.* **44**, 3100.

Mason, E. A., Weissman, S., and Wendt, R. P. (1964). *Phys. Fluids* **7**, 174.

Mason, E. A., Munn, R. J., and Smith, F. J. (1966). *Advan. At. Mol. Phys.* **2**, 33–91.

Mason, E. A., Malinauskas, A. P., and Evans, R. B., III (1967a). *J. Chem. Phys.* **46**, 3199.

Mason, E. A., Munn, R. J., and Smith, F. J. (1967b). *Phys. Fluids* **10**, 1827.

Mason, E. A., Miller, L., and Spurling, T. H. (1967c). *J. Chem. Phys.* **47**, 1669.

Mathur, B. P., and Saxena, S. C. (1968). *Appl. Sci. Res.* **18**, 325.

Maxwell, J. C. (1860). *Phil. Mag.* **20**, 21. Reprinted in "Scientific Papers," Vol. 1, pp. 392–409. Dover, New York (1962).

Maxwell, J. C. (1867). *Phil. Trans. Roy. Soc.* **157**, 49. Reprinted in "Scientific Papers," Vol. 2, pp. 26–78. Dover, New York (1962).

Maxwell, J. C. (1873). *Nature* **8**, 537. Reprinted in "Scientific Papers," Vol. 2, pp. 343–350. Dover, New York (1962).

Maxwell, J. C. (1879). *Phil. Trans. Roy. Soc.* **170**, 231. Reprinted in "Scientific Papers," Vol. 2, pp. 681–712. Dover, New York (1962).

Meyer, O. E. (1899). "Kinetic Theory of Gases," 2nd ed. Transl. by R. E. Baynes. Longmans, Green, New York.

Miller, L. (1966). *J. South African Chem. Inst.* **19**, 125.

Miller, L., and Carman, P. C. (1960). *Nature* **186**, 549.

Miller, L., and Mason, E. A. (1966). *Phys. Fluids* **9**, 711.

Miller, L., Spurling, T. H., and Mason, E. A. (1967). *Phys. Fluids* **10**, 1809.

Monchick, L. (1959). *Phys. Fluids* **2**, 695.

Monchick, L. (1962). *Phys. Fluids* **5**, 1393.

Monchick, L., and Mason, E. A. (1961). *J. Chem. Phys.* **35**, 1676.

Monchick, L., and Mason, E. A. (1967). *Phys. Fluids* **10**, 1377.

Monchick, L., Yun, K. S., and Mason, E. A. (1963). *J. Chem. Phys.* **39**, 654.

Monchick, L., Mason, E. A., Munn, R. J., and Smith, F. J. (1965). *Phys. Rev.* **139**, A1076.

Monchick, L., Munn, R. J., and Mason, E. A. (1966). *J. Chem. Phys.* **45**, 3051.

Monchick, L., Sandler, S. I., and Mason, E. A. (1968). *J. Chem. Phys.* **49**, 1178.

Morse, P. M. (1929). *Phys. Rev.* **34**, 57.

Munn, R. J., Mason, E. A., and Smith, F. J. (1964). *J. Chem. Phys.* **41**, 3978.

Munn, R. J., Smith, F. J., and Mason, E. A. (1965a). *J. Chem. Phys.* **42**, 537.

Munn, R. J., Mason, E. A., and Smith, F. J. (1965b). *Phys. Fluids* **8**, 1103.

Nelson, E. T. (1956). *J. Appl. Chem.* **6**, 286.

Nettley, P. T. (1954). *Proc. Phys. Soc. (London)* **B67**, 753.

Ney, E. P., and Armistead, F. C. (1947). *Phys. Rev.* **71**, 14.

O'Hara, H., and Smith, F. J. (1970). *J. Computational Phys.* **5**, 328.

Pakurar, T. A., and Ferron, J. R. (1966). *Ind. Eng. Chem., Fundamentals* **5**, 553.

Pal, A. K., and Barua, A. K. (1967). *J. Chem. Phys.* **47**, 216.

Partington, J. R. (1949). "An Advanced Treatise on Physical Chemistry," Vol. I: Fundamental Principles. The Properties of Gases, p. 903. Longmans, Green, New York.

Present, R. D. (1958). "Kinetic Theory of Gases." McGraw-Hill, New York.

Present, R. D., and de Bethune, A. J. (1949). *Phys. Rev.* **75**, 1050.

Reinganum, M. (1900). *Phys. Z.* **2**, 241.

Samoilov, E. V., and Tsitelauri, N. N. (1964). *High Temp. (English Transl.)* **2**, 509 [*Tepl. Vys. Temp.* **2**, 565 (1964)].

Sandler, S. I., and Dahler, J. S. (1967). *J. Chem. Phys.* **47**, 2621.

Sandler, S. I., and Mason, E. A. (1967). *J. Chem. Phys.* **47**, 4653.

Sandler, S. I., and Mason, E. A. (1968). *J. Chem. Phys.* **48**, 2873.

Saran, A. (1963). *Indian J. Phys.* **37**, 491.

Saxena, S. C., and Gambhir, R. S. (1963). *Mol. Phys.* **6**, 577.

Schäfer, K., and Moesta, H. (1954). *Z. Elektrochem.* **58**, 743.

Schäfer, K., Corte, H., and Moesta, H. (1951). *Z. Elektrochem.* **55**, 662.

Schmidt, R. (1904). *Ann. Phys.* **14**, 801.

Schramm, B. (1968). *Ber. Bunsenges. Phys. Chem.* **72**, 609.

Schramm, B. (1969). *Ber. Bunsenges. Phys. Chem.* **73**, 217.

Schwertz, F. A., and Brow, J. E. (1951). *J. Chem. Phys.* **19**, 640.
Scott, D. S., and Cox, K. E. (1960). *Can. J. Chem. Eng.* **38**, 201.
Seager, S. L., Geertson, L. R., and Giddings, J. C. (1963). *J. Chem. Eng. Data* **8**, 168.
She, R. S. C., and Sather, N. F., (1967). *J. Chem. Phys.* **47**, 4978.
Shih, K. T., and Ibele, W. E. (1968). *Trans. ASME (J. Heat Transfer)* **90C**, 413.
Smith, F. J., and Munn, R. J. (1964). *J. Chem. Phys.* **41**, 3560.
Smith, F. J., Mason, E. A., and Munn, R. J. (1965a). *J. Chem. Phys.* **42**, 1334.
Smith, F. J., Mason, E. A., and Munn, R. J. (1965b). *Phys. Fluids* **8**, 1907.
Smith, F. J., Munn, R. J., and Mason, E. A. (1967). *J. Chem. Phys.* **46**, 317.
Srivastava, B. N., and Saran, A. (1966). *Physica* **32**, 110.
Srivastava, K. P. (1958). *J. Chem. Phys.* **28**, 543.
Srivastava, K. P. (1959). *Physica* **25**, 571.
Stefan, J. (1871). *Sitzber Akad. Wiss. Wien* **63**, 63.
Stefan, J. (1872). *Sitzber. Akad. Wiss. Wien* **65**, 323.
Stefan, J. (1873). *Sitzber. Akad. Wiss. Wien* **68**, 385.
Storvick T. S., and Mason, E. A. (1966). *J. Chem. Phys.* **45**, 3752.
Strehlow, R. A. (1953). *J. Chem. Phys.* **21**, 2101.
Suetin, P. E., and Ivakin, B. A. (1961). *Soviet Phys.—Tech. Phys. (English Transl.)* **6**, 359 [*Zh. Tekh. Fiz.* **31**, 499 (1961)].
Suetin, P. E., and Volobuev, P. V. (1964). *Soviet Phys.—Tech. Phys. (English Transl.)* **9**, 859 [*Zh. Tekh. Fiz.* **34**, 1107 (1964)].
Sutherland, W. (1894). *Phil. Mag.* **38**, 1.
Taxman, N. (1958). *Phys. Rev.* **110**, 1235.
Toor, H. L. (1957). *Am. Inst. Chem. Engrs. J.* **3**, 198.
Umanskii, A. S., and Bogdanova, S. S. (1968). *High Temp. (English Transl.)* **6**, 518 [*Tepl. Vys. Temp.* **6**, 543 (1968)].
van Heijningen, R. J. J., Feberwee, A., van Oosten, A., and Beenakker, J. J. M. (1966). *Physica* **32**, 1649.
van Heijningen, R. J. J., Harpe, J. P., and Beenakker, J. J. M. (1968). *Physica* **38**, 1.
Volobuev, P. V., and Suetin, P. E. (1965). *Soviet Phys.—Tech. Phys. (English Transl.)* **10**, 269 [*Zh. Tekh. Fiz.* **35**, 336 (1965)].
Volobuev, P. V., and Suetin, P. E. (1967). *Soviet Phys.—Tech. Phys. (English Transl.)* **11**, 960 [*Zh. Tekh. Fiz.* **36**, 1292 (1966)].
von Obermayer, A. (1882). *Sitzber. Akad. Wiss. Wien* **85**, 147.
von Obermayer, A. (1887). *Sitzber. Akad. Wiss. Wien* **96**, 546.
Vyshenskaya, V. F., and Kosov, N. D. (1959). *Issled. Protsessov Perenosa. Vopr. Teorii Otnositel'nosti, Alma-Ata, Sb.* pp. 114–125.
Vyshenskaya, V. F., and Kosov, N. D. (1961). *Teplo i Massoperenos, Pervoe Vsesoyuznoe Soveschante, Minsk* pp. 181–187 [Transl. by A. L. Monks, *Oak Ridge Natl. Lab.* **ORNL-TR-506** (1965)].
Waitz, K. (1882). *Ann. Phys.* **17**, 201.
Waldmann, L. (1944). *Naturwissenschaften* **32**, 223.
Waldmann, L. (1947). *Z. Phys.* **124**, 2.
Waldmann, L. (1958). *In* "Handbuch der Physik" (S. Flugge, ed.), Vol. 12, pp. 295–514. Springer, Berlin.
Waldmann, L. (1963). *Z. Naturforsch.* **18a**, 1033.
Waldmann, L., and Schmitt, K. H. (1961). *Z. Naturforsch.* **16a**, 1343.
Walker, R E. (1958). Ph.D. Thesis, Univ. of Maryland, College Park, Maryland.
Walker, R. E. (1961). *J. Chem. Phys.* **34**, 2196.
Walker, R. E., and Westenberg, A. A. (1958a). *J. Chem. Phys.* **29**, 1139.

Walker, R. E., and Westenberg, A. A. (1958b). *J. Chem. Phys.* **29**, 1147.
Walker, R. E., and Westenberg, A. A. (1959). *J. Chem. Phys.* **31**, 519.
Walker, R. E., and Westenberg, A. A. (1960). *J. Chem. Phys.* **32**, 436.
Wang Chang, C. S., Uhlenbeck, G. E., and de Boer, J. (1964). *Studies Stat. Mech.* **2**, 241.
Wasik, S. P., and McCulloh, K. E. (1969). *J. Res. Natl. Bur. Std.* **A73**, 207.
Weijland, A., and van Leeuwen, J. M. J. (1968). *Physica* **38**, 35.
Weissman, S. (1964). *J. Chem. Phys.* **40**, 3397.
Weissman, S. (1965). *In* "Advances in Thermophysical Properties at Extreme Temperatures and Pressures" (S. Gratch, ed.), pp. 12–18. *Am. Soc. Mech. Engrs.* New York.
Weissman, S., and Mason, E. A. (1962). *J. Chem. Phys.* **37**, 1289.
Wendt, R. P. (1962). *J. Phys. Chem.* **66**, 1740.
Westenberg, A. A. (1966). *Advan. Heat Transfer* **3**, 253–302.
Westenberg, A. A., and Frazier, G. (1962). *J. Chem. Phys.* **36**, 3499.
Wicke, E., and Hugo, P. (1961). *Z. Physik. Chem. (Frankfurt)* **28**, 401.
Wicke, E., and Kallenbach, R. (1941). *Kolloid Z.* **97**, 135.
Wilke, C. R., and Lee, C. Y. (1955). *Ind. Eng. Chem.* **47**, 1253.
Williams, F. A. (1958). *Am. J. Phys.* **26**, 467.
Wilson, J. N. (1965). *J. Chem. Phys.* **43**, 2564.
Wilson, J. N. (1968). *J. Chem. Phys.* **49**, 3325.
Winn, E. B. (1950). *Phys. Rev.* **80**, 1024.
Wise, H. (1959). *J. Chem. Phys.* **31**, 1414.
Yao, Y. L. (1968). *J. Chem. Phys.* **48**, 537.
Yolles, R. S., and Wise, H. (1968). *J. Chem. Phys.* **48**, 5109.
Young, R. A. (1961). *J. Chem. Phys.* **34**, 1295.
Yun, K. S., Weissman, S., and Mason, E. A. (1962). *Phys. Fluids* **5**, 672.
Zhdanov, V. M. (1967). *Soviet Phys.—Tech. Phys. (English Transl.)* **12**, 134 [*Zh. Tekh. Fiz.* **37**, 192 (1967)].
Zhdanov, V. M. (1968). *Soviet Phys. JETP (English Transl.)* **26**, 1187 [*Zh. Eksperimen. Teor. Fiz.* **53**, 2099 (1967)].
Zhdanov, V., Kagan, Yu., and Sazykin, A. (1962). *Soviet Phys. JETP (English Transl.)* **15**, 596 [*Zh. Eksperim. Teor. Fiz.* **42**, 857 (1962)].

THEORY AND APPLICATION OF STURMIAN FUNCTIONS*

MANUEL ROTENBERG

Department of Applied Physics and
Institute for Pure and Applied Physical Sciences,
University of California, San Diego, La Jolla, California

I. Introduction

Although the eigenfunctions of the Schroedinger equation form a complete set, the fact that the continuum forms part of that set often reduces the completeness property to only formal importance. In expanding atomic wave functions of more than one particle in terms of Schroedinger functions, one

* This research was supported by the Advanced Research Projects Agency of the Department of Defense and was monitored by the U.S. Army Research Office, Durham under Contract DA-31-124-ARO-D-257.

invariably chooses to ignore the continuum terms—not because they are unimportant, but because the ensuing equations are impossible to deal with.

The analytical convenience offered by eigenfunction expansions have given way to less convenient but more accurate expansions (in the sense of convergence to the correct solution)—a trend made possible by use of digital computers. Even modern digital computers, however, do not allow all simplification due to analytical considerations to be abandoned, and so a compromise is sought. The use of Sturmian functions is one way to affect this compromise. [The term "Sturmian" was a whim of the author (Rotenberg, 1962) and has no historical significance other than the obvious one of recognizing that the functions are a solution of one form of the Sturm–Liouville problem.]

In the bound state problem Sturmians were used quite early by Holøien (1956) and Shull and Löwdin (1959), the latter showing considerable improvement in calculating the ground-state energy of helium over a hydrogen eigenfunction expansion carried out by Taylor and Parr (1952). The convergence difficulty in the two-electron problem seems to stem from the cusp in the two-electron wave function, for which a significant amount of continuum admixture seems necessary to adequately describe. This fact has been recognized by Hylleraas and Midtdal (1958), Schwartz (1961), and Pekeris (1962).

In this paper we have done little to review calculations of ground-state energies; noneigenfunction representations have been the rule in that field. We have instead concentrated on scattering problems which have made significant use of Sturmians, for better or for worse.

II. Mathematical Preliminaries

A. General Considerations

The formal definition of the Sturmian functions (Rotenberg, 1962) is the nontrivial, normalized solution of

$$\left[\frac{d^2}{dx^2} - \frac{l(l+1)}{x^2} + E_0 - \alpha_{nl} V(x) \right] S_{nl}(x) = 0 \tag{1}$$

$$S_{nl}(0) = 0; \quad S_{nl}(\infty) \to 0 \tag{2}$$

for $n = 1, 2, 3, \ldots$, and $l = 0, 1, \ldots, n - 1$, where E_0 is a fixed negative number and α_{nl} is an eigenvalue. Equation (1) can be regarded as a kind of Schroedinger equation, where, for a fixed total energy and angular momentum, the coupling constants α_{nl} are to be found such that Eqs. (2) are satisfied. In physical terms the potential energy is being increased as the α_{nl} increase [we assume $V(x)$ is attractive], so that the kinetic energy must also increase to keep E_0 constant. Therefore the number of oscillations of the Sturmian functions increases, which permits the possibility of orthogonality. A simple manipulation of Eq. (1) yields the orthogonality property

$$\int dx S_{nl}(x)V(x)S_{n'l}(x) = -\delta_{nn'}. \tag{3}$$

Equation (3) is strictly true provided $V(r)$ is of one sign: otherwise (3) may be zero for $n = n'$. The negative sign on the right of Eq. (3) results from the assumption that $V(x)$ is negative throughout, and from the desire to keep the S_{nl} real.

If $V(r)$ is the physically correct potential, then, obviously, there will be one Sturmian function which, apart from the normalization factor, is the same as the Schroedinger state of energy E_0. For this state, $\alpha = 1$. When E_0 is the ground-state energy, this function is of central importance in elastic scattering problems.

If $V(r)$ is purely attractive, then $1 \le \alpha_{nl} < \infty$. Should $V(r)$ be of both signs (an interatomic potential, for example) then the eigenvalue will range from plus to minus infinity, since both signs of the eigenvalue will provide binding. Thus, the functions fall into two categories: S_{nl}, corresponding to positive eigenvalues, and T_{nl}, corresponding to negative eigenvalues. The purpose served by distinguishing between them is the normalization integral (3): The negative sign is omitted in the case of the T_{nl}. The functions S_{nl} and T_{nl} are still mutually orthogonal with the weight $V(r)$. Note, however, that because $V(r)$ changes sign, the Sturm–Liouville theorem regarding different numbers of nodes in mutually orthogonal functions will not hold.

B. THE COULOMB STURMIANS

By far the most widely used Sturmian functions are those generated by the Coulomb potential. They satisfy

$$\left[-\frac{d^2}{dr^2} + \frac{l(l+1)}{r^2} - \frac{2\alpha_{nl}}{r} - E_0\right]S_{nl}(r) = 0 \tag{4}$$

with the associated boundary conditions that S_{nl} approach zero at both the origin and infinity. The unit of length and energy in Eq. (4) are the Bohr radius and the Rydberg, respectively. By comparing Eq. (4) with that for the Coulomb functions

$$\left[-\frac{d^2}{dr^2} + \frac{l(l+1)}{r^2} - \frac{2}{r} + \frac{1}{n^2}\right]u_{nl}(r) = 0 \tag{5}$$

where

$$u_{nl} = N'_{nl} e^{-r/n}(2r/n)^{l+1}L_{n+l}^{2l+1}(2r/n), \tag{6}$$

it is seen that

$$S_{nl}(r) = N_{nl} e^{-kr}(2kr)^{l+1}L_{n+l}^{2l+1}(2kr) \tag{7}$$

and

$$\alpha_{nl} = kn \tag{8}$$

where $k \equiv (-E_0)^{1/2}$. If the S_{nl} are normalized with respect to the weight $2/r$, then (see Section II, D)

$$N_{nl}^2 = \tfrac{1}{2}\{(n - l - 1)!/[(n + l)!]^3\}. \tag{9}$$

In Eq. (6) and following, $L_{n+l}^{2l+1}(x)$ is the Laguerre polynomial, defined as (Kemble, 1937)

$$L_{n+l}^{2l+1}(x) = \sum_{\kappa=0}^{n-l-1}(-1)^{\kappa+1}\frac{[(n + l)!]^2 x^{\kappa}}{(n - l - 1 - \kappa)!(2l + 1 + \kappa)!\kappa!}. \tag{10}$$

Because almost all bound-state hydrogenic wave functions are close to zero energy, they suffer from the peculiarity that the innermost zeros of the

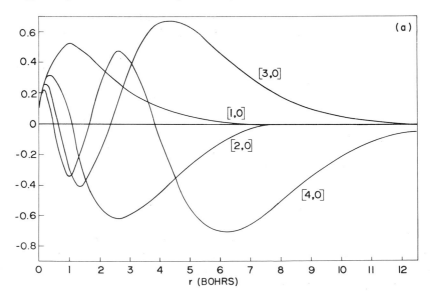

FIG. 1a. The $l = 0$ Sturmian functions for $n = 1, 2, 3$, and 4. Unlike hydrogen functions, the first node continues to move closer to the origin with increasing n, and the maximum antinode moves out linearly with n instead of quadratically.

functions are insensitive to the principal quantum number. For example, the first zero of u_{20} is at $r = 2.0$ Bohr radii; as n gets very large, the first zero approaches a value only slightly less than this: 1.88 Bohr radii. This accounts for the fact that the bound hydrogen functions do not form a complete set; the continuum is needed to analyze the region between the origin and the limiting first zero.

The maximum value of the bound Coulomb functions is at $r \sim n^2$, while that of the Sturmian functions extends linearly with n. Thus one should not expect rapid convergence with Sturmians when analyzing functions of large spatial extent. A few Sturmian functions for the $2/r$ potential are shown in Figs. 1a and 1b.

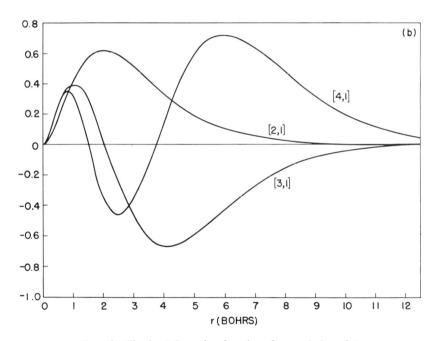

FIG. 1b. The $l = 1$ Sturmian functions for $n = 2$, 3, and 4.

C. THE PROJECTION OF THE STURMIANS ONTO COULOMB FUNCTIONS

It is of interest to examine how the Sturmians analyze the continuum Coulomb wavefunctions and vice versa. The Sturmians referred to are defined by Eq. (4) with $E_0 = -1$. The relevant expansion sare (Greek indices are summed)

$$u_{kl}(r) = \sum a(kl; vl)S_{vl}(r) \tag{11}$$

and

$$S_{nl}(r) = \int b(kl; nl)u_{kl}(r)\, dk, \tag{12}$$

where u_{kl} is the radial continuum hydrogenic function of momentum $k = [E(Ry)]^{1/2}$. The expansion coefficients, or Drang functions of the first and second kind, are given by

$$a(kl; vl) = \int dr S_{vl}(r)(2/r)u_{kl}(r) \tag{13}$$

$$b(kl; vl) = \int dr S_{vl}(r)u_{kl}(r) \tag{14}$$

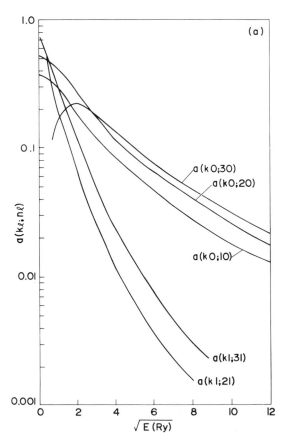

FIG. 2a. The Drang functions of the first kind for the first few $l = 0$ and $l = 1$ states. See Eq. (13).

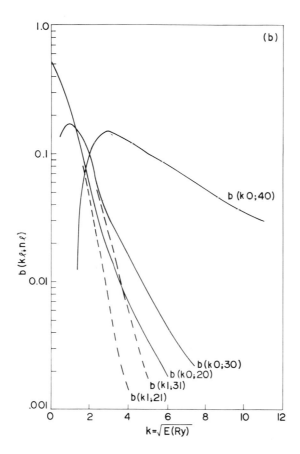

FIG. 2b. The Drang functions of the second kind for the first few $l = 0$ and $l = 1$ states. See Eq. (14).

and are plotted in Figs. 2a and 2b. For large k, we see the overlaps decrease approximately exponentially. For fixed k, the overlaps increase with n (sometimes dramatically), but decrease with l. The Drang functions are susceptible to nodes at low values of k for large values of n, although no more than one node has been found in the functions we have examined.

D. INTEGRALS INVOLVING STURMIAN FUNCTIONS

Integrals involving Sturmians and powers of the independent variable can be done by using expressions (7) and (10) and integrating term by term.

The result is

$$I_p(n'l'; nl) \equiv \int_0^\infty S_{n'l'}(x)x^p S_{nl}(x)\, dx$$

$$= \frac{(-1)^{n+n'+l+l'}}{2^{p+2}}(l+l'+p+2)!\left[\frac{(n-l-1)!(n'-l'-1)!}{(n+l)!(n'+l')!}\right]^{1/2}$$

$$\times \sum_{\tau=0}^{\alpha} \binom{l-l'+p+1}{n'-l'-1-\tau}\binom{l'-l+p+1}{n-l-1-\tau}\binom{l+l'+p+2+\tau}{\tau}$$

$$\tag{15}$$

were $p \geq -(l+l'+2)$, and α is the smaller of $(n-l-1)$ and $(n'-l'-1)$. The following relations among the binomial coefficients may be used:

$$\binom{s}{t} = \frac{s!}{t!(s-t)!}, \qquad s \geq t > 0 \tag{16}$$

$$\binom{-s}{t} = (-1)^t\binom{s+t-1}{t} \tag{17}$$

$$\binom{s}{t} = 0, \qquad t > s > 0. \tag{18}$$

In particular, some diagonal values of (15) are

$$I_0(nl; nl) = n/2 \tag{19}$$

$$I_{-1}(nl; nl) = 1/2 \tag{20}$$

$$I_{-2}(nl; nl) = 1/(2l+1). \tag{21}$$

Again, by integrating the explicit polynomial expressions for the Sturmians, the multipole integral appears as

$$W_p(nl, n'l'; r) \equiv 4\int_0^\infty S_{n'l'}(r')\frac{1}{r'}\frac{r_<^p}{r_>^{p+1}}S_{nl}(r')\, dr'$$

$$= 4\sum_{\kappa=0}^{n-l-1}\sum_{\kappa'=0}^{n'-l'-1}(-1)^{\kappa+\kappa'}C_\kappa(nl)C_{\kappa'}(n'l')(2r)^A$$

$$\times \left\{\frac{(A+p+1)!}{(2r)^{p+1}} - \sum_{\sigma=0}^{A+p+1}e^{-2r}\frac{(A+p+1)!}{(A+p+1-\sigma)!}(2r)^{-\sigma}\right.$$

$$\left. + \sum_{\sigma=0}^{A-p}e^{-2r}\frac{(A-p)!}{(A-p-\sigma)!}(2r)^{-\sigma}\right\} \tag{22}$$

where

$$C_\kappa(nl) \equiv \frac{[\tfrac{1}{2}(n+l)!(n-l-1)!]^{1/2}}{(n-l-1-\kappa)!(2l+1+\kappa)!\kappa!}$$

and $A \equiv l + l' + \kappa + \kappa'$. While such sums are usually performed by computer, it will be useful to note, for checking purposes, that

$$\sum_{\lambda'=0}^{n'-l'-1} \sum_{\lambda=0}^{n-l-1} (-1)^{\lambda+\lambda'}$$

$$\times \frac{(n-l-1)!(n+l)!(2l+1+\lambda+\lambda')!}{(n-l-1-\lambda)!(n-l-1-\lambda')!(2l+1+\lambda)!(2l+1+\lambda')!k!k'!} = 1.$$

$$(23)$$

E. STURMIAN FUNCTIONS IN MOMENTUM SPACE

Let the Sturmian function with three indices stand for the complete set in three dimensions

$$S_{nlm}(\mathbf{r}) = S_{nl}(r)Y_{lm}(\theta, \phi), \qquad (24)$$

where Y_{lm} is the conventional spherical harmonic. Further, denote by $\tilde{S}_{nlm}(\mathbf{p})$ the Fourier transform of $S_{nlm}(\mathbf{r})$:

$$\tilde{S}_{nlm}(\mathbf{p}) = (2\pi)^{-3/2} \int dr \, d\Omega_r \, S_{nlm}(\mathbf{r}) \exp(-i\mathbf{p} \cdot \mathbf{r}) \qquad (25)$$

where $d\Omega_r \equiv \sin \theta_r \, d\theta_r \, d\phi_r$. The Sturmian is now expressed in terms of Laguerre polynomials, Eq. (6), and the exponential is expanded in spherical harmonics. Equation (25) then becomes after integration

$$\tilde{S}_{nlm}(\mathbf{p}) = Y_{lm}(\theta_p \, \phi_p) \left[\frac{2^{4l+3} n(n-l-1)!}{(n+l)!} \right]^{1/2}$$

$$\times \frac{p^l l! (-E_0)^{(2l+3)/4}}{(p^2 - E_0)^{l+1}} C_{n-l-1}^{l+1} \left(\frac{p^2 + E_0}{p^2 - E_0} \right) \qquad (26)$$

where $C_t^s(x)$ is the Gegenbauer polynomial (Erdélyi, 1953). Because of the separation of p and Ω_p in Eq. (26), it is possible to put

$$\tilde{S}_{nlm}(p) = \Phi_{nl}(p)Y_{lm}(\theta_p, \phi_p). \qquad (27)$$

The $\Phi_{nl}(p)$ have the orthogonality property

$$\int_0^\infty dp^2 \, p\Phi_{nl}(p)(p^2 - E_0)^{-1}\Phi_{n'l}(p) = \pi\delta_{nn'}. \qquad (28)$$

The Gegenbauer polynomials in Eq. (26) are defined as (Erdélyi, 1953)

$$C_t^s(x) = \{[\Gamma(t+2s)]/\Gamma(t+1)\Gamma(2s)\}F[t+2s, -t; s+\tfrac{1}{2}; \tfrac{1}{2}(1-x)]$$

$$= \sum_{j=0}^t a_{j,t}^s [(x-1)/2]^j$$

where

$$a_{j,t}^{s} = \frac{(s+j-1)!(2s+t+j-1)!}{j!(s-1)!(2s+2j-1)!(t-j)!}.$$

III. The Close-Coupling Method for Scattering in the $(e^{+}-H)$ and $(e^{-}-H)$ Problems

The close-coupling method of Massey and Mohr (1932) has received widespread attention in recent years. [See Percival and Seaton (1957) and Burke and Smith (1962) for detailed discussions of close coupling.] In this method (as applied to the electron–hydrogen scattering problem) the two-electron wave function is expanded in terms of target eigenstates with unknown scattering coefficients. But only a small number of terms in this expansion can practicably be taken so that the rate of convergence of the expansion is of some concern. In terms of partial waves, the rate of convergence is fair (l^{-4}, Schwartz, 1963), but in terms of the principal quantum number, convergence seems to be poor. Indeed, without continuum states, convergence to the correct solution is impossible; Castillejo et al. (1960) have shown that almost 20% of the polarizability of the hydrogen atom in the ground state lies in the continuum.

The first attempt at applying Sturmian functions to the scattering problem was made by Rotenberg (1962), who substituted them for the usual hydrogenic expansion in the close-coupling method. The results for the cross section for elastic positron–hydrogen scattering differed from the extensive calculations of Schwartz (1961). A numerical error was found in the Sturmian calculation a few years later by both Rotenberg and Schwartz, but the full-scale calculation was never repeated: The question as to how fast the Sturmian expansion converges for the problem is still an open one. The use of Sturmians to include effects of the continuum has the appeal of a systematic development, but there are other ways to include the continuum. Burke and Taylor (1966) introduced the use of "correlation functions" within the close-coupling framework and recently, Burke et al. (1969) made use of a "pseudo-state" expansion with gratifying results. Pseudo-states are linear combinations of eigenfunctions of an operator which appears in the close-coupling formalism. One advantage of pseudo-states over Sturmians is that they are applicable to inelastic scattering problems, but by using linear combinations of Sturmians the same end can be achieved (see Section III, D).

A. THE MODIFIED CLOSE-COUPLING EQUATIONS

The differential equation governing the scattering of the Lth partial wave of positrons of energy k^2 from hydrogen in its ground state is

$$\{-\nabla_1{}^2 - \nabla_2{}^2 - [(2/r_1) - (2/r_2) + (2/r_{12})] - k^2\}\Psi_L(\mathbf{r}_1, \mathbf{r}_2) = 0 \quad (29)$$

where the subscript 1 refers to the electron.

If Ψ_L is expanded in terms of Sturmians for the electron (Rotenberg, 1962)

$$\Psi_L(\mathbf{r}_1, \mathbf{r}_2) = \sum S_{\nu_1\lambda_1}(r_1)Y_{\lambda_1\mu_1}(\Omega_1)\Phi_L(\nu_1\lambda_1\lambda_2 ; r_2)Y_{\lambda_2\mu_2}(\Omega_2)C_{\lambda_1\lambda_2}(LM; \mu_1\mu_2) \quad (30)$$

and substituted into Eq. (29), there results (Greek indices are summed)

$$-\left[\frac{d^2}{dr^2} + \frac{l_2(l_2 + 1)}{r^2} - (k^2 - 1)\right]\Phi_L(nl_1l_2 ; r)$$

$$+ 4\sum (\nu - 1)I_{-2}(nl_1 ; \nu l_2)\Phi_L(\nu l_1l_2 ; r)$$

$$+ \sum U_L(nl_1l_2 ; \nu\lambda_1\lambda_2 ; r)\Phi_L(\nu\lambda_1\lambda_2 ; r) = 0, \quad (31)$$

where $C_{l_1l_2}(LM; m_1m_2)$ is the Clebsch–Gordon coefficient (Rose, 1957), I_{-2} is defined in Eq. (15), and the shielded nuclear potential seen by the positron is

$$U_L(nl_1l_2, n'l_1'l_2'; r_2)$$

$$\equiv -\frac{2}{r_2} + \sum \int dr_1 \, d\Omega_1 \, d\Omega_2 \, S_{nl_1}(1)Y^*_{l_1\mu_1}(1)Y^*_{l_2\mu_2}(2)\frac{2}{r_1} \cdot \frac{2}{r_{12}}$$

$$\times S_{n_1'l_1'}(1)Y_{l_1'\mu_1'}(1)Y_{l_1'\mu_2'}(2)C_{l_1'l_2'}(LM; \mu_1'\mu_2')C_{l_1l_2}(LM; \mu_1\mu_2)$$

$$= -\frac{2}{r_2} + \sum W_\lambda(nl_1l_2, n'l_1'l_2'; r_2)(-1)^{L+l_1+l_1'}[l_1l_1'l_2l_2']^{1/2};$$

$$\times \begin{pmatrix} l_1 & l_1' & \lambda \\ 0 & 0 & 0 \end{pmatrix} \begin{pmatrix} l_2 & l_2' & \lambda \\ 0 & 0 & 0 \end{pmatrix} \begin{Bmatrix} l_1 & l_2 & L \\ l_2' & l_1' & \lambda \end{Bmatrix} \quad (32)$$

The symbol $[x] \equiv (2x + 1)$, the definition of W_λ is found in Eq. (22) and the factors at the end of Eq. (32) are the 3j and 6j symbols defined in Edmonds (1957) or Rotenberg et al. (1959). The reduction of Eq. (29) to Eq. (31) involves using the orthogonality of the Sturmians, the spherical harmonics, and the Clebsch–Gordon coefficients in the usual manner.

Since the first member of the set of Sturmians is proportional to the ground-state function of hydrogen, the boundary conditions remain simple in the elastic case,

$$\Phi_L(nl_1l_2 ; r) \xrightarrow[r \to \infty]{} 0 \quad (33)$$

for all quantum numbers except

$$\Phi_L(1, 0, l_2 ; r) \xrightarrow[r \to \infty]{} \sin(kr - L\pi/2 + \delta_L). \quad (34)$$

The subsequent interpretation of the phase shifts in terms of elastic scattering cross sections remains unaltered from the usual hydrogenic treatment (Burke and Smith, 1962).

To use Eq. (31) for electron–hydrogen scattering, the signs of the appropriate Coulomb potentials must be changed, and there must be appended to it the exchange potential $\pm X_L(n_1 l_1 l_2)$, where the positive or minus sign is used for the singlet or triplet case, respectively.

$$
X_L(r)
$$

$$
= 2 \Bigg\{ \sum (-1)^{L+l_1+l_2} \int dr_1 \Bigg[\left(\frac{2l_1}{r_1{}^3} + \frac{2(n-2)}{r_1{}^2} + 2\frac{(v-1)}{r_1 r} + \frac{(1-k^2)}{r_1} \right) S_{nl_1}(r_1)
$$

$$
+ \frac{2\sqrt{2}}{r_1^{5/2}(n-l_1-1)} S_{n-(1/2),\, l_1+(1/2)}(r_1) \Bigg] \Phi_L(v l_2\, l_1\,;\, r_1) \Bigg\} S_{v l_2}(r)
$$

$$
+ 2 \sum \int dr_1 S_{nl_1}(r_1) \frac{2r_<{}^\lambda}{r_1 r_>{}^{\lambda+1}} \Phi_L(v \lambda_1 \lambda_2\,;\, r_1)(-1)^{l_1+\lambda}
$$

$$
\times [l_1 \lambda_1 l_2 \lambda_2]^{1/2} \begin{pmatrix} \lambda_1 & l_2 & \lambda \\ 0 & 0 & 0 \end{pmatrix} \begin{pmatrix} l_1 & \lambda_2 & \lambda \\ 0 & 0 & 0 \end{pmatrix} \begin{Bmatrix} l_1 & l_2 & L \\ \lambda_1 & \lambda_2 & \lambda \end{Bmatrix} S_{v \lambda_1}(r). \qquad (35)
$$

The first term in Eq. (35) is the result of integration by parts.

B. USE OF STURMIANS ABOVE ELASTIC THRESHOLD

For a given physical value of E_0 in Eq. (4), there is only one Sturmian function which corresponds to a physical wave function. For this reason, it would seem inappropriate to use Sturmian expansions in scattering problems where more than one channel is open: the boundary condition for the channels in which the Sturmian function and the physical function do not coincide becomes intractable. The appeal of the Sturmians, however, lies in their ability to describe continuum admixtures to the closed channels, thus their usefulness can be retained if they are used only for this purpose. To illustrate the point, suppose N channels are open. Then in order to keep the boundary conditions simple, the total wave function can be expressed as

$$
\Psi(\mathbf{r}_1\mathbf{r}_2) = \sum u_\kappa(\mathbf{r}_1)\Phi_\kappa(\mathbf{r}_2) + \sum T_\mu(r_1)\Phi_\mu(r_2)
$$

where the u_κ are hydrogenic functions and the index $\kappa \equiv (v, \lambda, \mu)$ is summed over open channels only. The second term represents the contribution from the closed channels. The T_μ are linear combinations of the Sturmians corresponding to $E_0 = -1/n'^2$ where n' is the principal quantum number of first closed channel. This ensures that the Sturmian corresponding to a physical wave function does not belong to an open channel. Thus

$$T_1 = \sum a_{1\mu} S_\mu, \qquad T_2 = \sum a_{2\mu} S_\mu', \qquad T_3 = \sum a_{3\mu} S_\mu'', \quad \text{etc.,}$$

where the primes denote that the set of S_μ in each case is different, differing from the other sets by as little as one element. The number of terms in each sum is the same as the number of open channels. The coefficients are determined by requiring that each T_m is orthogonal to every u_k. [See Eq. (102) for the required overlap integrals between hydrogen functions and Sturmians.] Then, for convenience, the T_m can themselves be made mutually orthogonal through the Schmidt procedure.

IV. Application of Sturmians to the Faddeev Equations

The integral formulation of the three-body problem (Faddeev, 1963) was first applied by Ball *et al.* (1968) to the $(e^- - H)$ system. Like the differential methods, however, a choice of representation must be made. As will be seen, one of the important steps leading to tractability is the construction of a suitable expansion for the two-body off-shell t matrix. This problem has been worked out in some detail, and comparison between Coulomb expansion and Sturmian expansions have been made (Chen *et al.*, 1969b).

For the sake of completeness, an outline of the derivation of the Faddeev equations will be drawn, but for details other articles should be referred to, such as Faddeev (1963), Lovelace (1964a,b), or Watson and Nuttall (1967).

A. THE FADDEEV EQUATIONS

The scattering matrix for a three-particle system of total energy s is the solution of

$$T(s) = V + VG_0(s)T(s) \tag{36}$$

where V is the total interaction

$$V = \sum V_\alpha \tag{37}$$

V_α is the two-body interaction between particles β and γ, $G_0(s)$ is the free three-particle Green's function

$$G_0(s) = (s - H_0)^{-1} \tag{38}$$

The two-body off-shell scattering matrix is given by the solution of

$$T_i(s) = V_i + V_i G_0(s)T_i(s). \tag{39}$$

If the three-body scattering matrix is decomposed as

$$T(s) = T^{(1)}(s) + T^{(2)}(s) + T^{(3)}(s), \tag{40}$$

then each component satisfies

$$T^{(i)}(s) = V_i + V_i G_0(s) T(s). \tag{41}$$

The Faddeev equations are satisfied by the $T^{(i)}$ in terms of the two-body matrices T_i and are derived as follows (Ball *et al.*, 1968). Define the operator,

$$\Omega \equiv T^{(i)}(s) - T_i(s) - \sum_{j \neq i} T_i(s) G_0(s) T^{(j)}(s). \tag{42}$$

By using the last three equations one can verify that

$$\Omega = V_i + \sum_{j=1}^{3} V_i G_0 T^{(j)} - V_i - V_i G_0 T_i - \sum_{j \neq i} V_i G_0 T^{(j)}$$

$$- \sum_{j \neq i}^{3} V_i G_0 T_i G_0 T^{(j)} = V_i G_0 \Omega. \tag{43}$$

But $V_i G_0 \neq I$, the identity, which implies that $\Omega = 0$ for each i, so that Eq. (43) becomes the set of coupled equations

$$T^{(i)}(s) = T_i(s) + \sum_{j \neq i} T_i(s) G_0(s) T^{(j)}(s), \qquad i = 1, 2, 3. \tag{44}$$

These are the Faddeev equations.

Lovelace (1964b) has shown that the basis variables which are suitable for representing the operator $T^{(i)}$ in Eq. (44) are

$$\mathbf{p}_1 = [m_3 \mathbf{k}_2 - m_2 \mathbf{k}_3]/[2m_2 m_3(m_2 + m_3)]^{1/2} \tag{45}$$

$$\mathbf{q}_1 = [m_1(\mathbf{k}_2 + \mathbf{k}_3) - (m_2 + m_3)\mathbf{k}_1]/[2m_1(m_2 + m_3)(m_1 + m_2 + m_3)]^{1/2} \tag{46}$$

and cyclic for $T^{(2)}$ and $T^{(3)}$, where m_i and $\mathbf{k}_i$ are the mass and asymptotic momentum for the ith particle. The center-of-mass kinetic energy is

$$H_0 = p_1^2 + q_1^2 = p_2^2 + q_2^2 = p_3^2 + q_3^2. \tag{47}$$

The state vector $|\mathbf{k}_1 \mathbf{k}_2 \mathbf{k}_3\rangle$ can therefore be represented by any of the following

$$|\mathbf{k}_1 \mathbf{k}_2 \mathbf{k}_3\rangle = |\mathbf{p}_1 \mathbf{q}_1\rangle_1 = |\mathbf{p}_2 \mathbf{q}_2\rangle_2 = |\mathbf{p}_3 \mathbf{q}_3\rangle_3. \tag{48}$$

These basis momentum variables are linearly dependent; the relations among them are given by Ball *et al.* (1968).

We now proceed to separate out the angular momentum states in the Faddeev equations.

Let the projection of the ith momentum variable onto the spherical harmonics be

$$|plm_l, qLm_L\rangle_i = Y_{lml}(\hat{p}) Y_{LM_L}(\hat{q}) |p, q\rangle_i. \tag{49}$$

In terms of these partial wave states, the states diagonal in the total angular momentum J and its projection M are

$$|pqJMlL\rangle_i = (-1)^{L-l-M}(2J+1)^{1/2} \sum \begin{pmatrix} J & l & L \\ -M & \mu_l & \mu_L \end{pmatrix} |pl\mu_l, qL\mu_L\rangle \quad (50)$$

where the coefficient of the wave function on the right of Eq. (50) is the $3j$ symbol (Edmonds, 1957, or Rotenberg et al., 1959). The partial wave states in (49) are orthogonal:

$$_i\langle plm_l, qLm_L | p'l'm_l', q'L'm_L'\rangle_i$$
$$= (pq)^{-2}\,\delta(p-p')\,\delta(q-q')\,\delta_{ll'}\,\delta_{LL'}\,\delta_{m_lm_l'}\,\delta_{m_Lm_L'}. \quad (51)$$

Finally, in this representation, the Faddeev equations may be written as

$$\Psi_\alpha^{(i)}(p, q, s) = \Phi_\alpha^{(i)}(p, q, s)$$

$$-\tfrac{1}{4}\sum_{\alpha_j}\sum_{j\neq i}\int_0^\infty dp_j^2 \int_0^\infty dq_j^2\, K_j^{(i)}(pq\alpha \,|\, p_j q_j \alpha_j)$$

$$\times \left[\frac{p_j q_j}{p_j^2 + q_j^2 - s}\right]\Psi_{\alpha_j}(p_j, q_j, s) \quad (52)$$

where

$$\Psi_\alpha^{(i)}(p, q, s) \equiv {}_i\langle p, q, \alpha|\, T^{(i)}(s)\,|k_1 k_2 k_3\rangle \quad (53)$$

$$\Phi_\alpha^{(i)}(p, q, s) \equiv {}_i\langle p, q, \alpha|\, T_i(s)\,|k_1 k_2 k_3\rangle \quad (54)$$

$$K_j^{(i)}(pq\alpha \,|\, p_j q_j \alpha_j) \equiv {}_i\langle pq\alpha|\, T_i(s)\,|p_j q_j \alpha_j\rangle_j \quad (55)$$

and

$$\alpha \equiv (JMlL), \text{ not summed.} \quad (56)$$

To reduce Eq. (52) further, the kernel of Eq. (55) is decomposed as

$$K_j^{(i)}(pq\alpha \,|\, p_j q_j \alpha_j) = (-1)^{L+L'+l+l'} \sum \begin{pmatrix} J & l & L \\ -M & \mu_1 & \mu_2 \end{pmatrix}\begin{pmatrix} J & l' & L' \\ -M & \mu_l' & \mu_L' \end{pmatrix}$$

$$\times \int d\hat{p}_j\, d\hat{q}_j\, d\hat{p}_i\, d\hat{q}_i\, {}_i\langle \mathbf{pq}|\, T_i(s)\,|\mathbf{p}_j\mathbf{q}_j\rangle_j$$

$$\times (2J+1) Y_{l\mu_l}^*(\hat{p}_i) Y_{L\mu_L}^*(\hat{q}_i) Y_{l'\mu_l'}(\hat{p}_j) Y_{L'\mu_L'}(\hat{q}_j) \quad (57)$$

Further, from Eq. (39) it is seen that T_i involves only a two-body potential so that the matrix element in Eq. (57) becomes

$$_i\langle \mathbf{pq}|\, T_i(s)\,|\mathbf{p}_j\mathbf{q}_j\rangle_j = \delta(\mathbf{q}-\mathbf{q}_i)\langle \mathbf{p}|\, \tilde{T}_i(s-q^2)\,|\mathbf{p}_i\rangle \quad (58)$$

where Eqs. (47) and (48) have been used and where

$$\delta(\mathbf{q} - \mathbf{q}_i) \equiv 2q^{-1} \, \delta(q^2 - q_i^2) \, \delta(\cos \theta_q - \cos \theta_{q_i}) \, \delta(\phi_q - \phi_{q_i}). \qquad (59)$$

The two-particle scattering matrix can be decomposed as

$$\langle \mathbf{p} | \, \bar{T}_i(s - q^2) | \mathbf{p}_i \rangle = (1/2\pi^2) \sum (2\lambda + 1) P_l(\cos \theta_{pp_i}) t_l^{(i)}(p, p_i \, ; s - q^2). \qquad (60)$$

The two-particle amplitude for particles j and k is normalized on the energy shell

$$t_l^{(i)}(p, p; p^2) = e^{i\delta_l}(\sin \delta_l / p) \qquad (61)$$

where p^2 is the two-body center-of-mass energy.

Taking the special case $J = 0$, $\alpha = (00ll) \equiv l$, Eq. (52) becomes [with the help of Eqs. (57)–(60)]

$$\Psi_l^{(i)}(p, q, s) = \Phi_l^{(i)}(p, q, s) + \sum_{j \neq i} \sum_{l'=0} \int dq_j^2 \int_{a-}^{a+} dp_j^2$$

$$\times \frac{(-1)^{l+l'}[ll']^{1/2} P_l(\cos \theta_{p_i q_i}) P_{l'}(\cos \theta_{p_j q_j})}{4\pi \alpha_{ij} \beta_{ij}(p_j^2 + q_j^2 - s)}$$

$$\times t_l^{(i)}(p, p_i, s - q^2) \Psi_{l'}^{(j)}(p_j, q_j, s); \qquad i = 1, 2, 3 \qquad (62)$$

where

$$a_\pm \equiv (\alpha_{ij} q_j \pm q)^2 / \beta_{ij}^2$$

$$\alpha_{ij} \equiv \{m_i m_j / [(m_i + m_k)(m_j + m_k)]\}^{1/2}$$

$$\beta_{ij} \equiv (1 - \alpha_{ij}^2)^{1/2}$$

$$\cos \theta_{p_i q_i} = (ij)([\beta_{ij}^2 p_j^2 + \alpha_{ij}^2 q_j^2 - q^2]/2\alpha_{ij} \beta_{ij} p_j q_j)$$

and

$$(ij) = \begin{cases} +1 & \text{for (12), (23), (31)} \\ -1 & \text{otherwise.} \end{cases}$$

It is the representation of the off-shell amplitudes $t_l^{(i)}(p, p_i, s - q^2)$ by means of sums of separable terms that makes Eq. (62) tractable.

B. Expansion of the Off-Shell Amplitude

It is seen that the independent variable p appears only in the scattering amplitude $t_l^{(i)}(p, p', s - q^2)$ in Eq. (62), so that if a representation of $t_l^{(i)}$ were made in which a function of p were a factor, Eq. (62) could be made explicit.

In general, the partial wave components of the two-body amplitudes are obtained by expanding Eq. (41) in terms of spherical harmonics to obtain,

$$t_l^{(i)}(p, p'', E) = V_l^{(i)}(p) - \frac{1}{\pi} \int dp'^2 \, \frac{p'' V_l^{(i)}(p, p')}{p'^2 - E} \, t_l^{(i)}(p', p'', E) \qquad (63)$$

A natural expansion for the $t_l^{(i)}$ can be made by using the functions generated by making Eq. (63) homogeneous (Chen et al., 1969):

$$\lambda_{nl}^{(i)}(E)\Phi_{nl}^{(i)}(p, E) = -\frac{1}{\pi} \int dp'^2 \, \frac{p' V_l^{(i)}(p, p')}{p'^2 - E} \, \Phi_{nl}^{(i)}(p', E). \qquad (64)$$

Equation (64) is clearly the Fourier representation of the Sturmian functions $\Phi_{nl}(x, E)$ for arbitrary reduced mass and negative energy and

$$V^{(i)}(p, p') = \int e^{i(p - p')x} V^{(i)}(x) dx \qquad (65)$$

The solutions are given in terms of Gegenbauer polynomials (see Section II, E)

$$\Phi_{nl}^{(i)}(p, E) = N_{nl}(E) C_{n-l-1}^{l+1}[(p^2 + E)/(p^2 - E)] \qquad (66)$$

where $N_{nl}(E)$ is given in Eq. (26). In momentum space the Coulomb potential is

$$V_l^{(i)}(p, p') + (\mu_i/2)^{1/2}(Z_i/pp')Q_l[(p^2 + p'^2)/2pp'] \qquad (67)$$

where μ_i is the reduced mass of the pair (j, k), Z_i is the product of the charges of the pair (j, k), and Q_l is the Legendre function of the second kind. The normalization is taken as

$$\int dp^2 [p\Phi_{nl}^{(i)}(p, E)\Phi_{ml}^{(i)}(p, E)]/(p^2 - E) = \pi \, \delta_{mn} \qquad (68)$$

and the eigenvalues are

$$\lambda_n^{(i)}(E) = -(Z_i/n)[-(\mu_i/2E)]^{1/2}. \qquad (69)$$

The two-body scattering matrix appears as

$$t_l^{(i)}(p, p', E) = \sum_{v=0}^{\infty} \frac{\lambda_{vl}^{(i)}(E)}{1 - \lambda_{vl}^{(i)}(E)} \, \Phi_{vl}^{(i)}(p, E)\Phi_{vl}^{(i)}(p', E). \qquad (70)$$

The Faddeev equation for $J = 0$, Eq. (62), can now be written explicitly in p:

$$\Psi_l^{(i)}(p, q, s) = \Phi_l^{(i)}(p, q, s) + \sum \frac{\lambda_v^{(i)}(\varepsilon)}{1 - \lambda_{vl}^{(i)}(\varepsilon)} \, \Phi_{vl}(p, \varepsilon)\chi_{vl}^{(i)}(q, s) \qquad (71)$$

where

$$\varepsilon \equiv s - q^2, \quad \text{and} \quad \varepsilon_i \equiv s - q_i^2$$

$$\chi_{nl}^{(i)}(q, s) = \eta_{nl}^{(i)}(q, s) + \sum_{v\lambda} \sum_{j \neq i} \int_0^\infty dq_j{}^2 K_{nl, v\lambda}^{(i, j)}(q, q_j, s) \chi_{v\lambda}^{(j)}(q_j, s), \tag{72}$$

$$\eta_{nl}^{(i)}(q, s) = \sum_{\lambda} \sum_{j \neq i} \int_0^\infty dq_j{}^2 \int_{a-}^{a+} dp_j{}^2$$

$$\times \frac{(-1)^{l+\lambda}[ll']^{1/2} P_l(\cos \theta_{p_i q_i}) P_\lambda(\cos \theta_{p_j q_j})}{4\pi \alpha_{ij} \beta_{ij} q(p_j{}^2 + q_j{}^2 - s)}$$

$$\times \Phi_{nl}^{(i)}(p_i, \varepsilon) \Phi_{n\lambda}^{(j)}(p_j, \varepsilon_j), \tag{73}$$

and

$$K_{nl, n'l'}^{(i, j)}(q, q_j, s) = \int_{a-}^{a+} dp_j{}^2 \frac{(-1)^{l+l'}[ll']^{1/2} P_l(\cos \theta_{p_i q_i}) P_{j}(\cos \theta_{p_i q_i}) \lambda_{n'l'}^{(j)}(\varepsilon_j)}{4\pi \alpha_{ih} \beta_{ij} q(p_j{}^2 + q_j{}^2 - s)[1 - \lambda_{n'l'}^{(j)}(\varepsilon_j)]}$$

$$\times \Phi_{nl}^{(i)}(p_i, \varepsilon) \Phi_{n'l'}^{(j)}(p_j, s - q_j^2). \tag{74}$$

It is the kernel K that carries with it the two-body threshold information. For every two-body energy E below the three-body breakup threshold, the lowest Sturmian eigenvalue is unity and the corresponding eigenfunction is physical. The denominator of K vanishes for every projectile energy q^2 such that $q^2 = s - E$, creating a branch point. Elastic scattering from a two-particle system in its ground state occurs between the lowest two of these points; between the second and third points an inelastic channel is open, another inelastic channel opens between the third and fourth points, etc.

C. States of H^-

The ^1S state is the only bound configuration of H^-. For $J = 0$, the singlet multiplicity requires that $\chi_{nl}^{(1)} = (-1)^l \chi_{nl}^{(2)}$ and that for the electron–electron amplitude $\chi_{nl}^{(3)} = 0$ for odd l. Equations (71)–(74) can be written in matrix form

$$\chi(q, s) = \eta(q, s) + \int_0^\infty dq_j{}^2 K(q, q_j, s) \chi(q_j, s) \tag{75}$$

where the vector χ is

$$\chi^\dagger(q, s) = [\chi_0^{(1)}(q, s), \chi_0^{(3)}(q, s), \chi_1^{(1)}(q, s) \chi_1^{(3)}(q, s), \ldots]. \tag{76}$$

Each element $\chi_l^{(i)}(q, s)$ is a row whose dimension is the number of terms in the Sturmian expansion of t_l, Eq. (70). The subscripts refer to the partial waves. When the integral in Eq. (75) is put in terms of a weighted sum for

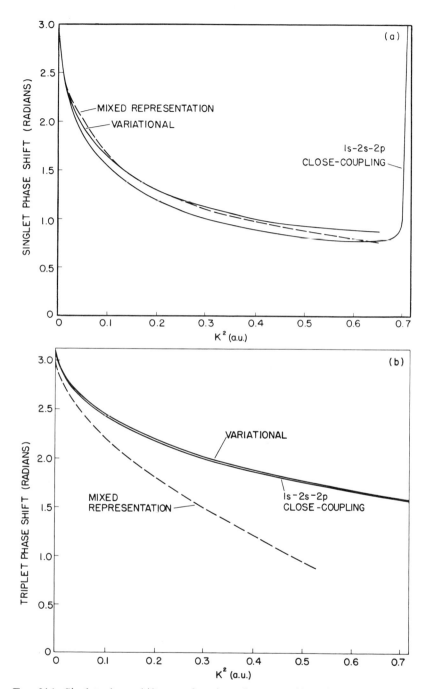

FIG. 3(a). Singlet phase shifts as a function of energy. (b). Triplet phase shifts as a function of energy. The close-coupling results are from Burke and Schey (1962); the curve marked "variational" is that of Schwartz (1961).

numerical purposes, the solution can be written formally as

$$\chi(s) = [I - K(s)]^{-1}\eta(s). \tag{77}$$

The bound H^- state is found at that s which makes $\text{Det}\{I - K(s)\} = 0$. Ball *et al.* (1968), using only the [1s] term in expansion of $t_i^{(i)}$, Eq. (70), found the bound state of H^- at -1.0516 Ry below the three-particle threshold. The best value available is -1.0555 Ry (Perkeris, 1959). (The symbol [nl] refers to Sturmian states.) As terms are added, the results oscillate and seem to converge to a result more positive than that of Pekeris (1959).

Again using only the [1s] Sturmian in the t-matrix expansion of Eq. (70), the singlet and triplet cross section have been calculated by Chen *et al.* (1969) throughout the elastic energy range, as shown in Figs. 3(a) and (b) (p. 251). The singlet phase shifts agree remarkably well with Schwartz' (1961) variational calculation, while the triplet phase shifts behave as one would expect from such a low-order approximation.

In the calculations of resonance widths, Ball *et al.* (1968) have plotted the elastic cross sections in the neighborhood of the lowest $J = 0$ singlet resonance in the [1s] + [2s] + [2p] + [3s] approximation (see Fig. 4). The calculated width is 0.034 eV while the measured width is 0.043 eV (McGowan, 1967).

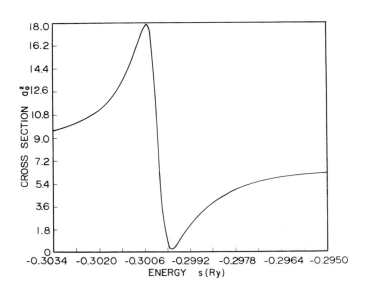

FIG. 4. Energy dependence of the elastic scattering cross section in the neighborhood of the lowest $J = 0$ resonance. The [1s] + [2s] + [2p] + [3s] approximation was used. (Ball *et al.*, 1968.)

By moving the total energy s above the elastic threshold, Ball *et al.* (1968) were able to observe the resonant H^- states. These are the complex poles of Eq. (77). The results are compiled in Table I. It is seen that the singlet resonance oscillates about the accepted value of 0.2973 Ry (see references next paragraph) and does not seem to converge rapidly.

For the triplet case, the symmetries

$$\chi_{nl}^{(1)} + (-1)^{l+1}\chi_{nl}^{(2)} \qquad \text{and} \qquad \chi_{nl}^{(3)} = 0 \qquad \text{for } l \text{ even}$$

are used. The triplet value shown in Table I is in reasonable agreement with calculations of Burke and Schey (1962), Gailitis and Damburg (1963), O'Malley and Geltman (1965), Burke (1965), Burke and Taylor (1966), Mittleman (1966), Holøien and Midtdal (1966), Bhatia *et al.* (1967), Chen (1967), and Chen and Rotenberg (1968).

TABLE I

$J = 0$ Resonance Levels below Three-Body Breakup Threshold (Ry)[a]

Sturmian states in expansion (70)	Singlet	Triplet
[1s]	None found	
[1s] + [2s]	0.286	
[1s] + [2p]	0.291	
[1s] + [2s] + [2p]	0.326	
[1s] + [2f] + [2p] + [3s] + [3p] + [3d]	0.298	
[1s] + [2s] + [2p] + [3s] + [3p]	—	0.257

[a] From Ball *et al.* (1968).

D. Comparison with the Coulomb Expansion

Instead of using Sturmian functions to expand the two-particle t matrix, as in Eq. (70), one can instead use a Coulomb representation (Chen and Ishihara, 1969a, b)

$$t_l^{(i)}(p, p', E) = V(p, p') + (\pi/2) \sum [(p^2 - \varepsilon_\nu^{(i)})(p'^2 - \varepsilon_\nu^{(i)})]/(E - \varepsilon_\nu^{(i)})$$

$$\times \psi_{\nu l}(np)\psi_{\nu l}(np') \tag{78}$$

where

$$V(pp') = -(\pi/4) \sum [(p^2 - \varepsilon_\nu) + (p'^2 - \varepsilon_\nu)]\psi_{\nu l}(np)\psi_{\nu l}(np') \tag{79}$$

$$\varepsilon_n^{(i)} = -(Z_i^2 \mu_i/2n^2)$$

and

$$\psi_{nl}(np) = \left[\frac{2^{4l+5}n(n-l-1)!}{\pi(n+l)!}\right]^{1/2} \frac{l!\,p^l(-\varepsilon_n)^{(2l+5)/4}}{(p^2-\varepsilon_n)^{l+2}}\,C_{n-l-1}^{+1}\left(\frac{p^2+\varepsilon_n}{p^2-\varepsilon_n}\right)$$

(see Section II, E). Equation (79) is the Fourier representation of the Coulomb function. In principle, the summations in Eqs. (78) and (79) include integrations over the continuum.

Both $V_l(p,p')$ and $t_l(p,p')$ have been calculated in terms of a Sturmian series and a Coulomb series by Chen and Ishihara (1969a, b). It is seen from Fig. 5a that the Sturmian expansion oscillates about the exact value of $V(p,p')$ while the Coulomb series converges rapidly to a slightly incorrect result due, presumably, to the absence of the continuum. The expansions for $t(p,p') - V(p,p')$ are rather smooth, the Sturmian expansion oscillating slightly (Fig. 5b). In an attempt to dampen the oscillations, Chen and Chung (1969) have rewritten the Faddeev equations in a mixed mode representation—Coulomb functions for attractive pairs and Sturmian functions for repulsive pairs where the continuum dominates.

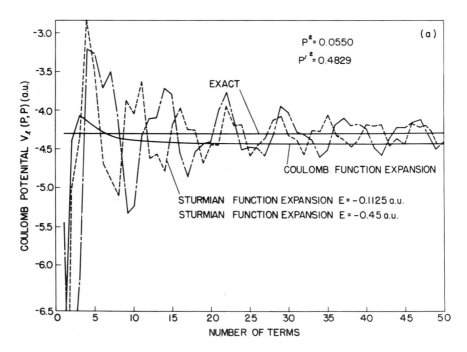

FIG. 5a. Sturmian function expansions of the Coulomb potential. (From Chen and Ishihara, 1969b.)

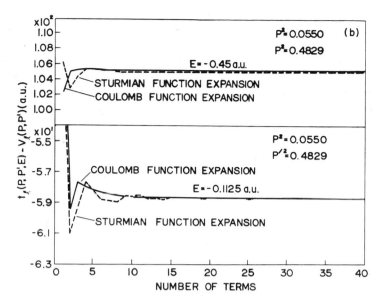

FIG. 5b. Sturmian function expansions of the difference between the two-body scattering matrix and the Coulomb potential. (From Chen and Ishihara, 1969b.)

E. ANOTHER INTEGRAL APPROACH TO THE THREE-BODY PROBLEM

A method reminiscent of Faddeev's, but not so general, has been devised by Eyges (1959, 1961, 1965) and carried out by Jasperse and Friedman (1967) for some heliumlike atoms using a Sturmian expansion. The method involves the same order of complexity as Faddeev's, but deals directly with bound wave functions rather than the more general t matrix. The two methods are quite similar for the on-shell case. An unfortunate degree of freedom appears in Jasperse's equations: when the one-particle wave functions are expanded in terms of Sturmian functions, the strength of the Coulomb potential defining them remains as a parameter. In the helium case, which involves two wave functions, two parameters appear of which the binding energy is a function. The energy chosen is the one for which small changes in the strength parameters is least sensitive.

V. Charge Transfer in (H–H$^+$) Collisions

Even simpler than the three-body problems discussed previously, is the (H–H$^+$) charge transfer problem. It involves only one electron, and in the impact-parameter formulation, is complicated only by the fact that the potential which the electron experiences is time dependent. The literature on this subject is vast, since, in spite of the problem's structural simplicity, the solution has been evasive. The major references in which Sturmian functions were not used are Bates and McCarroll (1958, 1962), Bates (1958), Ferguson (1961), McCarroll (1961), Bates and Williams (1964), Lovell and McElroy (1965), Wilets and Gallaher (1966), Cheshire (1968), and Schneiderman and Russek (1969). The appropriateness and applicability of the impact parameter formulation has been discussed by Mittleman (1961) and Wilets and Wallace (1967). Above collision energies of a few hundred volts, there seems to be little question but that the impact formulation introduces negligible error into charge transfer cross sections.

A. FORMULATION OF THE PROBLEM IN TERMS OF STURMIANS

The equation to be solved is

$$[-\nabla^2 - 2r_A^{-1} + r_B^{-1} - r_{AB}^{-1}]\Psi(\mathbf{r}) = 0 \tag{80}$$

where the variables are defined in Fig. 6. The wave function $\psi(\mathbf{r})$ is expanded about each of the nuclei A and B in a time-dependent linear combination of functions of the form

$$w_{nlm}(A, B) = \Phi(r_{A, B}) \exp(\mp ivz/2) \exp[-i\varepsilon_{nl} + \tfrac{1}{8}v^2 t] \tag{81}$$

where $\mathbf{v}$ is the relative velocity of the protons and ε_{nl} is the mean energy [see Eq. (86)]. The Φ's are usually expressed in terms of hydrogenic functions but that choice is a matter of convenience rather than rigor. The most extensive calculations have been carried out by Wilets and Gallaher (1966)

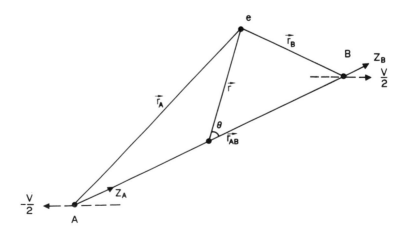

FIG. 6. Coordinate system used in (H–H$^+$) problem. Angles are measured relative to the line joining nuclei A and B.

who used a hydrogenic expansion, and Gallaher and Wilets (1968) who used Sturmian functions in Eq. (81).

The Sturmians used by Gallaher and Wilets (1968) are defined by Eq. (4) with

$$E_0 = -[1/(l+1)^2]. \tag{82}$$

This has the effect of making $S_{n,n-1}$ proportional to the hydrogen function of the same quantum numbers. Rather than using the normalization given by Eq. (9), it was found more convenient to use

$$\int S_{nl}^2(r)\, dr = 1 \tag{83}$$

which, in turn modifies the orthogonality integral, Eq. (3), to

$$\int dr S_{nl}(r)(2/r) S_{n'l}(r) = [2/n(l+1)]\, \delta_{nn'}. \tag{84}$$

The eigenvalues are, by Eq. (8),

$$\alpha_{nl} = n/(l+1). \tag{85}$$

In Eq. (81), the mean energy, ε_{nl}, is given by Eqs. (82) and (84)

$$\varepsilon_{nl} \equiv \int dr S_{nl}(r)[-\nabla^2 - (2/r)] S_{nl}(r) = [1/(l+1)^2] - [2/n(l+1)] \tag{86}$$

With these definitions, Ψ, in Eq. (80), can be expanded as

$$\psi(r, t) = \tfrac{1}{2} \sum \{ (b_{v\lambda}^+ + b_{v\lambda}^-)[S_{v\lambda}(r_A)/r_A] \mathscr{Y}_{v\lambda}(\theta_A, \phi_A)$$
$$+ (-1)^\lambda (b_{v\lambda}^+ - b_{v\lambda}^-)[S_{v\lambda}(r_B)/r_B] \mathscr{Y}_{v\lambda}(\theta_B, \phi_B) \} \tag{87}$$

where the b's are time-dependent coefficients and

$$\mathscr{Y}_{lm}(\theta_i, \phi_i) = \begin{cases} Y_{l0} & m = 0 \\ 2^{-1/2}[Y_{lm} + (-1)^l Y_{lm}^*] & m > 0. \end{cases} \tag{88}$$

Equation (88) insures invariance under reflections in the collision plane $(\phi_i \to -\phi_i)$, while the combination of coefficients in Eq. (87) gives the invariance of inversion through the center of mass $(r \to -r; r_{AB} \to -r_{BA})$.

The boundary conditions of the problem are most easily expressed through hydrogenic functions. Suppose that instead of Eq. (87), the wave functions were expressed as

$$\psi(r, t) = \tfrac{1}{2} \sum \{ (a_{v\lambda}^+ + a_{v\lambda}^-)[u_{v\lambda}(r_A)/r_A] \mathscr{Y}_{v\lambda}(\theta_A, \phi_A)$$
$$+ (-1)^\lambda (a_{v\lambda}^+ - a_{v\lambda}^-)[u_{v\lambda}(r_B)/r_B] \mathscr{Y}_{v\lambda}(\theta_B, \phi_B) \} \tag{89}$$

where the $u_{nl}(r)$ are the usual hydrogen functions. In both expansions (87) and (89), assuming that at $t = -\infty$ the hydrogen atom was in its ground state

$$a_{nl}^{\pm}(-\infty) = b_{nl}^{\pm}(-\infty) = \delta_{n1}\,\delta_{l0}\,. \qquad (90)$$

In the hydrogenic expansion,

$$D_{nl} = \tfrac{1}{2}[a_{nl}^{+}(+\infty) + a_{nl}^{-}(+\infty)] \qquad (91)$$

$$X_{nl} = \tfrac{1}{2}[a_{nl}^{+}(+\infty) - a_{nl}^{-}(+\infty)], \qquad (92)$$

where D_{nl} is the amplitude for the direct reaction and X_{nl} is that for the exchange reaction. These amplitudes will be obtained in terms of the Sturmian amplitudes [see Eq. (106)].

B. The Time-Dependent Coupled Equations

In this section is derived the equations for the b's in Eq. (87). We follow the method used by Gallaher and Wilets (1968).

We wish to solve the Schroedinger equation

$$i\dot{\Psi} = H\Psi \qquad (93)$$

with H give given by Eq. (80) for $-\infty < t < \infty$. Ψ is expressed as

$$\Psi(r, t) = \sum_{\pi\nu\lambda} b_{\nu\lambda}^{\pi}(t)W_{\nu\lambda}^{\pi}(r, t) \qquad (94)$$

where

$$W_{nl}^{\pi}(r, t) = 2^{-1/2}[w_{nl}(A) + \pi(-1)^{l}w_{nl}(B)] \qquad (95)$$

and where w_{nl} is defined in Eq. (81). The superscript π gives the parity upon interchange of A and B. Substitution into Eq. (93) gives

$$i\sum N_{nl,\,\nu\lambda}^{\pi}\,\dot{b}_{\nu\lambda}^{\pi} = \sum H_{nl,\,\nu\lambda}^{\pi}\,b_{\nu\lambda}^{\pi} \qquad (96)$$

where

$$N_{nln'l'}^{\pi} = \int d\mathbf{r}\,W_{nl}^{\pi*}\,W_{n'l'}^{\pi}$$

$$= \langle \Phi_{nl}|\Phi_{n'l'}\rangle + \pi(-1)^{l}\langle \Phi_{nl}(B)|\,e^{-i\nu z}\,|\Phi_{n'l'}(A)\rangle\,\exp i(\varepsilon_{nl} - \varepsilon_{n'l'})t$$

$$\equiv \hat{N}_{nl,\,n'l'}^{\pi}\,\exp i(\varepsilon_{n} - \varepsilon_{n}')t \qquad (97)$$

and

$$H_{nl,\,n'l'}^{\pi} = \int d\mathbf{r} W_{nl}^{\pi}\left(H - i\frac{\partial}{\partial t}\right)W_{n'l'}^{\pi}$$

$$= \left\{\frac{\hat{N}_{nl,\,n'l'}}{R} - \langle\Phi_{nl}(A)|\,2r_B^{-1}\,|\Phi_{n'l'}(A)\rangle\right.$$

$$+ \left[\frac{1}{n'(l'+1)} - \frac{1}{(l'+1)^2}\right](\hat{N}_{nl,\,n'l'} - \delta_{nn'}\,\delta_{ll'})$$

$$+ 2\pi(-1)^l\left[\left(\frac{n'}{l'+1} - 1\right)\langle\Phi_{nl}(B)|\,r_A^{-1}\,\exp(-ivz)\,|\Phi_{n'l'}(A)\rangle\right.$$

$$\left. - \langle\Phi_{nl}(B)|\,r_B^{-1}\,\exp(-ivz)\,|\Phi_{n'l'}(A)\rangle\right]$$

$$+ \frac{d\Theta}{dt}\left[\langle\Phi_{nlm}(A)|\,\mathcal{L}_A\,|\Phi_{n'l'm'}(A)\rangle\right.$$

$$\left.\left. + \pi(-1)^l\langle\Phi_{nl}(B)|\,e^{-ivz}\mathcal{L}_A\,|\Phi_{n'l'}(A)\rangle\right]\right\}\exp i(\varepsilon_{nl} - \varepsilon_{n'l'})t$$

$$\equiv \hat{H}_{nl',\,n'l'}\exp i(\varepsilon_{nl} - \varepsilon_{n'l'})t \tag{98}$$

with

$$\Phi_{nl}(A) = [S_{nl}(A)/r_A]\mathcal{Y}_{lm}(\theta_A, \phi_A) \tag{98a}$$

(the subscript m is suppressed unless the matrix element is nondiagonal in it) and

$$\mathcal{L}_A \equiv -i[z_A(\partial/\partial x_A) - x_A(\partial/\partial z_A)]. \tag{99}$$

The following matrix elements must be evaluated for the coefficients in the coupled equations (96):

$$\langle\Phi_{nl}|\Phi_{n'l'}\rangle \tag{100a}$$

$$\langle\Phi_{nl}(B)|\,e^{-ivz}\,|\Phi_{n'l'}(A)\rangle \tag{100b}$$

$$\langle\Phi_{nl}(B)|\,\frac{1}{r_{A,\,B}}\,e^{-ivz}\,|\Phi_{n'l'}(A)\rangle$$

$$= \langle\Phi_{n'l'}(B)|\,\frac{1}{r_{B,\,A}}\,e^{-ivz}\,|\Phi_{nl}(A)\rangle(-1)^{l+l'} \tag{100c}$$

$$\langle\Phi_{nl}(B)|\,\frac{1}{r_A}\,|\Phi_{n'l'}(B)\rangle \tag{100d}$$

$$\langle\Phi_{nl}(A)|\,\mathcal{L}_A\,|\Phi_{n'l'}(A)\rangle \tag{100e}$$

$$\langle\tilde{\Phi}_{nl}|\Phi_{n'l'}\rangle \tag{100f}$$

where $\tilde{\Phi}_{nl}$ is the same as Φ_{nl} [Eq. (98a)] except that the Sturmian is replaced by the hydrogen function of the same (n, l). The last matrix element is encountered in the problem of asymptotic amplitudes (Section V, C). The matrix element

$$\langle \Phi_{nl}(B) | e^{-ivz} \mathscr{L}_A | \Phi_{n'l'}(A) \rangle$$

can be reduced to the other types after $\mathscr{L}_A$ operates to the right.

Element (100a) can be evaluated by using Eq. (15) and observing the normalization (83)

$$\langle \Phi_{nl} | \Phi_{n'l'} \rangle = \delta_{ll'} \frac{(-1)^{n+n'}}{2} \left[\frac{(n-l-1)!(n'-l-1)!}{nn'(n+l)!(n'+l)!} \right]^{1/2}$$

$$\times \sum \frac{(2l+\sigma+2)!}{\sigma!(n-l-\sigma-1)!(n'-l-\sigma-1)!(\sigma+2-n+l)!(\sigma+2-n'+l)!}$$

$$\tag{101}$$

The index σ is summed until the argument of any factorial goes to zero. These are proportional to the Drang coefficients [Eq. (14)]. The overlap (100f) is given by Gallaher and Wilets (1968) as

$$\langle \tilde{\Phi}_{nl} | \Phi_{n'l'} \rangle = \frac{1}{2} \left[\frac{(n-l-1)!(n'-l-1)!}{nn'(n+l)!(n'+l)!} \right]^{1/2} \frac{(2l+2)!}{[n(l+1)]^{l+3/2}}$$

$$\times \sum \frac{\binom{n+l}{\lambda}\binom{n'+l}{\mu}\binom{-2l-3}{n'+n-2l-2-\lambda-\mu}\binom{n'+n-2l-2-\lambda-\mu}{n-l-1-\lambda}}{a^{1+n'+n-\lambda-\mu} n^{n-l-1-\lambda}(l+1)^{n'-l-1-\mu}}$$

$$\tag{102}$$

where

$$a = \frac{1}{2}\left(\frac{1}{n} + \frac{1}{l+1}\right).$$

In evaluating Eq. (102) observe the symmetries given by Eq. (17). Element (100f) is given by

$$\langle \Phi_{nlm} | \mathscr{L}_A | \Phi_{n'l'm'} \rangle \left[\frac{(l-m)(l+m+1)}{2-\delta_{m,0}} \right]^{1/2} \langle \Phi_{nlm} | \Phi_{n'l'm'+1} \rangle$$

$$- (1 - \delta_{m,0}) \left[\frac{(l+m)(l-m+1)}{2(2-\delta_{m,0})} \right]^{1/2} \langle \Phi_{nlm} | \Phi_{n'l'm'-1} \rangle. \tag{103}$$

The matrix elements including $\exp(-ivz)$ as a factor involve numerical integrals of a type which are not peculiar to Sturmians (Wilets and Gallaher, 1966); those of the form (100e) can be evaluated term by term, but there seems to be no general closed formula for them.

C. PROJECTION OF STURMIAN AMPLITUDE

Because the Sturmians are complete while the set of bound hydrogenic functions are not, the ionization amplitude can be obtained, in principle, from a Sturmian expansion but not from a bound hydrogenic expansion. In fact, however, only an indeterminate amount of ionization is described because the expansion is truncated. Quantitatively, when Sturmians are used, one has

$$P^{ion} + \sum_{nl} (P_{nl}^{D} + P_{nl}^{X}) = 1 \tag{104}$$

where $P_{nl}^{D,X}$ are the probabilities of direct and exchange reactions to the state (n, l). The ionization term in Eq. (104) is missing when hydrogenic functions are used. If one is willing to ignore the ionization amplitude in Eq. (104) and normalize only the summation, then

$$P_{total}^{D,X} = \tfrac{1}{4}[b_{nl}^{+} \pm b_{nl}^{-}]^{*}\langle \Phi_{nl} | \Phi_{n'l}\rangle[b_{n'l}^{+} \pm b_{n'l}^{-}] \exp[i(\varepsilon_{nl} - \varepsilon_{n'l})t]$$

where the b's are the solutions to Eq. (96), evaluated at a time when the two particles of the (H, H^{+}) system are no longer interacting.

D. NUMERICAL RESULTS

Some results of the calculation of Gallaher and Wilets (1968) are shown in Figs. 7 through 14. The curves shown in these figures are taken from Gallaher

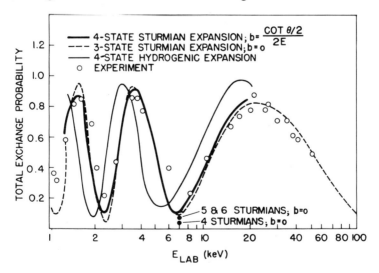

FIG. 7. Charge-transfer probability at 3° as a function of energy. The three-state Sturmian approximation for $b = 0$ includes [1s, 2s, 2p], $m = 0$ only. The four-state approximation includes both $m = 0, 1$ for the 2p state. The experimental points are from Helbig and Everhart (1965).

and Wilets (1968) and Wilets and Gallaher (1966). In Fig. 7 is shown the charge transfer probability at 3° as a function of energy. The Sturmian results agree with the data (Helbig and Everhart, 1965) much better than do hydrogenic results. Furthermore, adding Sturmians seems to have little effect on results at 7 keV. What effect on the phase this has is not known. The Sturmian expansion results seem to be converging also to the total cross-section data of McLure (1965), Fig. 8.

Less decisive than the total cross sections, however, are the results concerning excitation probabilities. As calculated using a hydrogenic expansion (Wilets and Gallaher, 1966), the direct and charge transfer cross sections with excitation to the 2s state shown in Fig. 9, seem to be in good agreement with experiment above 50 keV. The one experimental point at 45 keV agrees more with the 4-state Sturmian expansion than it does with the 6-state. The experimental points for charge transfer with 2p excitation, Fig. 10, also agree more with the 4-state Sturmian calculation than with the 8-state. In fact, the 8-state Sturmian expansion seems to agree with the 4-state hydrogenic. The direct 2p excitation cross sections are shown in Fig. 11. The 4-state Sturmian results are better than the 8-state Sturmian results, which leads one to wonder about convergence.

In Figs. 12 and 13 are shown results for direct and transfer probabilities with excitation to the 2p state as a function of impact parameter at 25 keV, of 4-, 8-, and 9-state Sturmian expansions, as well as a 4-state hydrogenic expansion. That the Sturmian expansion is converging, there seems little

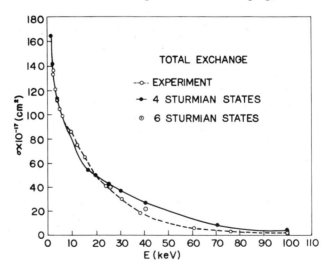

FIG. 8. Total cross sections for charge exchange. The four-state approximation includes the Sturmians [1s, 2s, 2p]. The six-state approximation includes [1s, 2s, 2p, 3s, 4d]. The experimental points are from McLure (1965).

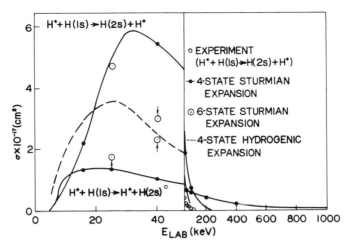

FIG. 9. Total cross sections for direct and charge exchange transfer to the 2s state. The experimental point is from Ryding *et al.* (1966).

doubt, and the results at impact parameters less than 6 Bohrs seem markedly different from the hydrogenic results. It would be interesting to see similar curves at lower energies to shed some light on the apparent lack of convergence in the excitation cross sections.

Finally, in Fig. 14 are compared the Sturmian and hydrogenic results for the resonant charge transfer cross section. The results agree above 15 keV. Below 15 keV there appears to be some structure in the hydrogenic treatment (Wilets and Gallaher, 1966) which is totally absent when Sturmians are used. [In the paper of Wilets and Gallaher (1966) the solid curve in Fig. 14 is shown

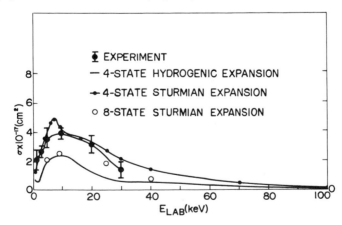

FIG. 10. Total cross sections for charge transfer to the 2p state. The experimental curve is from Stebbings *et al.* (1965).

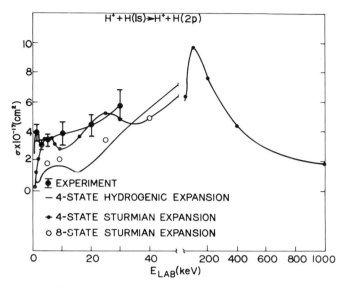

FIG. 11. Cross sections for direct 2p excitation. The experimental curve is from Stebbings *et al.* (1965).

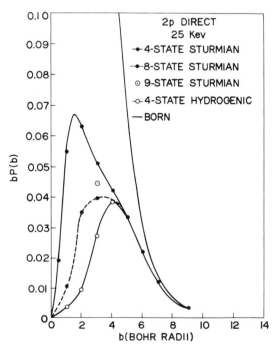

FIG. 12. Probability times impact parameter vs. impact parameter for 2p direct excitation at 25 keV. The Born curve is from Van den Bos (1966).

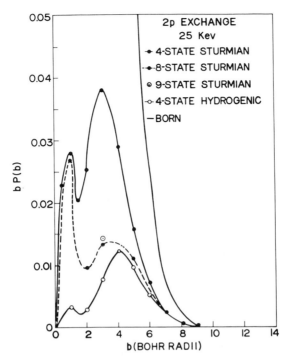

FIG. 13. Probability times impact parameter vs. impact parameter for 2p exchange excitation at 25 keV. The Born approximation is from Van den Bos (1966).

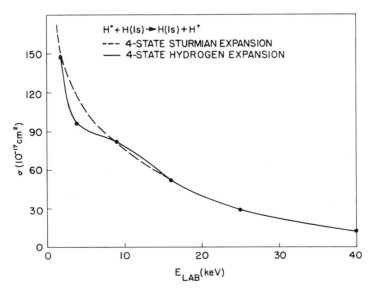

FIG. 14. Comparison of the Sturmian and hydrogenic results for the resonant charge exchange cross section. The curves are a superposition of curves taken from Wilets and Gallaher (1966) and Gallaher and Wilets (1968). See remark at the end of Section V, D regarding the low energy behavior of the hydrogenic results.

with a peak at about 2 keV. The author has been informed by Professor Wilets (private communication) that the point at 1 keV is in error and the peak probably does not exist. Because of the uncertainty, we have omitted the 1 keV point in our Fig. 14.]

VI. The Sturmian Functions with More Than One Potential

Up to this point we have defined the Sturmian functions with respect to the total energy and a single two-body potential. The definition can be generalized somewhat by including more than one two-body potential. Such a step is quite natural when one considers the Born–Oppenheimer method in the Sturmian representation (Lhagva and Zakhar'ev, 1968).

The Born–Oppenheimer Equations

We define a set of Sturmians with the equation

$$\{-(1/2m)(d^2/dr^2) + \alpha_n(R)[V_{12}(R) + V_{23}(r, R) + V_{13}(r, R)]\}$$
$$\times S_n(r, R) = E_0(R)S_n(r, R) \quad (105)$$

where we have, for simplicity, dropped all consideration of angular momentum. In the Sturmians thus defined, the variable R is considered a parameter. With $\alpha_1(R) = 1$, $E_0(R)$ is the lowest eigenvalue of Eq. (105) for every R. This defines $E_0(R)$. With $E_0(R)$ known, the Sturmian eigenvalue $\alpha_n(R)$ is obtained as a function of R for $n > 1$. Note that $S_1(r, R)$ will be the adiabatic atomic wave function, but all others will be nonphysical. If the expansion

$$\Psi(r, R) = \sum \phi_\nu(R)S_\nu(r, R) \quad (106)$$

is substituted into the Schroedinger equation

$$[-(1/2m)(\partial^2/\partial r^2) - (1/2M)(\partial^2/\partial R^2) + W(r, R)]\Psi = E\Psi, \quad (107)$$

where

$$W(r, R) \equiv V_{12}(R) + V_{23}(r, R) + V_{13}(r, R), \quad (108)$$

the results then multiplied on the left by $S_m(r, R)W(r, R)$ and integrated with respect to r, we have

$$-\frac{1}{2M}\frac{d^2}{dR^2}\phi_n(R) = \sum \left[V_{n\nu}(R) - \frac{1}{2M}J_{n\nu}(R) - \frac{1}{M}I_{n\nu}(R)\frac{d}{dR}\right]$$
$$\times \phi_\nu(R) = [E - E_0(R)]\phi_n(R) \quad (109)$$

where

$$V_{nn'} = [1 - \alpha_{n'}(R)] \int dr S_n(r\ R) W^2(r, R) S_{n'}(r, R) \tag{110}$$

$$J_{nn'} \equiv \int dr S_n(r, R) W(r, R) (d^2/dR^2) S_{n'}(r, R) \tag{111}$$

and

$$I_{nn'} \equiv \int dr S_n(r, R) W(r, R) (d/dR) S_{n'}(r, R). \tag{112}$$

Aside from the appearance of the Sturmians, the only formal difference between Eq. (109) and that derived in the normal way (Slater, 1968) is the appearance of extra factors of W. The effect of these modifications have not been calculated.

REFERENCES

Ball, J. S., Chen, J. C. Y., and Wong, D. Y. (1968). *Phys. Rev.* **173**, 202.

Bates, D. R. (1958). *Proc. Roy. Soc.* **A245**, 299.

Bates, D. R., and McCarroll, R. (1958). *Proc. Roy. Soc.* **A245**, 175.

Bates, D. R., and McCarroll, R. (1962). *Advan. Phys.* **11**, 39.

Bates, D. R., and Williams, D. A. (1964). *Proc. Phys. Soc. (London)* **83**, 425.

Bhatia, A. K., Temkin, A., and Perkins, J. F. (1967). *Phys. Rev.* **153**, 177.

Burke, P. G. (1965). *Advan. Phys.* **14**, 521.

Burke, P. G., and Schey, H. M. (1962). *Phys. Rev.* **126**, 149.

Burke, P. G., and Smith, K. (1962). *Rev. Mod. Phys.* **34**, 458.

Burke, P. G., and Taylor, A. J. (1966). *Proc. Phys. Soc. (London)* **88**, 549.

Burke, P. G., Gallaher, D. F., and Geltman, S. (1969). *6th Intern. Conf. Phys. Electron. Atomic Collisions*, p. 370. MIT Press, Cambridge, Massachusetts.

Castillejo, L., Percival, I. C., and Seaton, M. J. (1960). *Proc. Roy. Soc.* **A234**, 259.

Chen, J. C. Y. (1967). *Phys. Rev.* **156**, 150.

Chen, J. C. Y., and Chung, K. T., (1970). In press.

Chen, J. C. Y., and Ishihara, T. (1969a). *J. Phys. B. (At. Mol. Phys.)* **2**, 12.

Chen. J. C. Y., and Ishihara, T. (1969b). *Phys. Rev.* **186**, 25.

Chen, J. C. Y., and Rotenberg, M. (1968). *Phys. Rev.* **166**, 7.

Chen, J. C. Y., Chung, K. T., and Kramer, P. J. (1969). *Phys. Rev.* **184**, 64.

Cheshire, T. M. (1968). *Proc. Phys. Soc. (London)* **1B**, 428.

Edmonds, A. R. (1957). "Angular Momentum in Quantum Mechanics." Princeton Univ. Press, Princeton, New Jersey.

Erdélyi, A., ed. (1953). "Bateman Manuscript Project—Higher Transcendental Functions," Vol. I, p. 175. McGraw-Hill, New York.

Eyges, L. (1959). *Phys. Rev.* **115**, 1643.

Eyges, L. (1961). *Phys. Rev.* **121**, 1744.

Eyges, L. (1965). *J. Math. Phys.* **6**, 1320.

Faddeev, L. D. (1963). "Mathematical Problems of the Quantum Theory of Scattering for a Three-Particle System," Publ. Steklov Math. Inst., Leningrad. English Translation: H. M. Stationery Office, London, 1964.

Ferguson, A. F. (1961). *Proc. Roy. Soc.* **A264**, 540.

Gailitis, M., and Damburg, R. (1963). *Proc. Phys. Soc.* (*London*) **82**, 192.

Gallaher, D. F., and Wilets, L. (1968). *Phys. Rev.* **169**, 139.

Helbig, H. F., and Everhart, E. (1965). *Phys. Rev.* **140**, A715.

Holøien, E. (1956). *Phys. Rev.* **104**, 1301.

Holøien, E., and Midtdal, J. (1966). *J. Chem. Phys.* **45**, 2209.

Hylleraas, E. A., and Midtdal, J. (1958). *Phys. Rev.* **109**, 1013.

Jasperse, J. R., and Friedman, M. H. (1967). *Phys. Rev.* **159**, 69.

Kemble, E. C. (1937). "Fundamental Principles of Quantum Mechanics," Appendix G. McGraw-Hill, New York.

Lhagva, O., and Zakhar'ev, B. N. (1968). *Bull. Acad. Sci. USSR, Phys. Ser.* **32**, 239. (*Izv. Akad. Nauk. SSSR. Ser Fiz.* **32**, 264.)

Lovelace, C. (1964a). *In* "Strong Interactions of the Quantum Theory of Scattering" (R. G. Moorhouse, ed.). Oliver & Boyd, Edinburgh and London.

Lovelace, C. (1964b). *Phys. Rev.* **135**, B1225.

Lovell, S. E., and McElroy, M. B. (1965). *Proc. Roy. Soc.* **A283**, 100.

McCarroll, R. (1961). *Proc. Roy. Soc.* **A264**, 547.

McGowan, J. W. (1967). *Phys. Rev.* **156**, 165.

McLure, G. W. (1965). *Phys. Rev.* **140**, A769.

Massey, H. S. W., and Mohr, C. B. O. (1932). *Proc. Roy. Soc.* **A136**, 289.

Mittleman, M. H. (1961). *Phys. Rev.* **122**, 499.

Mittleman, M. H. (1966). *Phys. Rev.* **147**, 73.

O'Malley, T. F., and Geltman, S. (1965). *Phys. Rev.* **137**, A1344.

Pekeris, C. L. (1959). *Phys. Rev.* **115**, 1216.

Pekeris, C. L. (1962). *Phys. Rev.* **126**, 1470.

Percival, I. C., and Seaton, M. J. (1957). *Proc. Cambridge Phil. Soc.* **53**, 654.

Rose, M. E. (1957). "The Elementary Theory of Angular Momentum." Wiley, New York.

Rotenberg, M. (1962). *Ann. Phys.* (*N.Y.*) **19**, 262.

Rotenberg, M., Bivins, R., Metropolis, N., and Wooten, J. K., Jr. (1959). "The 3j and 6j Symbols." MIT Press, Cambridge, Massachusetts.

Ryding, G., Wittkower, A. B., and Gilbody, H. B. (1966). *Proc. Phys. Soc.* **89**, 547.

Schneiderman, S. B., and Russek, A. (1969). *Phys. Rev.* **181**, 311.

Schwartz, C. L. (1961). *Phys. Rev.* **124**, 1468.

Schwartz, C. L. (1963). *In* "Methods of Computational Physics" (B. Alder S. Fernbach, and M. Rotenberg, eds.), Vol. 2, p. 262. Academic Press, New York.

Schull, H., and Löwdin, P.-O. (1959). *J. Chem. Phys.* **30**, 617.

Slater, J. C. (1968). "Quantum Theory of Molecules and Solids," Vol. I. Chapter 1. McGraw-Hill, New York.

Stebbings, R. F., Young, R. A., Oxley, C. L., and Ehrhardt, H. (1965). *Phys. Rev.* **138**, A1312.

Taylor, G. R., and Parr, R. G. (1952). *Proc. Natl. Acad. Sci. U.S.* **38**, 154.

Van den Bos, J. (1966). "Foundation for Fundamental Research on Matter (Netherlands)," Rept. No. FOM22358 (unpublished).

Watson, K. M. and Nuttall, J. (1967). "Topics in Several Particle Dynamics," Chapter 4. Holden-Day, San Francisco, California.

Wilets, L., and Gallaher, D. F. (1966). *Phys. Rev.* **147**, 13.

Wilets, L., and Wallace, S. J. (1967). *Proc. 5th Intern. Conf. Phys. Electron. Atomic Collisions, Leningrad*, p. 62. Nauka, Leningrad.

USE OF CLASSICAL MECHANICS IN THE TREATMENT OF COLLISIONS BETWEEN MASSIVE SYSTEMS

D. R. BATES and A. E. KINGSTON

Department of Applied Mathematics and Theoretical Physics,
The Queen's University,
Belfast, Northern Ireland

I. Introduction

Just over 10 years ago Gryzinski (1959; see also 1965) revived interest in the application of classical mechanics to atomic scattering problems, which had lapsed since important early work by Thomson (1912), Thomas (1927a,b), and Williams (1931) [cf. reviews by Williams (1945) and Bohr (1948)]. The subject has since received considerable attention. A critical account of the basic theory has recently been given by Burgess and Percival (1968). Our prime objective here is to enable the worth of classical mechanics for atom–ion and atom–atom collisions to be assessed by presenting an extensive set of comparisons between the calculated and measured cross sections.

II. Fast Collisions

There are several factors which, taken together, would lead one to suppose that the classical impulse or binary encounter approximation (cf. Burgess and Percival, 1968) should achieve some success in treating fast collisions. First, a sufficient condition for the validity of the full impulse approximation is that the velocity of relative motion should be much greater than the orbital velocity of the active electron. Second, the classical version of this approximation uses, in the case of direct collisions, only permissible knowledge of the velocity distributions of the electrons and does not falsely assume that they move in definite trajectories. Third, classical mechanics happens to give the correct (Rutherford) differential cross section for two colliding distinguishable particles having a Coulomb interaction. Fourth, no ambiguity arises in the interpretation of the energy change in ionization.

The collision process for which the classical impulse approximation is best suited is clearly the removal of an electron from an atomic system by a fast projectile nucleus. Even for this process the classical impulse approximation would not of course be expected to yield precise results. Thus it ignores quantal interference effects. Again, it ignores the oscillator strengths associated with the cluster of charged particles comprising the target atom, and, in consequence, it fails to take proper account of distant encounters (which become relatively more important as the impact velocity is increased). The magnitude of the total error cannot be predicted with assurance.

Additional uncertainties occur in the cases of excitation, electron capture, and electron loss. These are due, respectively, to the difficulty of treating discrete energy and angular momentum changes classically, to the need for assuming that the electron follows a definite trajectory, and to a non-Coulomb interaction being involved.

Because of the ease with which the classical impulse approximation may be applied, it is of practical importance to obtain as much information as possible on its reliability. It is also of theoretical interest to find to what extent the errors are due to the use of the impulse approximation and to what extent they are due to the use of classical mechanics. As will be seen, much has to be done before all the questions which might be asked on these issues can be answered.

A. Removal of an Electron from an Atomic or Molecular System

A number of workers (cf. Gryzinski, 1959; Gerjuoy, 1966; Burgess and Percival, 1968; Vriens, 1970) have discussed classical collisions between protons, or other nuclei, and free electrons. We shall follow Vriens (1967), who has given the relevant basic relations in a particularly convenient form (and without misprints).

Let $\mathbf{V}$ and $\mathbf{v}$ be the respective velocities of an incident nucleus, charge Ze, and of a free electron, mass m_e, before an encounter. Averaging over the random distribution of angles between these two vectors, the differential cross section describing collisions in which the energy of the electron is increased by a positive amount between Δ and $\Delta + d\Delta$ is $\varsigma(V, v, \Delta)\, d\Delta$, where if

$$\Delta \leqslant 2m_e V(V - v) \tag{1}$$

then

$$\varsigma(V, v, \Delta) = (2\pi Z^2 e^4 / m_e V^2)\{\Delta^{-2} + (2m_e v^2/3)\,\Delta^{-3}\}; \tag{2}$$

if

$$2m_e V(V - v) \leqslant \Delta \leqslant 2m_e V(V + v) \tag{3}$$

then

$$\varsigma(V, v, \Delta) = (\pi Z^2 e^4 / 3V^2 v\,\Delta^3)\{4V^3 - \tfrac{1}{2}(v' - v)^3\} \tag{4}$$

with

$$v' \equiv [(2\Delta/m_e) + v^2]^{1/2}; \tag{5}$$

while if

$$2m_e V(V + v) \leqslant \Delta \tag{6}$$

then

$$\varsigma(V, v, \Delta) = 0. \tag{7}$$

In his pioneering work Gryzinski (1959) made an arbitrary approximation in averaging over the initial angular distribution. His formulas are hence not the same as those cited. However, the numerical results do not differ to any major extent.

The classical impulse approximation to the cross section for the removal of an electron from a shell of an atom is obtained by integrating the differential cross section over all Δ in excess of the ionization potential I of the shell, and then integrating over all v weighted according to the normalized velocity distribution $f(v)$ in the shell. In carrying through the first integration with the aid of formulas (1)–(7), it is found that the removal cross section per equivalent electron, $q_r(V)$, is given by

$$q_r(V) = \int_0^\infty f(v)\sigma(V, v, I)\, dv \tag{8}$$

in which

$$\sigma(V, v, I) \equiv (2\pi Z^2 e^4 / m_e V^2)\{(1/I) + (m_e v^2/3I^2) - [1/2m_e(V^2 - v^2)]\} \tag{9}$$

for

$$0 \leqslant v \leqslant V - (I/2m_e V); \tag{10}$$

$$\sigma(V, v, I) \equiv (\pi Z^2 e^4 / m_e V^2)\{(1/I) + (m_e/3vI^2)\{2V^3 + v^3 - [v^2 + (2I/m_e)]^{3/2}\}$$
$$+ [1/2m_e v(V + v)]\} \tag{11}$$

for

$$|V - (I/2m_e V)| \leqslant v; \tag{12}$$

and

$$\sigma(V, v, I) \equiv 0 \tag{13}$$

otherwise.

The shell removal cross section is usually taken to be

$$Q_r(V) = Nq_r(V), \tag{14}$$

where N is the number of equivalent electrons. To obtain the atom removal cross section it is, of course, necessary to sum the various shell removal cross sections.

The proportionality of $Q_r(V)$ to N in formula (14) arises from the implicit assumption that the chance that the projectile nucleus removes a given electron from the shell is independent of whether or not it has already removed another electron. This assumption is false. The removal of a second electron requires more energy, and is therefore less likely, than the removal of the first. An approximate correction may readily be introduced. Suppose for simplicity that the projectile can remove only a single electron from the shell; and suppose also that the electrons of the shell are a fixed distance p apart (which may conveniently be taken to be their root mean square distance apart). The extent to which the electrons shield each other from the projectile in this crude model may readily be calculated. It is found that the corrected removal cross section[1] is $\xi Q_r(V)$ where

$$\xi = 1 - [(N - 1)q_r(V)/4\pi p^2]. \tag{15}$$

Percival and Valentine (1966) have pointed out that while the condition,

$$\Delta \geqslant I, \tag{16}$$

defining the integration region used in obtaining the removal cross section, is a necessary condition for ionization, it is not a sufficient condition since the electron may become bound to the incident nucleus. Hence the removal cross section should not be identified with the ionization cross section. It is commonly taken to be the sum of the ionization and electron capture cross sections. However, this sum represents an upper limit to the removal cross section in that capture may occur without (16) being satisfied (cf. Section II, D).

During the past few years the classical impulse approximation has been used in the calculation of the cross sections describing electron removal from

[1] The quantal formula corresponding to (14) should similarly be corrected.

various atoms and molecules by fast protons (Percival and Valentine, 1966; McDowell, 1966; Catlow and McDowell, 1967; Vriens, 1967; Garcia *et al.*, 1968a; Tripathi *et al.*, 1969).

For simplicity the velocity distribution is commonly represented by a delta function:

$$f(v) = \delta(v - \bar{v}) \tag{17}$$

with $\bar{v}$ some mean value (Gryzinski, 1959). The usual choice is

$$\bar{v} = (2I/m_e)^{1/2}. \tag{18}$$

This choice is convenient but rather arbitrary. To be sure, if the active electron is in a Coulomb field, formula (18), by virtue of the virial theorem, yields the root mean square velocity. However, it is not obvious that the root mean square velocity is the best mean to adopt; and in any case the active electron is generally not in a Coulomb field. Though (17) combined with (18), which we shall refer to as the δ_I distribution, has in fact proved to be quite effective, we were anxious lest the uncertainty it introduces should obscure the main issue regarding the classical impulse approximation. We therefore carried out calculations in a number of representative cases using accurate velocity distributions (see Table I). These calculations are merely an extension, and in part repetition, of the earlier work cited in the preceding paragraph. Little computer time is required for a cross section curve.

TABLE I

WAVE FUNCTIONS USED TO CALCULATE ELECTRON VELOCITY DISTRIBUTIONS

Atom or molecule	Wave function
H	Exact
H$_2$	Two parameter variational (Weinbaum, 1933)
He, Li, Ne, Na, Ar, Kr	Hartree–Fock (Clementi, 1965)
Xe	Hartree–Fock (Synek and Sturgis, 1965)

The target systems considered are H, H$_2$, He, Li, Ne, Na, Ar, Kr, and Xe. In all cases the projectile is a proton. Removal cross section curves based on the classical impulse approximation are given in Figs. 1a–1i. Each figure in general contains three such curves. These are (*i*) the curve calculated using the δ_I distribution and the commonly adopted formula (14); the curves calculated using an accurate velocity distribution (Table I) and either (*ii*) again ignoring the shielding effect or (*iii*) multiplying the cross section obtained from formula (14) by the factor ξ of (15) to make approximate allowance for this effect. Curve (*i*) is always rather sharply peaked. For the lighter targets it lies close to the other two curves, but for the heavier targets it falls well below them in the high energy region. In a few cases (Figs. 1d, 1f–1i) the shell structure or the form of the velocity distribution in the important outer shell or both together cause a just discernible distortion in the cross section curve. The shielding effect does not occur for H and is so slight for H_2, He, Li, and Na that curves (*ii*) and (*iii*) are not distinguished. It is quite marked near the maxima of the cross section curves for some of the other targets. The correction applied is then unreliable.

FIGS. 1a–1i. [see pp. 274–278.] Electron removal cross sections of classical impulse approximation

Curve	Velocity distribution used	Screening effect
(*i*) Dash-dot	δ_l of (17) and (18)	Ignored
(*ii*) Long dash	Accurate (Table I)	Ignored
(*iii*) Solid line	Accurate (Table I)	Included by (15)

Related cross sections: short dash curve, first Born approximation to ionization cross section; ○ and □ measured ionization cross section, △ measured cross section for single-electron ionization by electron having same speed as proton; ● sum of measured ionization and capture cross sections. References and other information are given in the separate captions. In the case of multishell systems the symbol for a particular shell is affixed to a curve if the electrons of only that shell are involved.

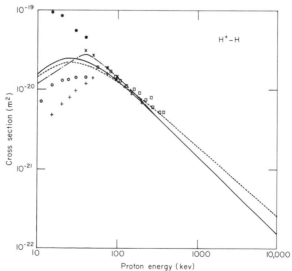

FIG. 1a. Atomic hydrogen. First Born approximation to ionization cross section curve from Bates and Griffing (1953). Ionization cross section data: <50 keV from Fite *et al.* (1960), >50 keV from Gilbody and Ireland (1963); capture cross section data from Fite *et al.* (1960). Results of exact classical calculations of Abrines and Percival (1966): + ionization cross section, × sum of ionization and capture cross sections.

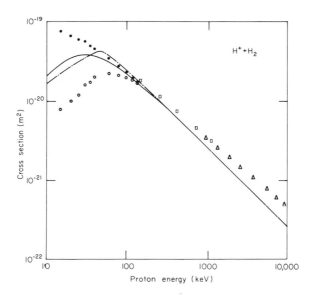

FIG. 1b. Molecular hydrogen. Ionization cross section data: <140 keV from de Heer *et al.* (1966), >150 keV from McDaniel (1964); equivalent electron impact cross section data from Schram (1966); capture cross section data from de Heer *et al.* (1966).

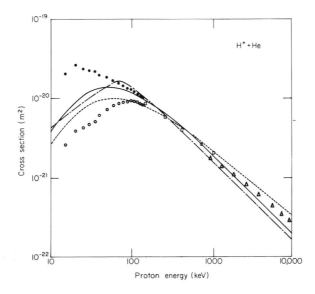

FIG. 1c. Helium. First Born approximation to ionization cross section curve from Bell and Kingston (1969). Ionization cross section data: <140 keV from de Heer *et al.* (1966), >150 keV from McDaniel (1964); equivalent electron impact cross section data from Schram (1966); capture cross section data from de Heer *et al.* (1966).

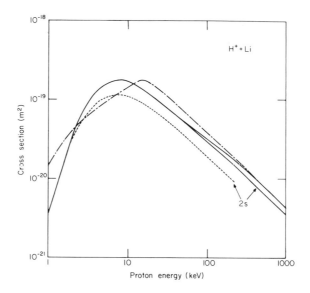

FIG. 1d. Lithium (valence electron and total). First Born approximation to 2*s* ionization cross section curve from Peach (1965).

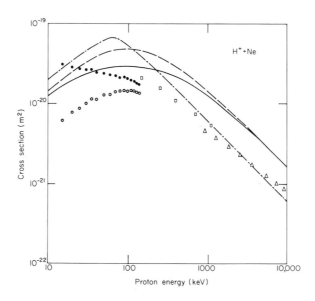

FIG. 1e. Neon. Ionization cross section data: <140 keV from de Heer *et al.* (1966), >150 keV from McDaniel (1964); equivalent electron impact cross section data from Schram (1966); capture cross section data from de Heer *et al.* (1966).

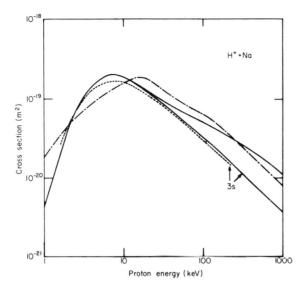

FIG. 1f. Sodium (valence electron and total). First Born approximation to $3s$ ionization cross section curve from Bates *et al.* (1965), Peach (1966).

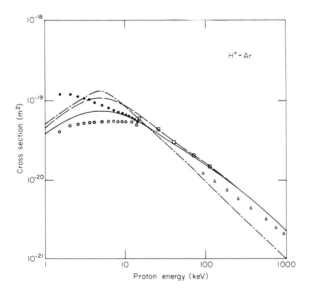

FIG. 1g. Argon. Ionization cross section data: <140 keV from de Heer *et al.* (1966), >150 keV from McDaniel (1964); equivalent electron impact cross section data from Schram (1966); capture cross section data from de Heer *et al.* (1966).

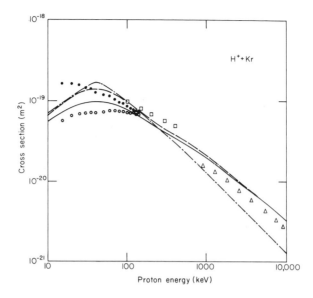

Fig. 1h. Krypton. Ionization cross section data: <140 keV from de Heer *et al.* (1966), >100 keV from Gilbody and Lee (1963); equivalent electron impact cross section data from Schram (1966); capture cross section data from de Heer *et al.* (1966).

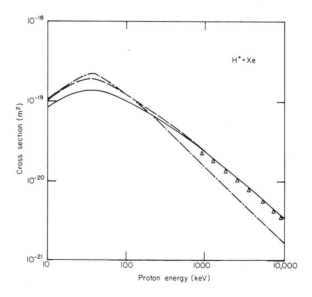

Fig. 1i. Xenon. Equivalent electron impact cross section data from Schram (1966).

Also marked in the figures are measured values of the cross sections for the ionization of the targets by protons (supplemented in the high energy region by corresponding data on ionization by equivalent electrons) and of the cross sections for electron removal (obtained by adding the ionization and capture cross sections). For H, He, Li, and Na (Figs. 1a, 1c, 1d, and 1f), the first Born approximation to the ionization curves for the outer electrons are included.

With the aid of the Monte Carlo method Abrines and Percival (1966) have simulated $H^+ - H(1s)$ scattering experiments on a digital computer without making any approximation in the classical equations of motion. Much computer time was needed. The results are of the utmost importance since they show unambiguously what may be achieved by classical mechanics (whereas in approximate calculations there is of course always the possibility of the chance cancellation of errors from different sources). The ionization and removal cross section curves that Abrines and Percival obtained are reproduced in Fig. 1a along with the other results on $H^+ - H(1s)$ collisions. They naturally join smoothly to the impulse approximation. Their good accord with the experimental data can scarcely be fortuitous.

Some comments will now be made on the pattern revealed by the set of figures (excluding those for Li, Na, and Ne (Figs. 1d, 1e, 1f) which are rather special cases and will be considered separately later).

We arbitrarily divide the energy range covered into three regions designated and defined thus:

I	II	III
$10\text{ keV} \leqslant E \leqslant 100\text{ KeV}$,	$100\text{ keV} \leqslant E \leqslant 1\text{ MeV}$	$1\text{ MeV} \leqslant E \leqslant 10\text{ MeV}$.

The so-called removal cross section of the classical impulse approximation should be greater than the ionization cross section (since its specifications include some captures), but it should be less than the true removal cross section (since its specifications do not cover all captures). In region I capture is important. Consequently, the calculated removal cross section is ill-defined, which prevents any exacting test being made of the classical impulse approximation. However, it can at least be seen that if allowance is made for shielding, the calculated removal cross sections lie between the required limits. A better check is possible in region II. There is good agreement. The range of incident velocities for which the classical impulse approximation gives useful results extends to lower values than would be expected from an inspection of the velocity distribution of the target electrons (see Fig. 2).

In region III the classical impulse approximation tends to underestimate the removal cross section for the lighter targets (Figs. 1a–1c). The ratio of the calculated cross section to the actual cross section decreases as the energy increases. It falls as low as 1 : 2 for H_2 (Fig. 1b), which is the worst case.

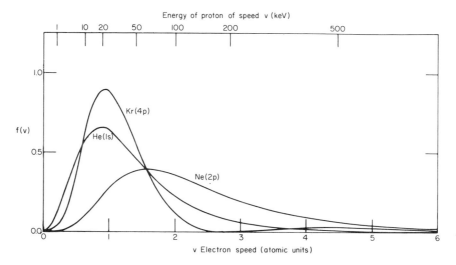

FIG. 2. Velocity distribution $f(v)$ of bound electrons in some representative cases. (The energy of a proton of speed v is indicated on top scale.)

For the heavier targets the position is quite different in that the calculated cross sections keep remarkably accurate. The underestimation for the lighter targets is presumably due to the failure of classical mechanics to take proper account of distant encounters. For the heavier targets (Figs. 1g–1i) these distant encounters remain relatively unimportant until region III is passed.

The simple pattern emerging does not fit Li, Na, and Ne in that the classical impulse approximation here overestimates the removal cross sections (moderately for Li, slightly for Na, and seriously for Ne). The seemingly anomalous behavior is associated with the outer electron in the cases of Li and Na and with the outer shell in the case of Ne. It is noteworthy that the corresponding photoionization cross sections are abnormally low (cf. Sampson, 1966; Stewart, 1967).

In the quantal description the total impact ionization cross section $Q_i(V, l)$ may be expressed as the sum of the partial impact ionization cross sections

---

FIG. 3a. Differential ionization cross sections for $H^+ + H_2$ collisions. Solid curves, classical impulse approximation with accurate velocity distribution (Table I); datum points ● from Rudd *et al.* (1966). The energy of the incident proton is indicated in each section of the figure. (Note: the insets show the regions near the origin with the energy scale magnified.)

FIG. 3b. Differential ionization cross sections for $H^+ + He$ collisions. Solid curves, classical impulse approximation with accurate velocity distribution (Table I); dashed curves, first Born approximation from Bell and Kingston (1969); datum points ■ from Rudd and Jorgensen (1963) and ● from Rudd *et al.* (1966). The energy of the incident proton is indicated in each section of the figure. (Note: the insets show the regions near the origin with the energy scales magnified.)

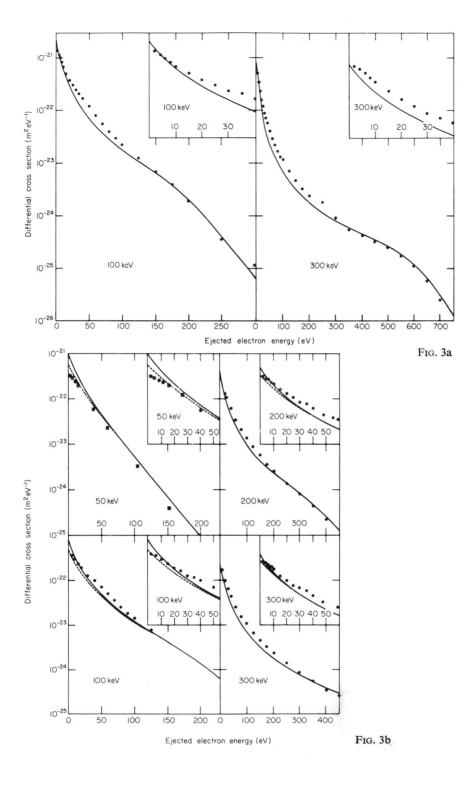

FIG. 3a

FIG. 3b

$q_i(V, l, l')$ associated with continua of different angular momenta. Thus

$$Q_i(V, l) = \sum_{l'} q_i(V, l, l') \qquad (19)$$

where l and l' are the azimuthal quantum numbers of the active electron in its initial and final states. There are two features of immediate relevance. First, the partial cross section $q_i(V, l, l + 1)$ which is normally dominant, is very small for Li and Na because of severe cancellation within the transition matrix element (Peach, 1965, 1966) and is very small for Ne because of the scanty overlap between the initial and final wave functions (Cooper, 1962). Second, as l' is increased beyond $l + 1$ the partial cross sections for Ne, Li, and Na fall rapidly because of the diminution of the overlap between the compact initial wave functions and the final wave functions (the form of which naturally reflects the increase in the angular momentum). Taken together, these two features make the overestimation of the cross sections by the classical impulse approximation qualitatively understandable.

Rudd and his associates (Rudd and Jorgensen 1963; Rudd *et al.*, 1966) have investigated the position regarding differential ionization cross sections. They measured these cross sections in molecular hydrogen and helium and made an interesting comparison with results they obtained using Gryzinski's form of the classical impulse approximation and the δ_l distribution and also with results they obtained by scaling the first Born approximation to the differential ionization cross sections for atomic hydrogen. Figures 3a and 3b show some of the experimental data together with the differential ionization cross sections calculated from the true classical impulse approximation. Results obtained by Bell and Kingston (1969) for He using the first Born approximation, with accurate wave functions, are included. It may be seen that for high energies of ejection the classical impulse approximation is in very close accord with the first Born approximation. The extent of the agreement with the measured cross sections over an extremely wide range of values is striking.

An indirect check is provided by K-shell ionization by proton impact. Measurements have only been made in the low energy region where the cross section for a particular process is rising rapidly towards its maximum. The first Born approximation (Henneberg, 1933) overestimates the cross section in this region. Bang and Hansteen (1959), however, obtained good agreement with experiment by allowing for the deflection of the incident protons in the field of the target while continuing to treat the electronic transition by first-order perturbation theory. It is therefore reasonable to suppose that, should the classical impulse approximation give results close to those of the first Born approximation, then a more elaborate classical approximation which makes allowance for the deflection of the incident protons would also give

Figs. 4a–4e. [see pp. 284–286.] *K*-shell ionization cross sections of classical impulse approximation

Curve	Velocity distribution used	Screening effect
(*i*) Dash-dot	δ_I of (17) and (18)	Ignored
(*ii*) Solid line	Accurate (Table I)	Included by Eq. (15)

The δ_I distribution is unsatisfactory for low energy collisions. Results obtained using it are therefore only shown in one representative case (oxygen). Related cross sections: dashed curve, first Born approximation to ionization cross section from Freeston and Kingston (1970a) for hydrogen, Merzbacher and Lewis (1958) for all other cases; ○, △, ☐, ▽ and ● measured cross section for K-shell ionization by protons; ■ quarter of measured cross section for K-shell ionization by α particles having same speed as protons of the indicated energy.

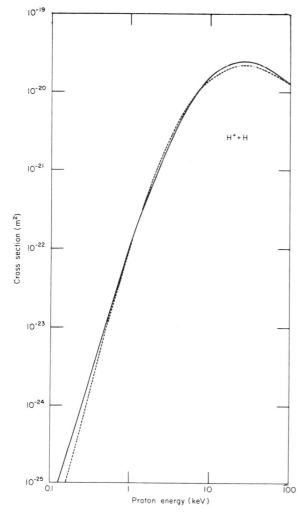

$H^+ + H$

FIG. 4a.
Hydrogen.

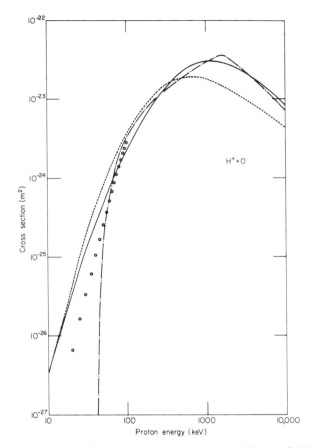

FIG. 4b. Oxygen. Ionization cross section data: ☉ Hart *et al.* (1969).

good agreement with experiment. Cross section curves calculated from the classical impulse approximation and from the first Born approximation are compared in Figs. 4a–4e. They lie above the measured cross section curves for small proton energies, but perhaps more significantly they are indeed in fair accord with each other. The absolute values of the measured cross sections are rather uncertain because the fluorescent yields, on which they depend, are poorly determined. The theoretical cross sections are quite sensitive to the assumed velocity distributions and ionization energies. Recently a paper by Garcia (1970) appeared giving the results of calculations based on the classical impulse approximation modified to allow for the effect of the nuclear repulsion on the motion of the proton. The effect is in the expected sense but for the light elements studied (C, O, Mg, Al) it is small in magnitude except at very low energies.

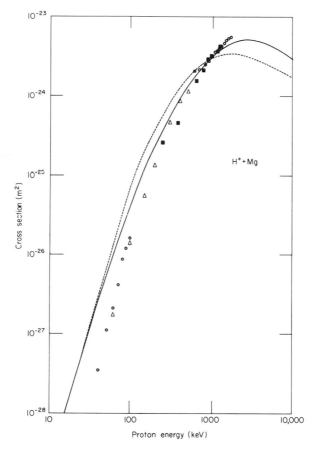

FIG. 4c. Magnesium. Ionization cross section data: ⊙ Khan *et al.* (1965); △ Khan and Potter (1964). Equivalent α particle impact cross section data: ■ Sellers *et al.* (1969).

B. EXCITATION

The classical impulse approximation does not lend itself as readily to the treatment of excitation as it does to the treatment of ionization. For it to be useful, a simple procedure is needed which takes cognisance of the fact that a bound–bound transition in general involves discrete changes both in the energy and in the orbital angular momentum. Such a procedure has not yet been devised.

In some cases the problem presented by the discrete changes in the orbital angular momentum may be ignored and attention may be concentrated on the more amenable problem presented by the discrete changes in the energy. For example, this is possible if the total cross section for the excitation of a

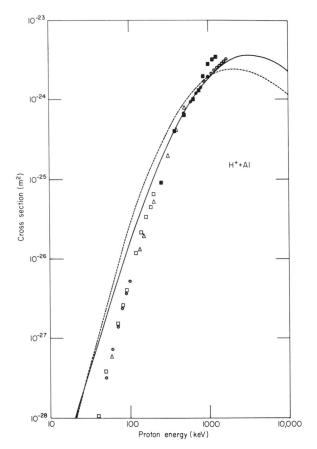

Fig. 4d. Aluminium. Ionization cross section data: ⊙ Khan *et al.* (1965); △ Khan and Potter (1964); ⊡ Brandt and Laubert (1969). Equivalent α particle impact cross section data: ■ Sellers *et al.* (1969).

hydrogen atom to any of the states of the level of principal quantum number n is sought:

$$H^+ + H(1s) \rightarrow H^+ + H(nl[l = 0 \rightarrow n - 1]). \tag{20}$$

Gryzinski (1959) has suggested a prescription—proceed as for electron removal but take the interval over which the differential cross section is integrated to be from the excitation energy Δ_n of level n to the excitation energy Δ_{n+1} of the next level. Because of lack of adequate comparison data it is scarcely worth carrying out calculations on process (20) at present.

Freeston and Kingston (1970b) have investigated

$$H^+ + He(1s^2) \rightarrow H^+ + He(1snl[l = 0 \rightarrow n - 1]) \tag{21}$$

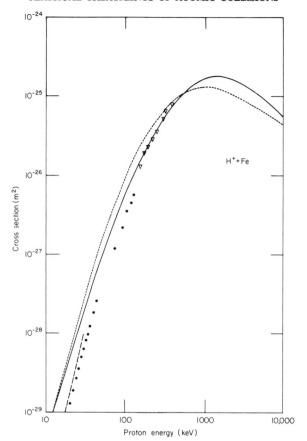

FIG. 4e. Iron. Ionization cross section data: ▽ Merzbacher and Lewis (1958); ● Messelt (1958). Long-dash curve, quantal calculations by Bang and Hansteen (1959) with nonrectilinear trajectories.

neglecting the dependence of the final energy on l and adopting Gryzinski's prescription. Since the optically allowed transitions are the strongest, they judged it best to set Δ_n equal to the excitation energy of the 1snp singlet level. The derived total excitation cross section curves are not sensitive to this choice. In Figs. 5a–5c they are compared with experiment and with the corresponding curves obtained from the first Born approximation [accurate initial and final wave functions being used (Bell, *et al.*, 1969b)]. The accord is poorer than in the case of ionization (Fig. 1c) but is not so much poorer as to suggest that Gryzinski's prescription is seriously at fault. Moreover, some increase in the error might be expected from the trend of the error in the differential ionization cross section (Fig. 3b) as the energy of ejection is reduced.

FIGS. 5a–5c. Excitation of helium by proton impact: solid curve, classical impulse approximation with accurate velocity distribution; dashed curve, first Born approximation (Bell *et al.*, 1969b); ● and ☐ measured excitation cross sections from van den Bos *et al.* (1968) and Thomas and Bent (1967), respectively; △ measured cross section [Moustafa Moussa (1967)] for excitation by electron having same speed as proton. (After Freeston and Kingston, 1970b).

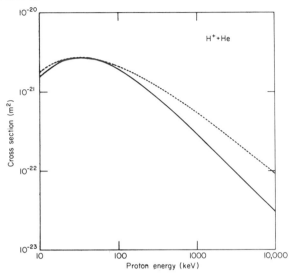

FIG. 5a. Excitation to 1s 2s or 2p levels.

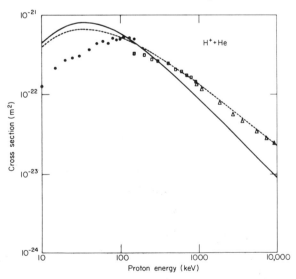

FIG. 5b. Excitation to 1s3s, 3p, or 3d levels. Thomas and Bent (1967) only measured the cross section for excitation to the 3p level. The cross sections for excitation to the 3s and 3d levels are small at large impact energies.

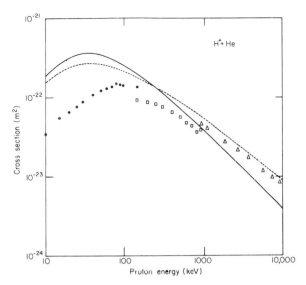

FIG. 5c. Excitation to 1s4s, 4p, 4d, or 4f levels. The cross section for excitation to the 4f level is small and is not included in the calculated or measured cross sections.

Percival and Richards (1970a,b,c) have recently made important advances based on correspondence principle arguments. They extended Bohr's correspondence principle for spontaneous transitions to include transitions induced by external fields in those circumstances in which both quantal perturbation theory and classical perturbation theory hold. Contrary to what has usually been done, they specifically recognized that the probability of a collision-induced quantum transition for small $|\Delta_n|$ is not directly obtainable from the classical energy transfer but that it is instead obtainable from partial classical energy transfers associated with the Fourier coefficients of the classical motion of the atom with respect to time (Percival and Richards, 1970a). A consequence of this, which was first pointed out by Williams (1931), is that if n is high, the total energy transfer, summed over all final states, is the same in classical and quantum mechanics. Percival and Richards (1970a) were able to demonstrate that the classical excitation cross section has the correct form in the high energy limit. Using a generalization of Heisenberg's correspondence principle between matrix elements and classical Fourier components they later (Percival and Richards, 1970b,c) obtained what they believe to be accurate cross sections for transitions induced in highly excited hydrogen atoms by incident charged particles.

C. ELECTRON LOSS

Bates *et al.* (1969) have examined the use of the classical impulse approximation in the treatment of electron loss from a projectile system A in collision with a target system B:

$$A + B \rightarrow A^+ + e + B(S). \tag{22}^2$$

They expressed the total loss cross section $Q_l(V)$ at impact velocity $\mathbf{V}$ as the sum of an elastic loss cross section $Q_{lo}(V)$ and an inelastic loss cross section $Q_{lx}(V)$ arising, respectively, from collisions in which B is left in its initial state and from collisions in which B is excited or ionized. Loss is presumed to occur as a consequence of the active electron gaining sufficient energy relative to A by, in the elastic case, being simply deflected through a large enough angle in the static field of B or by, in the inelastic case, making a hard enough collision with one of the electrons of B, this electron being, like the active electron, regarded as free except in that the least energy which may be given to it is Δ_e, the first excitation energy.

Assume in the first instance that the active electron initially has a fixed speed u relative to its parent nucleus. Let $S_o(v, \theta) \sin \theta \, d\theta \, d\phi$ be the differential cross section of the target describing elastic scattering of electrons of velocity $\mathbf{v}$ into the solid angle $\sin \theta \, d\theta \, d\phi$ about the direction (θ, ϕ); let

$$S_{xi}(v, \theta, \Delta) \sin \theta \, d\theta \, d\phi$$

be the differential cross section of shell i of the target describing similar inelastic scattering with diminution of energy between Δ and $\Delta + d\Delta$; and let H be the ionization energy of the projectile. In the classical impulse approximation the elastic and inelastic loss cross sections are then given by

$$Q_{lo}(V, u) = (1/Vu) \int_{|V-u|}^{V+u} \sum \int_{\alpha}^{\beta} S_o(v, \theta)\phi_l v \sin \theta \, d\theta \, dv, \tag{23}$$

and

$$Q_{lx}(V, u) = (1/Vu) \sum_i \int_{v_{oi}}^{V+u} \int_{\Delta_{ei}}^{m_e v^2/2} \int_{\alpha}^{\beta} S_{xi}(v, \theta, \Delta)\phi_l v \sin \theta \, d\theta \, dv \, d\Delta, \tag{24}$$

where

$$\phi_l = \begin{cases} 0 & \text{for} \quad \mu \geqslant 1 \\ \cos^{-1}\mu, \, (0 \leqslant \phi_l \leqslant \pi) & \text{for} \quad 1 \geqslant \mu \geqslant -1 \\ \pi & \text{for} \quad \mu \leqslant -1 \end{cases} \tag{25}$$

in which

$$\mu \equiv [(a - b) - (a + b) \tan^2(\theta/2)]/2c \tan(\theta/2), \tag{26}$$

with

$$a \equiv m_e[1 - (2\Delta/m_e v^2)]^{1/2}(V^2 + v^2 - u^2), \tag{27}$$

$$b \equiv m_e(V^2 + v^2 - u^2) - 2H - 2\Delta, \tag{28}$$

$$c \equiv m_e[1 - (2\Delta/m_e v^2)]^{1/2}\{4V^2v^2 - (V^2 + v^2 - u^2)^2\}^{1/2}, \tag{29}$$

---

[2] Here **S** indicates the totality of bound and free states.

(Δ being of course zero in the elastic case); where α and β are scattering angles between which ϕ_l is non-zero (see Bates et al., 1969); where v_{oi} is the lesser of $|V - u|$ and $(2 \Delta_{ei}/m_e)^{1/2}$; and where the summation in (23) is over all such pairs of angles (of which, because of orbiting, there may be a number).

When evaluating (23) Bates et al. (1969) found it convenient to use the relation

$$S_o(v, \theta) \sin \theta \, d\theta = \rho \, d\rho, \tag{30}$$

and to follow the procedure described by Smith (1964) for the treatment of elastic scattering in a given central potential. In (24) they used the approximation for $S_{xi}(v, \theta, \Delta)$ which was derived by Gryzinski (1959). They obtained $Q_{lo}(V)$ from $Q_{lo}(V, u)$ by averaging over the exact distribution $f(u)$ appropriate to the projectile (a normal hydrogen atom) but in the inelastic case they were content to adopt the δ_H distribution which leads immediately to

$$Q_{lx}(V) = Q_{lx}(V, \bar{u}), \qquad \bar{u} = (2H/m_e)^{1/2}. \tag{31}$$

Formula (23) for $Q_{lo}(V, u)$ depends on treating elastic scattering in a non-Coulomb field by classical mechanics. Its validity is therefore open to special doubt. Because of the uncertainty principle a necessary condition for the classical description of elastic scattering to be satisfactory is that

$$d \gg \hbar/2m_e \tag{32}$$

where

$$d \equiv \rho \theta v. \tag{33}$$

Bates et al. (1969) argued that the classical description would be unjustified should condition (32) not be fulfilled for impact parameters up to about

$$\rho_o = \{Q_{lo}(v)/\pi\}^{1/2}. \tag{34}$$

This criterion indicates that formula (23) is untrustworthy in the case of light targets like helium but provides no evidence against its applicability in the case of heavy targets like argon.

In Figs. 6a, 6b, and 7a, 7b the H–H and H–He elastic and inelastic loss cross sections obtained from the classical impulse approximation are compared with the corresponding cross sections obtained from the first Born approximation. These first Born cross sections are probably quite accurate at impact energies above about 200 keV since their sums are in good agreement with the H–H total loss cross sections measured by Wittkower et al. (1967) and by McClure (1968), and with the H–He total loss cross sections measured by Stier and Barnett (1956) and by Barnett and Reynolds (1958). As would be expected with such light target systems, the classical impulse approximation is rather poor for elastic loss (especially at high impact energies). It is remarkably successful for inelastic loss.

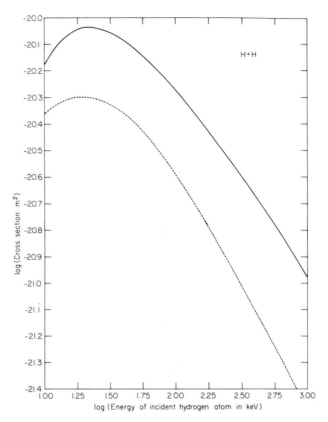

FIG. 6a. Elastic contributions to electron loss from hydrogen atoms passing through atomic hydrogen. Solid curves, classical impulse approximation (Bates *et al.*, 1969); dashed curves, first Born approximation (for H–H, Bates and Griffing 1953, 1955; for H–He, Bell *et al.*, 1969a).

The only comparison results available in the case of the heavier target systems are the measured total loss cross sections. It may be seen from Figs. 8a–8d that the classical impulse approximation reproduces these quite well. As in the case of removal it gives useful results down to surprisingly low velocities of relative motion.

Bates and Walker (1966, 1967) have developed a version of the classical impulse approximation which uses little computer time but which is unjustified if the electron being stripped from the projectile system is strongly bound. In this version the target system is replaced by a structureless model which is assigned an elastic cross section equal to the total electron scattering cross section, and the scattering is assumed to be isotropic. The loss cross section

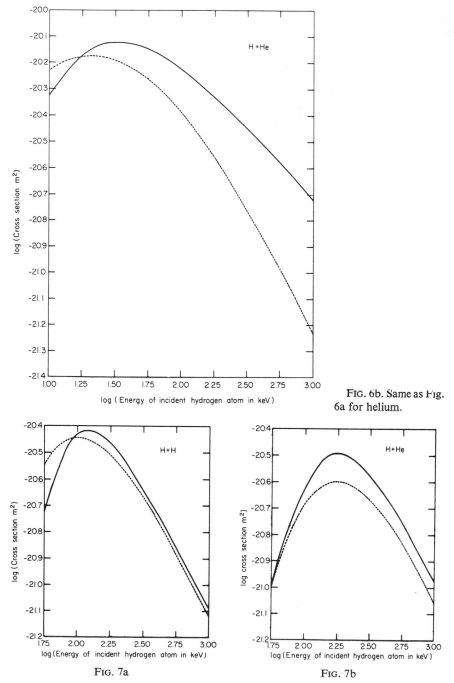

FIG. 6b. Same as Fig. 6a for helium.

FIG. 7a

FIG. 7b

FIG. 7. Inelastic contributions to electron loss from hydrogen atoms passing through (a) atomic hydrogen, (b) helium. Curves and references same as Fig. 6.

Fig. 8. Total loss cross section from hydrogen atoms passing through (a) molecular nitrogen, (b) neon, (c) argon, (d) krypton. Solid curves, classical impulse approximation; chain curves, experiment (SB, Stier and Barnett, 1956; BR, Barnett and Reynolds, 1958; FAPT, Fogel *et al.*, 1958; SIOF, Solov'ev *et al.*, 1962; W, Williams, 1967; TNL, Toburen *et al.*, 1968) (after Bates *et al.*, 1969).

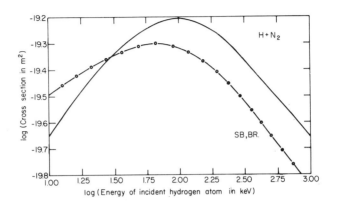

Fig. 8a

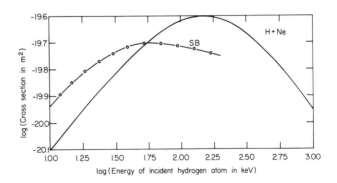

Fig. 8b

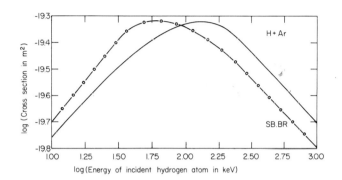

FIG. 8c

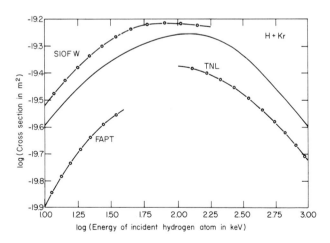

FIG. 8d

$Q_l(V, \bar{u})$ is found to be given by

$$Q_l(V, \bar{u}) = \frac{1}{3\pi v^2}$$

$$\times \int_{\tau/4v}^{\infty} \left\{ \frac{8vx - 1 - (v - x)^2 - 2\tau}{[1 + (v - x)^2]^3} + \frac{1}{[1 + (v + x)^2 - \tau]^2} \right\} s_T(x\bar{u}) \, dx \tag{35}$$

where

$$v \equiv V/\bar{u}, \qquad \tau \equiv H/\tfrac{1}{2}m_e \bar{u}^2 \tag{36}$$

and $s_T(x\bar{u})$ is the total electron scattering cross section at speed $x\bar{u}$.

The calculated and measured cross sections for loss from He(2^3S) in collision with H_2, Ne, Ar, and Kr are compared in Fig. 9. They are in very

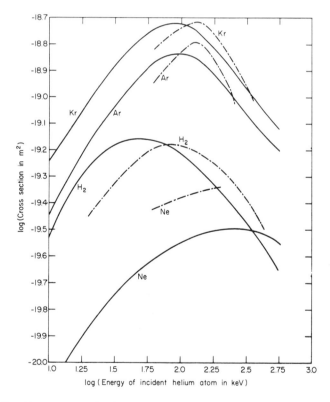

FIG. 9. Total loss cross section from metastable helium atoms passing through molecular hydrogen, neon, argon, and krypton: dash-dot curves represent the experimental results of Gilbody *et al.*, 1968, 1970; solid curves, the classical impulse approximation with the target system replaced by a structureless model (Bates *et al.*, 1969).

good accord. Again, recent experiments by Thomas *et al.* (1969) gives the loss cross section from 168 keV H($n = 3$) atoms to be $5(\pm 1) \times 10^{-20} \text{m}^2$ which is close to the predicted (Bates and Walker, 1966) value ($6.9 \times 10^{-20} \text{ m}^2$). If allowance is made for the shielding of one of the electrons of the projectile by the other, satisfactory agreement is also obtained for the cross sections for detachment in H^-–H, He, Ne, Ar, H_2, N_2, and O_2 encounters (Bates and Walker, 1967).

Drawin (1968) has proposed a remarkably simple procedure for calculating ionization cross sections in atom–atom collisions, and Russell (1969) has extended it to cover excitation cross sections. Some quite striking successes have been achieved. The procedure is commonly but misleadingly described as classical. It is actually empirical and therefore outside the scope of this article. The origin of the misnomer is simply that the formula involved was obtained by making certain arbitrary adjustments to the classical formula of Thomson (1912) for ionization in electron–atom collisons.

D. ELECTRON CAPTURE

There is one tractable problem concerning electron capture. This is the determination of the asymptotic form that the cross section for

$$H^+ + H(1s) \rightarrow H(\Sigma) + H^+ \tag{37}$$

would have if the protons could be regarded as distinguishable particles.[3]

The quantal solution in the case of capture into the 1s state has been investigated by Mapleton (1964) using the (nonrelativistic) first Born and distorted wave approximations, and by Coleman and McDowell (1964) using the impulse approximation. The asymptotic form of the cross section which they obtained is

$$Q_c(1s - 1s)|\text{quantal} = (2M/3m_e E^3)\pi a_0^2 \tag{38}$$

where E is the energy of the incident proton in atomic units and M is its mass.

Bates and Mapleton (1965) have applied the classical impulse approximation to the problem. They supposed that capture occurs in close knock-on collisions in which the target proton B, which is initially stationary at a fixed point O, acquires a high speed and the projectile A is nearly brought to rest. In carrying out the detailed calculations they assumed (*i*) that the velocity and position of the electron do not change appreciably during the encounter; (*ii*) that the motion of the two protons is controlled by the Coulomb

---

[3] Here Σ indicates the totality of bound states.

interaction between them; (*iii*) that capture, which takes place only in large angle scattering, requires

$$\tfrac{1}{2}m_e(\mathbf{V}' - \mathbf{v})^2 - (e^2/r) \leqslant 0 \qquad (39)$$

where $\mathbf{V}'$ is the velocity of A after scattering by B and where $\mathbf{v}$ is the velocity of the electron and r is its distance from O so that classically

$$\tfrac{1}{2}m_e v^2 - e^2/r = -(e^2/2a_0). \qquad (40)$$

Using the exact velocity distribution for the electron, they found the asymptotic form of the cross section for process (37) to be

$$Q_c(1s - \Sigma)|\,\text{classical} = (5M/6m_e\,E^3)\pi a_0{}^2. \qquad (41)$$

Comparing formula (38) with (41) it is seen that classical mechanics is here very successful in that

$$Q_c(1s - \Sigma)|\,\text{classical} = 1.25 Q_c(1s - 1s)|\,\text{quantal}. \qquad (42)$$

In the case of small angle scattering the first Born approximation gives the high velocity limit to the cross section for capture into all states to be 1.23 times that for capture into the 1s state (Jackson and Schiff, 1953).

The problem just discussed is not typical. The treatment of electron capture is, in general, a formidable task.

The simplest prescription for estimating the cross section is due to Gryzinski (1965). According to it when the impact speed is V, the capture cross section per equivalent electron of speed v is

$$Q_c(V, v) = \int_{\Delta_l}^{\Delta_u} \varsigma(V, v, \Delta)\, d\Delta, \qquad (43)$$

in which $\varsigma(V, v, \Delta)$ is defined by (1)–(7) and in which

$$\Delta_l = I + \tfrac{1}{2}m_e\,V^2 - H \qquad (44)$$

and

$$\Delta_u = I + \tfrac{1}{2}m_e\,V^2 + H, \qquad (45)$$

I being the ionization energy of the target atom and H being that of the system formed by capture. The sum of the first two terms of Eqs. (44) and (45) give the energy Δ_0 which must be transferred to the electron to enable it to escape from the target atom and travel at the same speed as the projectile; the third term is intended to be a measure of the amount by which the energy transferred may differ from Δ_0 and yet leave binding of the electron to the projectile possible. However, as Gryzinski recognized, the course of the collision is not fully determined by the amount of energy transferred: capture may occur even if Δ is outside the limits defined by (44) and (45), and it does not necessarily occur if Δ is within these limits. Formula (43) is really only empirical.

If I is less than H, then Δ_l becomes zero for sufficiently small V in which case the predicted cross section becomes infinite. Gryzinski argued that this unacceptable feature is due to failure to take into account the fact that in its initial and final states the electron must be within certain distances r_B and r_A from B and A, respectively. He suggested that a geometrical upper limit

$$q_{geom} = \pi(r_A + r_B)^2 \qquad (46)$$

should be imposed on the cross sections.

Garcia *et al.* (1968b) have carried out extensive computations on capture cross sections using Gryzinski's prescription (which we shall refer to by the letter G alone) with the δ_I distribution (17) and (18) as an approximation to the velocity distribution of the atomic electrons. They also considered two further ways of avoiding the divergence mentioned in the preceding paragraph. One way (GGWI) is to replace the lower limit Δ_l of (43) by

$$\Delta_l' = I + \tfrac{1}{2}m_e V^2. \qquad (47)$$

Another (GGWII) is to find the cross section for any exothermic capture process by treating the endothermic inverse process and then applying the principle of detailed balance.

In Figs. 10a–10e the results of some of the calculations of Garcia *et al.* are compared with experiment. Clearly, all prescriptions are very unreliable in the energy region up to about where a cross section curve passes through its maximum. Prescription GGWI is also poor in the region above this, but prescriptions G and GGWII are here remarkably successful. Conclusions regarding the applicability of classical mechanics naturally cannot be drawn.

To decide if the electron and projectile become bound together, it is necessary to know their separation and relative velocities after the encounter. These quantities appear in the capture condition (39) for the knock-on problem considered earlier. They may readily be determined in that special case because capture may be regarded as due to a binary collision between the two nuclei in which the positions of the two nuclei are effectively interchanged and by which the electron is not affected. In the general case there is a binary collision between the electron and the projectile nucleus. Their separation and relative velocity after they have left the field of the target obviously depend on the trajectory followed by the electron.

The problem in classical mechanics this presents is very complicated. Thomas (1927b) succeeded in carrying through the mathematical analyses involved in an approximation that is appropriate when the speed is high and where the target is a heavy atom. In these circumstances it may be assumed (*i*) that the impact parameter of the A–e binary encounter is very much smaller than the magnitude of the position vector **r** of the electron relative to the nucleus of the target, (ii) that **r** does not change appreciably during this encounter, (iii) that the subsequent trajectory of the electron is controlled by the potential of the target so that there is in effect a second binary encounter, (iv) that this potential $\mathscr{V}(r)$ may be represented by

$$\mathscr{V}(r) = m_e \lambda/r^2, \tag{48}$$

λ being a constant.

Let **V** be the (constant) velocity of the projectile relative to the target; let **v**, **v′**, and **v″**, respectively, be the corresponding velocities of the electron before and after the first binary collision and after the second binary collision; let w be its speed of escape from the target so that

$$(v')^2 = w^2 + (v'')^2 \tag{49}$$

Figs. 10a–10e. [see pp. 300–302.] Electron capture cross sections. The letters G, GGWI, and GGWII labeling various curves indicate the prescription followed (see text). The cross section curve obtained using prescription GGWII appears only in Fig. 10e since this is the only figure referring to an endothermic process. Gryzinski's upper limit (46) to the cross section is off-scale in Figs. 10a–c and is indicated by the broken horizontal line in Figs. 10d and 10e. The references from which the datum points were taken are as follows: × de Heer *et al.* (1966) in a–d, Afrosimov *et al.* (1960) in e; ○ Stedeford and Hasted (1955); ● Stier and Barnett (1956) or Barnett and Reynolds (1958). The calculations and the comparisons with experiment are due to Garcia *et al.* (1968b).

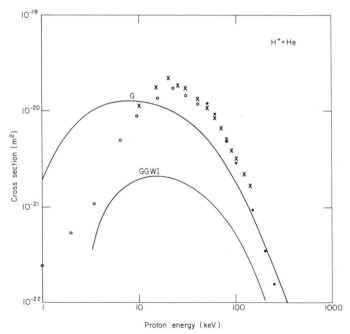

Fig. 10a.
Helium.

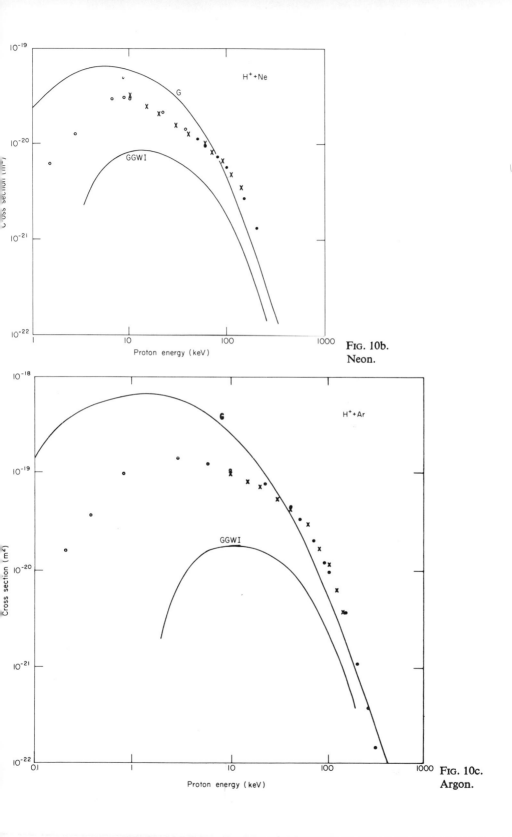

FIG. 10b.
Neon.

FIG. 10c.
Argon.

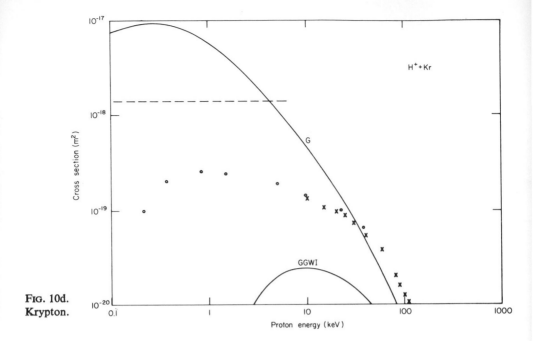

**Fig. 10d.
Krypton.**

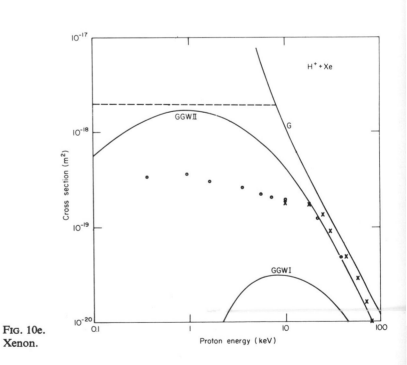

**Fig. 10e.
Xenon.**

and let

$$u = v - V, \qquad u' = v' - V, \qquad u'' = v'' - V, \tag{50}$$

$$\eta' = \widehat{rv'}, \qquad \eta'' = \widehat{rv''}, \qquad \chi' = \widehat{Vv'}, \tag{51}$$

$$\phi = \widehat{(Vv')}\widehat{(rv')}. \tag{52}$$

On the assumptions indicated in the previous paragraph Thomas (1927b) showed that the capture cross section for a distribution $X(v, r)$ of target electrons is

$$Q_c(V) = \frac{2^{3/2}\pi\mu^{7/2}}{3V^{3/2}} \iiint_{\mathscr{C}} \frac{X(v, r)t(u, v, V)}{vr^{3/2}u^{11/2}} \, dv \, dr \, d(\cos \eta') \tag{53}$$

where

$$t(u, v, V) = \frac{2uV(w^2 + V^2)^{1/2}\{4u^2V^2 - (u^2 - w^2)(V^2 + u^2 - v^2)\}}{|(v^2 - V^2 - w^2)|^3} \tag{54}$$

$$\mu = Ze^2/m_e \tag{55}$$

and $\mathscr{C}$ signifies that the integration is taken over all accessible parts of multi-dimensional space in which the condition for capture is met. This condition is

$$(u'')^2 < 2\mu/x \tag{56}$$

x being the A–e separation after the second binary encounter. It requires

$$\eta' + \eta'' \simeq x', \qquad \phi \simeq 0, \qquad v'' \simeq V. \tag{57}$$

Thomas found that

$$x = ru/V \tag{58}$$

and that

$$\eta'' = \frac{(1 + y^2)^{1/2}}{(1 - y^2 \cot^2 \eta')^{1/2}} \left[\tfrac{1}{2}\pi \pm \cos^{-1}\{(1 - y^2 \cot^2 \eta')^{1/2} \sin \eta'\} \right] - \pi \tag{59}$$

$$+ \text{ sign for } 0 \leqslant \eta' \leqslant \tfrac{1}{2}\pi$$

$$- \text{ sign for } \tfrac{1}{2}\pi \leqslant \eta' \leqslant \pi$$

where

$$y \equiv w/V = (2\lambda)^{1/2}/rV. \tag{60}$$

The simplicity of the expressions for x and y is a special feature of the inverse square potential (48) assumed. Without it the calculation of $Q_c(V)$ would be a very formidable task.

In the earlier studies the distribution function for the electrons of a Fermi–Thomas atom (Landau and Lifshitz, 1959) was adopted:

$$X(v, r) = (32\pi^2 m_e^3/h^3) r^2 v^2. \tag{61}$$

Formula (53) may then be reduced to

$$Q_c(V) = \frac{m_e^3 (2\mu)^{7/2}}{6h^3 V^{9/2}} \iint_{\mathscr{C}} \{4V^2 w^2 - (V^2 + w^2 - u^2)^2\}$$

$$\times \{w^2 + V^2\}^{1/2} r^{1/2} u^{-9/2} \, dr \, d(\cos \eta'). \tag{62}$$

It is convenient to change from the variable r to y the minimum value of which may be taken to be

$$y_1 = (2I/m_e V^2)^{1/2} \tag{63}$$

where I is the ionization potential of the target atom (Bates and Mapleton, 1966). Formula (62) becomes

$$Q_c(V) = (2m_e^3/3h^3)(2\mu)^{7/2}(2\lambda)^{3/4} V^{-11/2} C(y_1) \tag{64}$$

where

$$C(y_1) \equiv \int_{y_1}^{\infty} \int_{-1}^{+1} \frac{(1 + y^2)^{1/2} \{y^2 - \frac{1}{4}(1 + y^2 - \varepsilon^2)^2\}}{y^{5/2} \varepsilon^{9/2}} \, dy \, d(\cos \eta') \tag{65}$$

and

$$\varepsilon^2 \equiv u^2/V^2 = (V^2 + v'^2 - 2Vv' \cos \chi')/V^2. \tag{66}$$

Use of approximation (57) yields

$$\varepsilon^2 = 2 + y^2 - 2(1 + y^2)^{1/2} \cos(\eta' + \eta''). \tag{67}$$

Because of (59) the right-hand side of this depends on y and η' only.

Examination of (63) and (64) shows that according to the theory there should be a universal curve which expresses

$$\mathscr{R} \equiv Q_c E^{11/4}/M^{11/4} Z^{7/2} \lambda^{3/4} \tag{68}$$

as a function of $m_e E/MI$ for any capture process, E being the energy and M the mass of the projectile nucleus. The computed curve (Bates and Mapleton, 1966) is given in Fig. 11 with the independent variable changed to E/MI as is of course permissible since attention is here confined to *electron* capture.

Stier and Barnett (1956) and Barnett and Reynolds (1958) have measured the cross sections for capture by protons from neon, argon, krypton, xenon, nitrogen, and oxygen. The corresponding values of $\mathscr{R}$ were calculated from (68), the parameters λ being obtained with the aid of the tables of Herman and Skillman (1963). They are marked as datum points in Fig. 11. Their

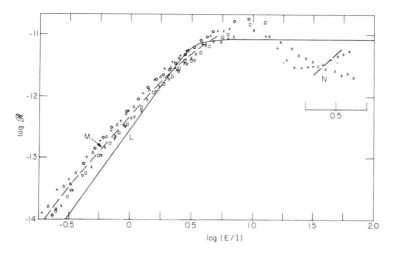

FIG. 11. Electron capture parameter $\mathscr{R}$ of Eq. (68) for protons with Q_c in m², M in units of the mass of the proton, λ in atomic units, E in keV, and I in eV. The full curve L is the prediction of the classical theory. The broken curve M is an empirical curve through the datum points obtained from measurements in various gases: O, Ne; +, Ar; △, Kr; ▽, Xe; ×, N_2 (per atom); □, O_2 (per atom). The raised horizontal scale on the right refers to the argon points + only; for it I is taken to be the ionization potential of the $2p$ shell; and the broken curve N is identical with the broken curve M (after Bates and Mapleton, 1966).

scatter about the empirical curve which has been drawn to represent them is remarkably small. This suggests that $\mathscr{R}$ is indeed very nearly a function of $m_e E/MI$ only.

The predicted curve lies unexpectedly close to the datum points in the moderate-to-low energy region where some of the assumptions made are quite unjustified. However, good accord is not achieved in the high energy region. Pointing out that the second rise exhibited by the datum points for argon may reasonably be attributed to the $2p$ shell (Fig. 11), Bates and Mapleton (1966) suggested that the failure is due in part at least to the adoption of distribution (61) which takes no account of the shell structure of the target atom.

To check this they later (1967) carried out further calculations with the Fermi–Thomas description of the target atom replaced by the Hartree–Fock–Slater description. The necessary modification of the theory is straightforward.

Consider each shell i separately. Let I_i be the ionization potential of the shell, N_i be the number of electrons it contains, w_i be the escape velocity of one of its electrons when distant r from the nucleus, and $P_i(r)/r$ be the Hartree–Fock–Slater radial wave function normalized so that

$$\int_0^{r_i} P_i(r)^2 \, dr = 1, \tag{69}$$

r_i being the classical radius of the shell.[4] The classical velocity–radial–distance distribution function is then

$$X_i(v, r) = N_i \, \delta(v - v_i) P_i(r)^2 \tag{70}$$

where

$$v_i{}^2 \equiv w_i{}^2 - (2I_i/m_e). \tag{71}$$

On substituting from (70) into (53) it is found that the contribution to the capture cross section from shell i is given by

$$Q_{ci}(V) = \frac{2^{5/2} N_i \pi}{3V^7} \int_0^{r_i} \int_{-1(\mathscr{C})}^{+1}$$

$$\frac{P_i(r)^2 (1 + y^2)^{1/2} \{4\varepsilon^2 - (\varepsilon^2 - y^2)(1 + \varepsilon^2 + a^2 - y^2)\}}{r^{3/2} \varepsilon^{9/2} (1 + a^2)^3 (y^2 - a^2)^{1/2}} \, dr \, d(\cos \eta') \tag{72}$$

where

$$y^2 \equiv 2|\mathscr{V}(r)|/m_e \, V^2 \tag{73}$$

$\mathscr{V}(r)$ being the potential function, where

$$a^2 \equiv 2I_i/m_e \, V^2 \tag{74}$$

and where ($\mathscr{C}$) indicates that the integrations are not over the entire region defined by the limits but are confined to the subregions where the integrand is real and positive and in which

$$|1 - \varepsilon| < (y^2 - a^2)^{1/2}. \tag{75}$$

The computational labor required to obtain the total capture cross section

$$Q_c(V) = \sum_i Q_{ci}(V) \tag{76}$$

is heavy.

Bates and Mapleton (1967) have treated

$$\mathrm{H}^+ + \mathrm{Ne} \rightarrow \mathrm{H}(\Sigma) + \mathrm{Ne}^+ \tag{77}$$

and

$$\mathrm{H}^+ + \mathrm{Ar} \rightarrow \mathrm{H}(\Sigma) + \mathrm{Ar}^+; \tag{78}$$

[4] The cutoff at r_i is, in fact, unimportant.

and Mapleton (1967) and Mapleton and Grossbard (1969) have treated

$$H^+ + N \rightarrow H(\Sigma) + N^+, \tag{79}$$
$$H^+ + O \rightarrow H(\Sigma) + O^+, \tag{80}$$
$$H^+ + Mg \rightarrow H(\Sigma) + Mg^+, \tag{81}$$

and

$$He^{2+} + N \rightarrow He^+(\Sigma) + N^+. \tag{82}$$

Figure 12 shows the calculated contributions from particular shells. As would be expected, the importance of inner shells relative to outer shells tends to increase as the impact energy E is raised. Moreover, the secondary loop of some of the wave functions causes the rate of fall of Q_{ci} with E to exhibit a quite sharp diminution. Structural effects are, however, almost obscured in the total capture cross section curves. These are compared with laboratory data in Figs. 13a–13d. Results obtained using the Fermi–Thomas distribution are included. The refinement of the theory makes the agreement

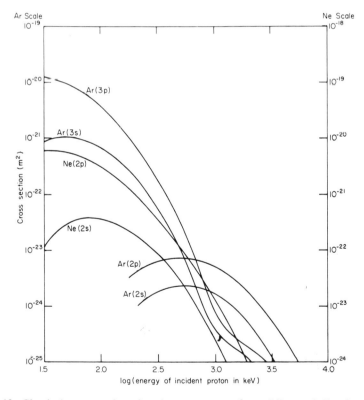

FIG. 12. Classical cross sections for electron capture from different shells of neon and argon (after Bates and Mapleton, 1967).

with experiment less good in the low energy region, but this is not significant since some of the basic assumptions are here invalid. It improves the accord in the high energy region where the success achieved is quite impressive. Serious error does not seem to arise either on account of the uncertainty principle or on account of the potential involved in the second binary encounter being non-coulombic.

Thomas (1927b) also considered capture from light target atoms. The mathematical analysis is very complicated, and to make progress he was obliged to ignore the initial orbital velocity v of the electron, which he judged to be permissible if the velocity V of the projectile is high enough. He argued that capture occurs through two binary encounters as in the case of a heavy target atom. Taking the active electron to be initially in a circular orbit of radius r_0, he found that the capture cross section is given by

$$\tilde{Q}_c(1s - \Sigma \,|\, r_0) = \frac{4 \times 2^{1/2}\pi}{3} \frac{S_{0A}(V, 60°)S_{0B}(V, 60°)}{m_e^{3/2} r_0^3 V^3} |\mathcal{U}(r_0)|^{3/2} \quad (83)$$

where $S_{0A}(V, 60°)$ and $S_{0B}(V, 60°)$ are the 60° differential elastic cross sections presented to the electron in the two binary encounters, and where $\mathcal{U}(r_0)$ is

FIGS. 13a–13d. [see pp. 308–310.] Electron capture cross sections. Classical calculations: dashed curves are based on the Fermi–Thomas electron distribution, solid curves on the Hartree–Fock–Slater distributions. The incident nuclei are protons except where otherwise stated.

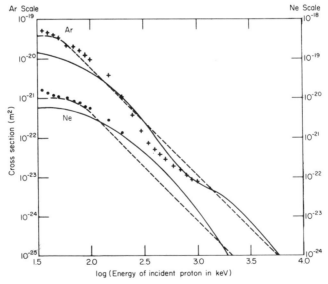

FIG. 13a. Neon and argon. Datum points from Stier and Barnett (1956) and Barnett and Reynolds (1958), ● Ne, + Ar (after Bates and Mapleton, 1967).

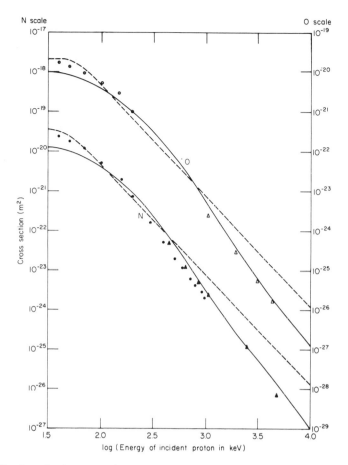

FIG. 13b. Atomic nitrogen and oxygen. Datum points from Stier and Barnett (1956) and Barnett and Reynolds (1958), ● N_2, ○ O_2; from Welsh *et al.* (1967) ▲ N_2; from Schryber (1967), △ O_2; they refer to the electron capture cross section per gas atom (after Mapleton, 1967).

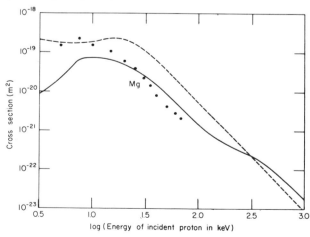

FIG. 13c. Atomic magnesium. Datum points from Berkner *et al.* (1969) ● Mg. (after Mapleton and Grossbard, 1969).

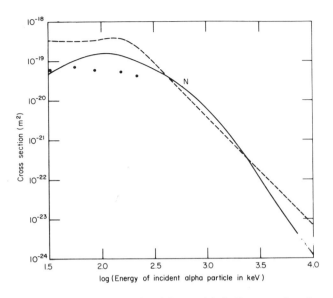

Fig. 13d. Atomic nitrogen (capture by alpha particles). Datum points from Berkner *et al.* (1969), ● N_2; they refer to the electron capture cross section per gas atom (after Mapleton, 1967).

the potential energy of the electron in the field of the projectile when they are distance r_0 apart.

In the H^+–$H(1s)$ case formula (83) reduces to

$$\tilde{Q}_c(1s - \Sigma \,|\, r_0) = \frac{64 \times 2^{1/2}\pi e^{11}}{3m_e^{11/2} r_0^{7/2} V^{11}}. \qquad (84)$$

Thomas identified r_0 with a_0 the radius of the first Bohr orbit. It is instructive to compare $\tilde{Q}_c(1s - \Sigma \,|\, a_0)$ with the high energy limit to the cross section $Q_c(1s - 1s \,|\, 2^{nd}B)$ for

$$H^+ + H(1s) \rightarrow H(1s) + H^+ \qquad (85)$$

as given by the second Born approximation (cf. Bransden, 1965). If the impact energy E is expressed in keV then

$$Q_c(1s - 1s \,|\, 2^{nd}B)/\tilde{Q}_c(1s - \Sigma \,|\, a_0) \simeq 6.7 + (2.6 \times 10^3/E^{1/2}). \qquad (86)$$

The ratio of the cross sections is thus much greater than unity—indeed, unless E is extremely high it is very much greater than unity. This does not signify that classical mechanics is necessarily inapplicable; it merely signifies that at least one of the approximations made is bad. Bates and Mapleton

(1966) pointed out that Thomas in effect adopted $\delta(r - a_0)$ for the radial distribution function of the electron in the hydrogen atom. They pointed out also that the use of the correct distribution function leads to a divergent integral. Collisions in which the electron is initially very close to its parent nucleus are not properly covered by Thomas's theory of capture. A classical theory without this serious defect is needed.

Exact classical calculations on process (85) have been carried out by Abrines and Percival (1966) using the Monte Carlo method already mentioned. Their results are compared in Fig. 14 with those of McCarroll (1961) and McElroy (1963), who carried out quantal calculations taking into account the nonorthogonality of the wave functions describing the initial and final states, and with the laboratory data of Gilbody and Ryding (1966). Almost a one-hundredfold range of cross sections is covered. The remarkable success achieved by classical mechanics can scarcely be fortuitous.

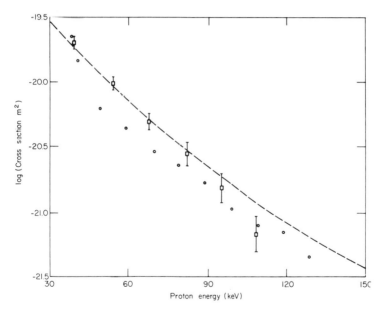

Fig. 14. Electron capture in H^+–H collisions. Broken curve, calculation (McCarroll, 1961; McElroy, 1963); O measurement (Gilbody and Ryding, 1966); □ Monte Carlo classical calculation with bars to indicate the statistical r.m.s. standard deviation errors (after Abrines and Percival, 1966).

The Monte Carlo method has also been employed by Banks and Valentine (1969) in an investigation on the dependence of charge transfer on the incident proton energy at a fixed angle of scattering. They discovered that the Everhart effect (cf. Lockwood and Everhart, 1962; Helbig and Everhart,

1965) is not entirely of a quantal nature. Though the exact classical calculations greatly underestimate the amplitude of the oscillations, they give correctly the energies at which the maxima and minima of the charge transfer probabilities occur.

III. Slow Collisions

Little progress has been made with slow collisions. These are, in general, very difficult to treat by classical mechanics (just as they are by quantum mechanics). The only process that has yet proved at all amenable is symmetrical resonance charge transfer

$$X^+ + X \rightarrow X + X^+. \tag{87}$$

This is rather surprising since the mechanism concerned here is seemingly peculiarly quantal, involving as it does the change in the relative phase of a pair of the eigenfunctions of the $X^+ - X$ quasimolecules (cf. Bates, 1962).

If X^+ and X are at a large distance Ra_0 apart, the total energy of the active electron is approximately

$$\varepsilon = -[(1/2n_s^2) + (1/R)](e^2/a_0) \tag{88}$$

n_s being the effective principal quantum number in the separated atoms limit. Bates and Mapleton (1966) argued that if the electron is to have sufficient energy to cross over the potential energy col at the midpoint O of the internuclear line, it is, from a classical viewpoint, necessary that

$$\varepsilon \geqslant -4e^2/Ra_0. \tag{89}$$

This condition is satisfied provided

$$R \leqslant R_0 \tag{90}$$

where

$$R_0 = 6n_s^2. \tag{91}$$

Assuming that if the electron is energetically able to transverse the col at any stage during a slow encounter it has equal chances of being attached to either system after the encounter, Bates and Mapleton deduced that the low velocity limit to the classical cross section for symmetrical resonance charge transfer is

$$Q_c(0) = 18n_s^4 \pi a_0^2. \tag{92}$$

If the velocity of relative motion V is nonzero, the time available for transfer is finite, and therefore (92) overestimates the classical cross section. The required modification has not yet been made (though it has been made in a related case to be discussed later).

As the velocity tends towards zero the cross section should not tend towards a finite limit. If the influence of the polarization attraction on the relative motion is neglected it should instead tend logarithmically to infinity (cf. Demkov, 1952; Dalgarno and McDowell, 1956). This behavior cannot be reproduced on a pure classical theory. Bates and Mapleton (1966) attributed it to the quantal tunnel effect. Taking the tunnel effect crudely into account they estimated that the probability of the electron being transferred in a collision of impact parameter

$$\rho \equiv \zeta R_0 a_0 \qquad (\zeta \geqslant 1) \tag{93}$$

is

$$\mathscr{P} = \frac{2^{1/2}\zeta^{9/4}e^2}{5 \times 3^{1/4}\pi n_s^{5/2}(3\zeta - 1)^{1/2}V\hbar} \exp\left\{\frac{-3^{3/2}\pi\zeta^{1/2}(\zeta - 1)n_s}{2}\right\}. \tag{94}$$

By the approximation of Firsov (1951) the cross section may be taken to be

$$Q_c(V) = \tfrac{1}{2}\pi\rho^{\dagger 2}a_0^2 \tag{95}$$

where $\rho^\dagger$ is the greatest impact parameter for which

$$\mathscr{P} = \sin^2 (1/\pi). \tag{96}$$

The procedure indicated is quite successful (cf. Fig. 15), though for the reason already specified it does not give a cross section which falls off sufficiently rapidly as the velocity is increased.

Using the quantal two-state approximation, Bates and Reid (1969) have carried out calculations on

$$H^+ + H(n_s) \rightarrow H(n_s) + H^+, \qquad 1 \leqslant n_s \leqslant 5. \tag{97}$$

In the low velocity region the derived cross sections $Q_c(n_s \,|\, V)$ may be closely represented by

$$Q_c(n_s \,|\, V) = n_s^3 A(V), \qquad n_s \gtrsim 1, \tag{98}$$

where $A(V)$ is a function of V only. Noting that the powers of n_s occurring in the classical formula (92) and in the quantal formula (98) are different, Bates and Reid inferred that the physical models do not correspond; and they showed how the original classical description of symmetrical resonance charge transfer may be modified to bring it into satisfactory accord with the two-state quantal description.

The quantal two-state approximation involves a gerade and an ungerade state of $H_2{}^+$ both characterized by the trio (n_s, m, I), where n_s is the principal quantum number in the separated atoms limit, where m is the magnetic quantum number, and where

$$I \equiv n_u - l \tag{99}$$

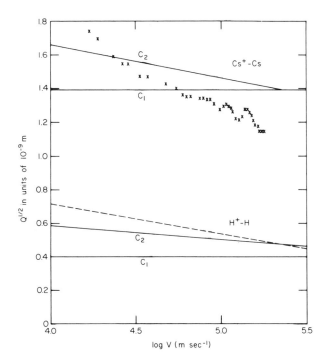

FIG. 15. Symmetrical resonance charge transfer in H^+–H and Cs^+–Cs collisions. Solid curves C_1 are based on the pure classical formula (92); solid curves C_2 are based on the modified classical formulas (93)–(96) which make allowance for the tunnel effect (Bates and Mapleton, 1966); the broken curve is as calculated by Dalgarno and Yadav (1953) using the quantal perturbed stationary state approximation; datum points represent the measurements of Perel *et al.* (1965).

n_u and l being the principal and azimuthal quantum numbers in the united atom limit. In specifying the states, the quantum numbers n_s and I may be replaced by the eigenenergy E and the separation constant C that arises in solving the Schrödinger equation expressed in prolate spheroidal (λ, μ, ϕ) coordinates.

In the derivation of (92) the energy is the only assumed constant of the classical motion of the electron in the static field of the protons. To obtain a classical approximation simulating the quantal two-state approximation, Bates and Reid took as constants of the motion the classical dynamical variables corresponding to the quantal operators which have m and C as eigenvalues. They then found that

$$(1 - \mu^2)p_\mu^2 = C - (R^2E/2)\mu^2 - [m^2/(1 - \mu^2)] \tag{100}$$

in which p_μ is the momentum conjugate with μ and in which atomic units are used. Transfer of the electron is possible classically if p_μ is real at the midplane. From (100) it is hence possible provided

$$R \leqslant \bar{R}(n_s, m, I) \tag{101}$$

where $\bar{R}$ is the root of the equation

$$C - m^2 = 0. \tag{102}$$

Using an expansion for C in inverse powers of R this equation may be solved approximately to yield

$$\bar{R}(n_s, m, I) = n_s^2 \left(3 + \frac{1}{n_s} [-4I - 2|m| + 2 \right.$$

$$\left. + \{6n_s^2 - 5n_s(2I + |m| - 1) + 8I(I + |m| - 1) - 4|m|\}^{1/2}] \right).$$

$$\tag{103}$$

It may be seen that

$$\bar{R}(n_s, m, I) < R_0(n_s). \tag{104}$$

The low velocity limit to the classical symmetrical resonance charge transfer cross section for state (n_s, m, I) clearly is

$$Q_c(n_s, m, I \,|\, 0) = \tfrac{1}{2}\pi \bar{R}(n_s, m, I)^2 a_0^2. \tag{105}$$

Bates and Reid also treated symmetrical resonance charge transfer collisions having a non-zero velocity of relative motion. Letting $n_A(t)$ and $n_B(t)$ be the probabilities that the electron is on either side of the midplane at time t and letting $v_\mu(R)$ be the frequency of the μ motion (the calculation of which is a straightforward problem in classical mechanics), they observed that

$$dn_B(t)/dt = -\gamma(R)\{n_B(t) - n_A(t)\} \tag{106}$$

$$= -\gamma(R)\{2n_B(t) - 1\}, \tag{107}$$

where

$$\gamma(R) = 2v_\mu(R), \qquad R \leqslant \bar{R} \tag{108}$$

$$= 0 \quad, \qquad R > \bar{R}. \tag{109}$$

Solving Eq. (107) with the appropriate boundary condition, they found that the probability of electron transfer occurring in an encounter at impact parameter ρ is

$$\mathscr{P} = \tfrac{1}{2}[1 - \exp\{-\mathscr{L}(\rho/\bar{R})\}] \qquad (110)$$

where

$$\mathscr{L}(\rho/\bar{R}) \equiv 2 \int_{-\infty}^{\infty} \gamma(R)\, dt, \qquad (111)$$

and therefore that the cross section is

$$Q_c(n_s, m, I \,|\, V) = \tfrac{1}{2}\pi \bar{R}(n_s, m, I)^2 \int_0^1 [1 - \exp\{\mathscr{L}(s)\}]\, ds^2. \qquad (112)$$

All that need be done to make approximate allowance for tunneling is to extend the upper limit to the integration in (112) to infinity and to replace (109) by

$$\gamma(R) = v_\mu(R)T(R), \qquad R \geqslant \bar{R}, \qquad (113)$$

where

$$T(R) \equiv \exp\left(-4 \int_0^{|\mu_1|} |p_\mu|\, d\mu\right) \qquad (114)$$

is the transmission coefficient, $|\mu_1|$ being the magnitude of μ at the classical turning points.

Figures 16a and 16b compare the symmetrical resonance charge transfer cross sections obtained from the pure classical theory outlined, from the modified version of this which allows for tunneling, and from the quantal two-state approximation. It is seen that tunneling is very important if n_s and V are low. The cross sections obtained when this effect is grafted onto the classical theory and those obtained from the quantal two-state approximation present patterns which are in remarkable harmony. The corresponding cross sections for process (97), found by averaging over m and I, are naturally also in good accord (Fig. 17).

Neither the classical nor the quantal calculations take into account the fact that the expectation value of the momentum of the electron in its orbit around A is not the same as that in its orbit around B because of the relative velocities of the two protons. In consequence the cross sections are over-estimated.

The success achieved in treating symmetrical resonance charge transfer in slow collisions is especially interesting as it is quite unrelated to the success of the classical impulse approximation.

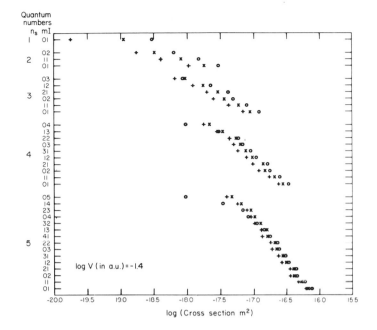

FIG. 16a

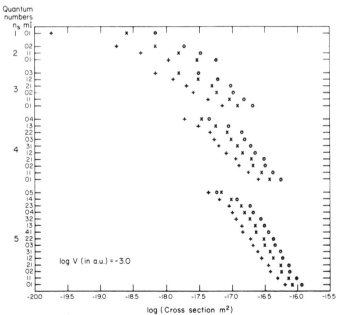

FIG. 16b

FIGS. 16a and 16b. Symmetrical resonance charge transfer in H^+–$H(n_s, m, I)$ collisions: pure and modified classical cross sections denoted by $+$ and $\times$, respectively; quantal (two-state approximation) cross sections denoted by O. The quantum numbers n_s, m, and I are given on the ordinate (after Bates and Reid, 1969).

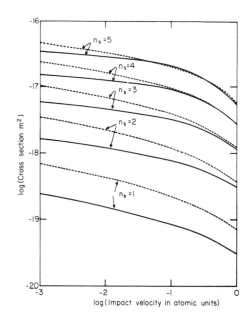

FIG. 17. Symmetrical resonance charge transfer in $H^+-H(n)$ collisions: average modified classical cross sections are denoted by solid curves and average quantal (two-state approximation) cross sections are denoted by dashed curves (after Bates and Reid 1969).

ACKNOWLEDGMENT

This research was supported by the Advanced Research Projects Agency of the Department of Defense and was monitored by the Office of Naval Research under Contract N00014-69-C-0035.

REFERENCES

Abrines, R., and Percival, I. C. (1966). *Proc. Phys. Soc. (London)* **88**, 861, 873.
Afrosimov, V. V., Il'in, R. N. and Solov'ev, E. S. (1960). *Soviet Phys.—Tech. Phys. (English Transl.)* **5**, 661.
Bang, J., and Hansteen, J. M. (1959). *Kgl. Danske Videnskab. Selskab Mat. Fys. Medd.* **31**, No. 13.
Banks, D., and Valentine, N. A. (1969). *Abstr. 6th Intern. Conf. Phys. Electron. At. Collisions* p. 123. M.I.T. Press, Cambridge, Massachusetts.
Barnett, C. F., and Reynolds, H. K. (1958). *Phys. Rev.* **109**, 355.
Bates, D. R. (1962). "Atomic and Molecular Processes," p. 601. Academic Press, New York.
Bates, D. R., and Griffing, G. W. (1953). *Proc. Phys. Soc. (London)* **A66** 961.

Bates, D. R., and Griffing, G. W. (1955). *Proc. Phys. Soc.* (*London*) **A68**, 90.
Bates, D. R., and Mapleton, R. A. (1965). *Proc. Phys. Soc.* (*London*) **85**, 605.
Bates, D. R., and Mapleton, R. A. (1966). *Proc. Phys. Soc.* (*London*) **87**, 657.
Bates, D. R., and Mapleton, R. A. (1967). *Proc. Phys. Soc.* (*London*) **90**, 909.
Bates, D. R., and Reid, R. H. G. (1969). *J. Phys. B* (*At. Mol. Phys.*) **2**, 851, 857.
Bates, D. R., and Walker, J. C. G. (1966). *Planetary Space Sci.* **14**, 1367.
Bates, D. R., and Walker, J. C. G. (1967). *Proc. Phys. Soc.* (*London*) **90**, 333.
Bates, D. R., Boyd, A. H., and Prasad, S. S. (1965). *Proc. Phys. Soc.* (*London*) **85**, 1121.
Bates, D. R., Dose, V., and Young, N. A. (1969). *J. Phys. B* (*At. Mol. Phys.*) **2**, 930.
Bell, K. L., and Kingston, A. E. (1969). *J. Phys. B.* (*At. Mol. Phys.*) **2**, 653.
Bell, K. L., Dose, V., and Kingston, A. E. (1969a). *J. Phys. B* (*At. Mol. Phys.*) **2**, 831.
Bell, K. L., Kennedy, D. J., and Kingston, A. E. (1969b). *J. Phys. B* (*At. Mol. Phys.*) **1**, 1037.
Berkner, K. H., Pyle, R. V., and Stearns, J. W. (1969). *Phys. Rev.* **178**, 248.
Bohr, N. (1948). *Kgl. Danske Videnskab. Selskab Mat. Fys. Medd.* **18**, No. 8.
Brandt, W., and Laubert, R. (1969). *Phys. Rev.*, **178**, 226.
Bransden, B. H. (1965). *Advan. At. Mol. Phys.* **1**, 85.
Burgess, A., and Percival, I. C. (1968). *Advan. At. Mol. Phys.* **4**, 109.
Catlow, G. W., and McDowell, M. R. C. (1967). *Proc. Phys. Soc.* (*London*) **92**, 875.
Clementi, E. (1965). *IBM J. Res. Devel.* **9**, 2.
Coleman, J., and McDowell, M. R. C. (1964). *Proc. Phys. Soc.* (*London*) **83**, 334 and 907.
Cooper, J. W. (1962). *Phys. Rev.* **128**, 681.
Dalgarno, A., and McDowell, M. R. C. (1956). *Proc. Phys. Soc.* (*London*) **A69**, 615
Dalgarno, A., and Yadav, H. N. (1953). *Proc. Phys. Soc.* (*London*) **A66**, 173.
de Heer, F. J., Schutten, J., and Moustafa Moussa, H. R. (1966). *Physica* **32**, 1766.
Demkov, Yu. N. (1952). *Uch. Zap. Leningr. Gos. Univ.* **146**, 74.
Drawin, H. W. (1968). *Z. Phys.* **211**, 404.
Firsov, O. B. (1951). *Zh. Eksper. Teor. Fiz.* **21**, 1001.
Fite, W. L. Stebbings, R. F., Hummer, D. G.. and Brackmann, R. T. (1960). *Phys. Rev.* **119**, 663.
Fogel, Ya. M., Ankudinov, V. A., Philipenko, D. V., and Topolia, N. V. (1958). *Soviet Phys. JETP* (*English Transl.*) **7**, 400.
Freeston, M. J., and Kingston, A. E. (1970a). In preparation.
Freeston, M. J., and Kingston, A. E. (1970b). In preparation.
Garcia, J. D. (1970). *Phys. Rev.* **A1**, 280.
Garcia, J. D., Gerjuoy, E., and Welker, J. E. (1968a). *Phys. Rev.* **165**, 66.
Garcia, J. D., Gerjuoy, E., and Welker, J. E. (1968b). *Phys. Rev.* **165**, 72.
Gerjuoy, E. (1966). *Phys. Rev.* **148**, 54.
Gilbody, H. B., and Ireland, J. V. (1963). *Proc. Roy. Soc.* **A277**, 137.
Gilbody, H. B., and Lee, A. R. (1963). *Proc. Roy. Soc.* **A274**, 365.
Gilbody, H. B., and Ryding, G. (1966). *Proc. Roy. Soc.* **A291**, 438.
Gilbody, H. B., Browning, R., Levy, G., McIntosh, A. I., and Dunn, K. F. (1968). *J. Phys. B* (*At. Mol. Phys.*) **1**, 863.
Gilbody, H. B., Browning, R., Dunn, K. F., and Latimer, C. (1970). *J. Phys. B* (*At. Mol. Phys.*) In Press.
Gryzinski, M. (1959). *Phys. Rev.* **115**, 374.
Gryzinski, M. (1965). *Phys. Rev.* **138**, A 305, 322, 336.
Hart, R. R., Reuter, F. W., III, Smith, H. P., Jr. and Khan, J. M. (1969). *Phys. Rev.*, **179**, 4.
Helbig, H. F., and Everhart, E. (1965). *Phys. Rev.* **140**, A715.

320   *D. R. Bates and A. E. Kingston*

Henneberg, W. (1933). *Z. Phys.* **86**, 592.
Herman, F., and Skillman, S. (1963). "Atomic Structure Calculations." Prentice-Hall, Englewood Cliffs, New Jersey.
Jackson, J. D., and Schiff, H. (1953). *Phys. Rev.* **89**, 359.
Khan, J. M., and Potter, D. L. (1964). *Phys. Rev.* **133**, A890.
Khan, J. M., Potter, D. L., and Worley, R. D. (1965). *Phys. Rev.* **139**, 1735.
Landau, L. D., and Lifshitz, E. M. (1959). "Quantum Mechanics," p. 235. Pergamon, Oxford.
Lockwood, G. L., and Everhart, E. (1962). *Phys. Rev.* **125**, 567.
McCarroll, R. (1961). *Proc. Roy. Soc.* **A264**, 547.
McClure, G. W. (1968). *Phys. Rev.* **116**, 22.
McDaniel, E. W. (1964). "Collision Phenomena in Ionized Gases," Chapter 6. Wiley, New York.
McDowell, M. R. C. (1966). *Proc. Phys. Soc. (London)* **89**, 23.
McElroy, M. B. (1963). *Proc. Roy. Soc.* **A272**, 542.
Mapleton, R. A. (1964). *Proc. Phys. Soc. (London)* **83**, 895.
Mapleton, R. A. (1967). *Phys. Rev.* **164**, 51.
Mapleton, R. A. and Grossbard, N. (1969). *Phys. Rev.* **188**, 228.
Merzbacher, E., and Lewis, H. W. (1958). *In* "Handbuch der Physik" (S. Flugge, ed.), Vol. 34, p. 166. Springer, Berlin.
Messelt, S. (1958). *Nucl. Phys.* **5**, 435.
Moustafa Moussa, H. R. (1967). Ph.D. Thesis, Univ. of Leiden, Holland.
Peach, G. (1965). *Proc. Phys. Soc. (London)* **85**, 709.
Peach, G. (1966). *Proc. Phys. Soc. (London)* **87**, 375.
Percival, I. C., and Valentine, N. A. (1966). *Proc. Phys. Soc. (London)* **88**, 885.
Percival, I. C. and Richards, D. (1970a) *J. Phys. B. (At. Mol. Phys.)* **3**, 315.
Percival, I. C. and Richards, D. (1970b) *J. Phys. B. (At. Mol. Phys.)* In press.
Percival, I. C. and Richards, D. (1970c) *Astrophys. Letters* **4**, 235.
Perel, J., Vernon, R. H., and Daley, H. L. (1965). *Phys. Rev.* **138**, A 937.
Rudd, M. E., Sautter, C. A., and Bailey, C. L. (1966). *Phys. Rev.* **151**, 20.
Rudd, M. E., and Jorgensen, T., Jr. (1963). *Phys. Rev.* **131**, 666.
Russell, G. R. (1969). *J. Chem. Phys.* **50**, 4597.
Sampson, J. A. R. (1966). *Advan. Atom. Molec. Phys.* **2**, 177.
Schram, B. L. (1966). Ph.D. Thesis, Univ. of Amsterdam, Holland.
Schryber, U. (1967). *Helv. Phys. Acta* **40**, 1023.
Sellers, B., Hansen, F. A., and Wilson, H. H. (1969). *Phys. Rev.* **182**, 90.
Smith, F. J. (1964). *Physica* **30**, 497.
Solov'ev, E. S., Il'in, R. N., Oparin, V. A., and Fedorenko, N. V. (1962). *Soviet Phys.— JETP (Eng. Transl.)* **15**, 459.
Stedeford, J. B. H., and Hasted, J. B. (1955). *Proc. Roy. Soc.* **A227**, 466.
Stewart, A. L. (1967). *Advan. Atom. Molec. Phys.* **3**, 1.
Stier, P. M., and Barnett, C. F. (1956). *Phys. Rev.* **103**, 896.
Synek, M., and Sturgis, G. E. (1965). *J. Chem. Phys.* **42**, 3068.
Thomas, E. W., and Bent, G. (1967). *Phys. Rev.* **164**, 143.
Thomas, E. W., Edwards, J. L., and Ford, J. (1969). *Abstr. 6th Intern. Conf. Phys. Electron. At. Collisions* p. 462. M.I.T. Press, Cambridge, Massachusetts.
Thomas, L. H. (1927a). *Proc. Cambridge Phil. Soc.* **23**, 714, 828.
Thomas, L. H. (1927b). *Proc. Roy. Soc.* **A114**, 561.
Thomson, J. J. (1912). *Phil. Mag.* **23**, 449.
Toburen, L. H., Nakai, M. Y., and Langley, R. A. (1968). *Phys. Rev.* **171**, 114.

Tripathi, A. N., Mathur, K. C., and Joshi, S. K. (1969). *J. Phys. B (At. Mol. Phys.)* **2**, 155.

Van den Bos, J., Winter, G. J., and De Heer, F. J. (1968). *Physica* **40**, 357.

Vriens, L. (1967). *Proc. Phys. Soc. (London)* **90**, 935.

Vriens, L. (1970). "Case Studies in Atomic Collision Physics. I" (E. W. McDaniel and M. R. C. McDowell, eds.), p. 335, North-Holland Publ., Amsterdam.

Weinbaum, S. (1933) *J. Chem. Phys.* **1**, 593.

Welsh, L. M., Berkner, K. H., Kaplan, S. N., and Pyle, R. V. (1967). *Phys. Rev.* **158**, 85.

Williams, E. J. (1931). *Proc. Roy. Soc.* **A130**, 328.

Williams, E. J. (1945). *Rev. Mod. Phys.* **17**, 217.

Williams, J. F. (1967). *Phys. Rev.* **153**, 116.

Wittkower, A. B., Levy, G., and Gilbody, H. B. (1967). *Proc. Phys. Soc. (London)* **91**, 306.

Author Index

Numbers in italics refer to the pages on which the complete references are listed.

A

Abram, R. A., 148, *151*
Abrines, R., 274, 279, 311, *318*
Accardo, C. A., 50, *57*
Afanas'er, A. M., 185, *228*
Afrosimov, V. V., 66, *100*, 300, *318*
Ahearne, J. F., 64, *101*
Aikin, A. C., 49, 51, *54*
Akhiezer, A. I., 64, 66, *100*
Akhiezer, I. A., 64, 66, *100*
Albritton, D. L., 25, *54*
Alexeff, I., 64, 66, 94, *100, 102*
Alievskiĭ, M. Ya., 185, 187, *226*
Allen, C. W., 98, *101*
Allison, A. C., 138, 147, *151*
Altshuler, S., 120, 133, *151*
Amdur, I., 190, 200, 206, 213, 223, 225, *226*
Anderson, V. E., 145, *153*
Andō, K., 62, *103*
Ankudinov, V. A., 294, *319*
Annis, B. K., 201, 203, 213, *226*
Appleton, E. V., 50, *54*
Ardill, R. W. B., 136, 140, *151*
Armistead, F. C., 205, 207, *230*
Armstrong, B. H., 62, *101*
Arnold, J. H., 205, 209, *226*
Arthurs, A. M., 113, 114, 134, *151*
Artsimovich, L. A., 67, 68, *101*

B

Bailey, C. L., 280, 282, 283, *320*
Bakanov, S. P., 203, *227*
Balescu, R., 64, 66, 79, 95, *101*
Ball, J. S., 245, 246, 252, 253, *267*
Bang, J., 282, 286, *318*
Banks, D., 311, *318*
Bardsley, J. N., 36, 38, 40, 41, 43, 44, 45, 46, *54*, 115, *151*
Barker, J. A., 197, 200, *226*
Barnett, C. F., 291, 294, 300, 304, 308, 309, *318, 320*
Barrell, H., 62, *101*

Barua, A. K., 224, *230*
Bates, D. R., 2, 3, 6, 11, *54*, 256, *267*, 274, 277, 289, 291, 292, 294, 296, 297, 304, 305, 306, 307, 308, 310, 312, 313, 314, 317, 318, *318*
Bauer, E., 41, *55*
Bearden, A. J., 62, 77, 91, *103*
Beatty, J. W., Jr., 206, *226*
Beckers, J. M., 76, *101, 103*
Beenakker, J. J. M., 191, 207, 219, 220, 221, 222, 224, *231*
Bell, B., 75, *101*
Bell, K. L., 275, 280, 282, 283, 287, 288, 292, *319*
Belyaev, Yu. N., 225, *226*
Belyaeva, V. A., 62, 88, *104*
Bendt, P. J., 205, 210, *226*
Bent, G., 288, *320*
Berezin, A. B., 62, 70, 88, *101*
Berkner, K. H., 309, 310, *319, 321*
Berlande, J., 16, 17, *55*
Berning, W. W., 51, *57*
Bernstein, I. B., 64, *101*
Bernstein, M. J., 64, 84, 93, *101*
Bernstein, R. B., 146, *152*, 197, *228*
Berry, R. S., 47, *55*
Beth, M.-U., 84, *101*
Bertoncini, P. J., 44, *57*
Bhatia, A. K., 253, *267*
Biedenharn, L. C., 113, 115, *151*
Bielski, A., 84, *101*
Billings, D. E., 68, *101*
Biondi, M. A., 3, 5, 7, 8, 11, 12, 14, 15, 16, 17, 18, 19, 20, 21, 22, 23, 24, 25, 26, 27, 28, 29, 31, 33, 34, 42, 43, 47, 53, *56*
Bird, R. B., 156, 169, 188, 193, 196, 197, 200, 219, 221, *228*
Birkebak, R. C., 84, *101*
Bivins, R., 243, 247, *268*
Blatt, J. M., 113, 115, *151*
Boardman, L. E., 206, 224, *226*
Bogdanova, S. S., 198, *231*
Bogen, P., 82, 98, *101*

323

Subject Index

335